T0221731

Advances in Information Security

Volume 103

Series Editors
Sushil Jajodia, George Mason University, Fairfax, VA, USA
Pierangela Samarati, Milano, Italy
Javier Lopez, Malaga, Spain
Jaideep Vaidya, East Brunswick, NJ, USA

The purpose of the *Advances in Information Security* book series is to establish the state of the art and set the course for future research in information security. The scope of this series includes not only all aspects of computer, network security, and cryptography, but related areas, such as fault tolerance and software assurance. The series serves as a central source of reference for information security research and developments. The series aims to publish thorough and cohesive overviews on specific topics in Information Security, as well as works that are larger in scope than survey articles and that will contain more detailed background information. The series also provides a single point of coverage of advanced and timely topics and a forum for topics that may not have reached a level of maturity to warrant a comprehensive textbook.

Dietmar P. F. Möller

Guide to Cybersecurity in Digital Transformation

Trends, Methods, Technologies, Applications and Best Practices

 Springer

Dietmar P. F. Möller
Clausthal University of Technology
Clausthal-Zellerfeld, Germany

ISSN 1568-2633   ISSN 2512-2193   (electronic)
Advances in Information Security
ISBN 978-3-031-26844-1   ISBN 978-3-031-26845-8   (eBook)
https://doi.org/10.1007/978-3-031-26845-8

This Springer imprint is published by the registered company Springer Nature Switzerland AG
The registered company address is: Gewerbestrasse 11, 6330 Cham, Switzerland

Foreword

Changing market dynamics are reviving up industrial, public and private organizations through digital transformation. Digital transformation is the incorporation of computer-based technologies and practices into organizations products, processes, strategies and services. Organizations also use digital transformation to better engage and serve their customers and employees, and thus improve their ability to compete in the market. In this sense, successful digital transformations yield ongoing business benefits. Therefore, a digital transformation strategy deeply positions organizations to survive and thrive today and in the future because digital transformation is the key economic driver. However, there is still a risk in using the latest digital technologies and their applications, meaning a bigger attack surface, from the perspective of cybersecurity and resilience. Therefore, the book outlines internal in-depth research and experience gained over multiple engagements of digitization in the context of digital transformation and the need for cybersecurity, which has achieved great interest in industrial, public and private organizations. In this context, the book provides a systematic overview of the latest development of core research areas in methods and technologies in cybersecurity in digital transformation. The focus is on threats, threat intelligence, intrusion detection and prevention, cyberattack models and scenarios, cybersecurity ontology, NIST cybersecurity framework and MITRE cybersecurity criteria, ransomware attack risks with regard to ransom cost and loss of reputation, cybersecurity maturity models and SWOT analysis as well as machine learning and deep learning in the application of defending the cybersecurity risk. Thus, the book *Guide to Cybersecurity in Digital Transformation – Trends, Methodologies, Technologies, Applications and Best Practices* is a showcase of creative ideas of ongoing research work and best practices. The book shows how to analyze the intrinsic complexity of cybersecurity in digital transformation accurately and under varying operational conditions and scenarios to predict its behavior for engineering and planning purposes to provide adequate academic answers for today's emerging technology management questions in cybersecurity in digital transformation. The chapters are well written, showing academic rigor and professionalism of the author. Therefore, the book can be stated as an important reading for new researchers entering the fields of cybersecurity and digital transformation research. It also offers new perspectives and documents important methods and progress in cybersecurity and digital transformation related analysis and development. I strongly recommend Prof. Dr. Dietmar P.F. Moeller's scholarly writing to students, academicians and industrialists who are keen to learn advance methodologies in cybersecurity in digital transformation. I can say without reservation that this book, and, more specifically the method it espouses, will change the fundamental ideas for better understanding of innovations and opportunities required todays in industry. I failed to mention how much more I enjoy reading and reviewing the book. I think that the author can be confident that there will be many grateful readers, who will have gained a broader perspective of cybersecurity and digital transformation, as result of their efforts.

Indian Institute of Information   Roland E. Hass
Technology, Bangalore, India

Preface

Digital transformation and cybersecurity is a global phenomenon, capturing the attention in every industry and spurring major investment. However, digital transformation and cybersecurity is not a single objective; it is multifaceted depending on the respective industrial goals and their digital maturity. Therefore, the goal of this book is to provide a comprehensive, in-depth, and state-of-the-art summary of trends, methods, technologies, applications, and best practices in cybersecurity in digital transformation. It describes the digital transformation paradigm clearly, showing the essential activities required and applied to fit within the overall effort of digital transformation. However, the intrinsic complexity of digital transformation also makes cybersecurity awareness and cybersecurity strategies a sine qua non objective because cybersecurity deals with the presence of cyberattackers with their cybercriminal cyberattack repertoire. Thus, the book provides an ideal framework for understanding the complexity of cybersecurity in digital transformation. For this reason, some choices have been made in selecting the material for this book. A top-down approach taken into consideration introduces the fundamentals of digital transformation and cybersecurity, threat and threat intelligence, intrusion detection and prevention, as well as cyberattack models, scenarios, and ontologies as most important subject areas. Furthermore, the NIST cybersecurity framework and MITRE cybersecurity criteria, cybersecurity maturity models and SWOT analysis as well as machine learning and deep learning describing how current methods relate to and support a greater scope in cybersecurity in digital transformation. This provides a framework within which the reader can assimilate the associated requirements. Without such a reference, the practitioner is left to ponder the plethora of terms, standards, and practices that have been developed independently and that often lack cohesion, particularly in nomenclature and emphasis. Therefore, this book intends both to cover all aspects of cybersecurity in digital transformation and to provide a framework for the consideration of the many objectives associated with cybersecurity in digital transformation. These subjects are discussed with regard to trends, methods, technologies, applications, and best practices.

This book can serve as textbook or as reference book for college courses on information security and cybersecurity and can be offered in computer science, electrical and computer engineering, information technology and information systems, applied mathematics and operations research, business informatics, and management departments. The contents of the book are also very useful to researchers who are interested in the development of cybersecurity in the digital transformation era. Company engineers can use the principles described in the book for their product design work.

The material in the book can be difficult to comprehend if the reader is new to such an approach because cybersecurity in digital transformation is a multidisciplinary domain, founded in computer science, engineering, mathematics, intelligent manufacturing, and other. However, the material of the book may not be read and comprehended quickly and easily. Therefore, specific best practices examples have been included with related topics to help the reader mastering the material. It is assumed that the reader has some knowledge of basic calculus-based probability and statistics and some experience with systems and software engineering.

The use of the book can be as primary text in a course in various ways. It contains more material than covered in detail in a quarter-long (30-h) or semester-long (45-h) course. Instructors may elect to choose their own topics and add their own case studies. The book is also suitable for self-study as a reference for engineers, scientists, computer scientists, and CIOs for on-the-job training, for study in graduate schools, and as a reference for cybersecurity in digital transformation practitioners and researchers.

However, a textbook cannot describe all of the innovative aspects of cybersecurity in digital transformation. For this reason, the reader is referred to specific supplemental material, such as references, other textbooks, case studies, and others as well as Internet-based information that address to topics selected for the book.

The book contains eight chapters, which can be read independently or consecutively.

Chapter 1, "Cybersecurity in Digital Transformation," introduces relevant definitions on digital transformation and their essential technological background. It covers the confluence of emerging technologies applied in an increasing number of industrial applications with unprecedented results. Digital transformation has received a great attention in all industrial sectors, which will fundamentally change operational business approaches in all sectors. However, the intrinsic complexity of digital transformation also makes cybersecurity awareness an essential objective fundamental to both protecting secret data and business assets and enabling their defense. Therefore, Chap. 1 focuses in Sect. 1.1 on the manifold aspects of digital transformation with their great impact on industrial, public, and private organizations, enhancing their business success. Section 1.2 introduces the interaction of digitization and emerging technologies required in digital transformation, while Sect. 1.3 reflects the challenges in digital transformation. Section 1.4 discusses about applications in digital transformation, while Sect. 1.5 refers to gain leadership. Section 1.6 focuses on cybersecurity, described in detail in three subsections, referring on knowledge about cybersecurity risks, the CIA Triad, and explain why cybersecurity is still paramount. Section 1.7 focuses on the approach of circular economy, referring to the challenges of digital transformation to build up a circular economy. Section 1.8 contains comprehensive questions from the topics such as digital transformation, cybersecurity, and circular economy, followed by references for further reading in "References" section.

Chapter 2, "Threats and Threat Events," introduces threats with their intension to inflict harm because threat event attacks tries, e.g., to shut down targeted systems, networks, infrastructure resources, and others, making it inaccessible to regular operation or users. Threat intelligence is evidence-based knowledge, including context, mechanisms, indicators, implications, and action-oriented advice, about an existing or emerging menace or hazard to assets. Therefore, a solid understanding of the impact and potential consequences of threat event attacks and/or threat intelligence is required, to cyber-secure critical computer systems, networks, infrastructure resources, and others. This requires a detailed analysis of known and documented threat event attacks to avoid that they cause a loss of confidentiality, integrity, and availability, as described in the CIA Triad (see Sect. 1.6.2). Therefore, Sect. 2.1 discusses about threats, threat events, threat intensions, threat types, cybersecurity residual risk rating, likelihood and consequence level, and cybersecurity risk management and quantifying cybersecurity risk in four subsections. In Sect. 2.2, threat intelligence is introduced, taking into account the problem of known-known, known-unknown, and unknown-unknown; digital forensic and threat intelligence platforms; threat event profiling; threat intelligence and threat lifecycle; and threat intelligence sharing and management platforms, described in detail in four subsections. Section 2.3 contains comprehensive questions from the topics threat event attacks, threat intelligence, threat risk level and security model, threat event lifecycle, and threat event platforms, followed by "References" section with references for further reading.

Chapter 3, "Intrusion Detection and Prevention," summarizes the knowledge to detect, respond, and prevent cybersecurity risks for computer systems, networks, and infrastructure resources, using different methods to impede successful execution of cybersecurity risks. In this context, the signature-based approach that corresponds to known threat event attacks is used, or the anomaly-based detection that compares definitions of what activity is considered normal against observed threat event attacks, to identify significant deviations. Other methods are the stateful protocol analysis, which compares predetermined profiles of general accepted definitions of benign protocol activities for each protocol state against observed events, to identify deviations, or the hybrid system approach that combines some or all of the other methodologies to detect and respond to cybersecurity risks. However, the need of intrusion detection and prevention systems architectures requires distinguished decisions. Against this background, Chap. 3 introduces intrusion detection in Sect. 3.1, which is organized into seven subsections that cover the subject's characteristics and capabilities of the most important intrusion detection methods, their advantages and disadvantages, and finally the architecture of intrusion detection systems. Thereafter, in Sect. 3.2, the focus lies on pre-processing in intrusion detection systems, while Sect. 3.3 covers the metric of intrusion detection capability.

Section 3.4 refers to intrusion prevention, organized into two subsections that cover intrusion prevention methods and the architecture of the intrusion prevention systems. Section 3.5 finally focuses on the intrusion detection and prevention system architecture. Section 3.6 summarizes the intrusion detection and intrusion prevention methods concerning a stable and resilient system operation. Comprehensive questions from the topics intrusion detection and intrusion prevention methodologies and architectures presented in Sect. 3.7, followed by "References" section with references for further reading.

Chapter 4, "Cyberattack Models, Scenarios, and Ontology," refers to the number and the sophistication of cyberattacks and cyberattack scenarios, which mainly cause resource disruption for example through a Denial of Service (DoS) attack that affect system operation, or disclosure of resources for example by an eavesdropping attack that alone cannot disrupt the system operation. Thus, cyberattacks cause serious problems to all industries and public and private organizations because a safe and reliable operation of computer systems, networks, infrastructure resources, and others is a major concern. Therefore, methods and tools are necessary to detect data breaches to respond immediately to the identified cyberattack. In this context, ontology specifies some sort of shared understanding based on description logic by making use of semantic web languages and ontologies for cybersecurity awareness. Therefore, Chap. 4 focuses in Sect. 4.1 on cyberattacks, cyberattack measures, and cyberattacker types in the era of digital transformation and covers the topic cyberattacker profiles in a subsection. Section 4.2 focuses on cyberattack models with subsections on modeling formalism and generic cyberattack models, applicable to several cyberattack scenarios. Section 4.3 refers to cyberattacker behavior modeling with subsections on generic cyberattacker behaviors modeling and cyberattacker simulation model. Furthermore, Sect. 4.4 considers the topic cybersecurity ontology as a formal specification of a shared conceptualization within a knowledge model, which introduces ontology and ontology types in Sect. 4.4.1. Section 4.4.2 refers to cybersecurity ontology, and in two subsections in detail in a generic cybersecurity data space ontology framework and a cyberattack ontology model. Section 4.5 contains comprehensive questions from the topics cyberattack models, cybersecurity ontology, and cyberattack scenarios, followed by "References" section with references for further reading.

Chapter 5, "NIST Cybersecurity Framework and MITRE Cybersecurity Criteria," refers to the need to institute policies and procedures that enforce the way their users access information and interact with network or system resources. Here the NIST Cybersecurity Framework and the MITRE Cybersecurity Criteria come into place. The NIST Cybersecurity Framework is a set of best practices, standards, and recommendations that support organizations to improve their cybersecurity measures. The MITRE Cybersecurity Criteria enable a collective response against cybersecurity threat events, worked out in conjunction with industry and government authorities. It describes the common tactics, techniques, and procedures of advanced persistent threats against organizations' computer systems and networks. NIST's and MITRE's goal is to develop cyber-resiliency approaches and controls to mitigate malicious cyberattacks. Therefore, Chap. 5 discusses the NIST

Cybersecurity Framework (NIST CSF) in Sect. 5.1 with their manifold of possible applications, to improve industrial, public, and private organizations' cybersecurity. Thus, Sect. 5.1 is divided into five subsections that cover the CIS CSC, the ISA/IEC 62443, the MITRE adversarial tactics, techniques and common knowledge, and the NIST 800-53 and the NIST cybersecurity framework. Section 5.2 refers to the NIST cybersecurity framework for critical infrastructure, embedding a best practice use case, applying the NIST Cybersecurity Framework in a critical infrastructure domain. Section 5.3 focuses on the MITRE Cybersecurity Criteria that provides a common taxonomy of tactics, techniques, and procedures, applicable to defend cyberattacks to withstand cyberattackers' activities like unauthorized access, whereas Sect. 5.4 focuses the MITRE taxonomy of cybersecurity criteria. Section 5.5 contains comprehensive questions of the topics NIST Cybersecurity Framework and MITRE Cybersecurity Criteria, followed by "References" section with references for further reading.

Chapter 6, "Ransomware Attacks: Costs Factors and Loss of Reputation," introduces ransomware, a type of malware that typically lock the data on a targeted computer system or user's files by encryption. This cyberattack threat event demands a payment (ransom) before the ransomed data is decrypted and access returned to the targeted user, but ransomware comes in many forms. In this cyberattack, the tools used by cybercriminals are developed and adjusted with new attack pattern, so cybercriminals hit big in terms of payout and notoriety. Such ransomware attacks led to an evolution capitalizing on a growing number of cybercriminals who want to get in. These successful cybercriminals started as cybercriminal entrepreneurs offering ransomware as a service (RaaS) by carrying out ransomware cyberattacks much easier by other cybercriminals, lowering the barrier to entry, and expanding the reach of ransomware. Against this background, Chap. 6 covers the general aspects of cyberattackers and the rapid growth of cyberattacks in Sect. 6.1. Section 6.2 refers to ransomware attacks, which allows unprecedented opportunities gaining ransom payment by potential ransomware attack scenarios and ransomware attacks on OT systems. Section 6.3 refers to cost factors of ransomware attacks and introduces two important metrics to evaluate and enhance the maturity level in disaster recovery in ransomware prevention, to diminish ransom cost. These metrics are the recovery point objective and the recovery time objective, described in two subsections, as well as a powerful design of the recovery point objective and the recovery time objective. The focus in Sect. 6.4 lies on loss of reputation and the question of how to prevent it. Section 6.5 contains comprehensive questions of the topics ransomware, cost of ransomware attacks, and loss of reputation through ransomware attacks, followed by "References" section with references for further reading.

Chapter 7, "Cybersecurity Maturity Models and SWOT Analysis," introduces cybersecurity maturity skills and SWOT analysis. Cybersecurity maturity enables organizations to investigate their cybersecurity awareness, one of the most relevant skills to defend cyberattacks besides a powerful cybersecurity plan. Therefore, measures required quantifying the actual state in cybersecurity awareness and/or cybersecurity defense strategy to identify the essential actions to achieve a target state in cybersecurity. This is where cybersecurity maturity models come into play because cybersecurity maturity models describe an anticipated desired or necessary development path of criteria in consecutive discrete ranks. The prerequisite for this is that the characteristics of the individual maturity development stages are clearly defined beforehand, so user gets an overview of what actually is necessary to achieve the next maturity level. Besides the cybersecurity maturity model, there is another important method available, recording the economic and technical initial situation in an organization, the strength-weakness-opportunities-threats (SWOT) analysis. SWOT is a pragmatic approach capturing the current state of specific and relevant organizational characteristics, to initiate further improvement. In this context, Chap. 7 covers organizations' ability to defend their critical data and business assets from cybersecurity risks in Sect. 7.1. Section 7.2 refers to the maturity index, the Maturity Models, and Maturity Models after ISO 9004:2008 in three subsections. Section 7.3 covers cybersecurity maturity models, while Sect. 7.4 refers to a cybersecurity maturity best practice model. Thereafter, Sect. 7.5 discusses the SWOT analysis, a framework to identify and analyze organizations' strengths, weaknesses, opportunities, and threats. Section 7.5.1 covers SWOT analysis. Section 7.5.2 refers to SWOT analysis best practice in two subsections focusing on a company analysis and a cybersecurity SWOT analysis. Section 7.6 contains comprehensive questions from the topics cybersecurity maturity and SWOT analysis, followed by "References" section, which covers references for further reading.

Chapter 8, "Machine Learning and Deep Learning," introduces machine learning as a broader category of algorithms using data sets to identify patterns, discover insights and/or enhance understanding, and make decisions or predictions. Compared with machine learning, deep learning is a particular branch of machine learning that uses machine learning functionality, and move beyond these capabilities. Machine learning and deep learning algorithms are used in applications that identify and respond to cybercriminals manifold cyberattacks, by analyzing big data sets of cybersecurity incidents to identify patterns of malicious activities. Against this background, Chap. 8 covers machine learning in Sect. 8.1, which consists of four subsections that cover supervised, unsupervised, and reinforcement machine learning, and a comparison of machine learning methods. Section 8.2 discusses machine learning and cybersecurity. This section is divided into three subsections with the topics machine learning-based intrusion detection in industrial applications and machine learning-based intrusion detection of unknown cyberattacks. Section 8.3 introduces the deep learning method, subdivided into three subsections that cover the classification of deep learning methods, the deep Bayesian neural network, and the deep learning-based intrusion detection system. Finally, Sect. 8.4 contains the subject deep learning methods in cybersecurity. Section 8.5 presents comprehensive questions from the topics machine learning and deep learning, followed by "References" section with references for further reading.

Besides the methodological and technical content, all chapters in the book contain chapter-specific comprehensive questions to help students and readers to determine if they have gained the required knowledge. This also allows to identify possible knowledge gaps, which need to close. Moreover, all chapters include references that also act as suggestions for further reading.

I would like to thank Wayne Wheeler, Springer, for his help with the organizational procedures between the author and the publishing house. Furthermore, I sincerely thank all of the authors who have published material about cybersecurity and/or digital transformation, which contribute to this book through citation.

Most notably, I would deeply thank my wife Angelika for her encouragement, patience, and understanding without which this book would have never been written.

Clausthal-Zellerfeld, Germany   Dietmar P. F. Möller

Contents

Chapter 1
Cybersecurity in Digital Transformation

Abstract Digital transformation is a global phenomenon, capturing the attention in every industry and spurring major investment. However, digital transformation is not a single objective; it is a multifaceted approach depending on the goals of the respective industry and digital maturity. Thus, digital transformation is the way of change from a monolithic business approach to fully digitized business concepts. Therefore, Chap. 1 introduces us to relevant definitions on digital transformation and their main technological background. One major concern in digital transformation is the confluence of emerging technologies such as Artificial Intelligence, Big Data and Analytics, Blockchain, Cloud Computing and Services, Internet of Things and Industrial Internet of Things, and others. In this regard, Cloud Computing and Services achieve a new generation of Artificial Intelligence, applied in an increasing number of industrial applications with unprecedented results. Besides this, the Internet of Things connects industrial devices and entities within industrial value chains and infrastructure, however, generating terabytes of data every day, which requires Big Data and Analytics to deal with them. Therefore, the digital transformation has a huge impact on industrial control systems and processes, as well as on the Key Performance Indicator (KPI). Moreover, emerging technologies also will have a huge impact on the realization of circular economy in the industrial sectors, a paradigm in today's digital transformation in industrial, public, and private discussion to reduce the greenhouse effect. Against this background, digital transformation has received a great attention in all industrial, public, and private organizations, which will be fundamentally changing operational business approaches in all sectors. Hence, the transformation of organizations through digitization and emerging technologies can be referred to as a technological wave, such as the third and now the fourth industrial transformation. In this context, the fourth industrial transformation optimizes computerization of the third through digitization, wireless infrastructure networks, intelligent algorithms, and others, which today is known as digital transformation paradigm. Thus, digitization and technologies in digital transformation create, and consequently change, market offerings, business processes, or business models, which result from the use of both. However, the intrinsic complexity of digital transformation also makes cybersecurity awareness a sine qua non. Cybersecurity deals with the presence of cyber attackers with their cybercriminal attack repertoire. Cybercrime causes $ 100 billion and more in damage annually, which means cybercriminal attack prevention has a clear need to address it. However, cybersecurity spans many areas, including, but not limited to, data security, information security, operational security, and others. Moreover, cybersecurity is fundamental to both protecting secret data and information and enabling their defense. Therefore, cybersecurity is understood as a body of knowledge with regard to technologies, processes, and practices designed to protect computer systems, networks, infrastructure resources, and others, from cyber-criminal attacks, damage, manipulation, or unauthorized access. In this context, Chap. 1 introduce in Sect. 1.1 into the manifold aspects of digital transformation with their great impact on industrial, public, and private organizations development, enhancing their business success. Section 1.2 focusses on emerging technologies required and used in digital transformation, presented in twenty Subsections. Against this background, Section 1.3 discusses the challenges of digital transformation in industrial businesses, while Sect. 1.4 introduces applications in digital transformation. Section 1.5 answers the question how to become a leader in digital transformation. Section 1.6 introduces the topic cybersecurity, taking into account knowledge about the interaction of digitization and emerging technologies to gain knowledge about cybersecurity risks through threat event attacks. In this sense, Section 1.6 also focusses on the CIA Triad and the actionable knowledge in cybersecurity that is still paramount. Finally, Section 1.7 describes the challenges of digital transformation to build up a circular economy. Section 1.8 contains comprehensive questions from the topics digital transformation, cybersecurity, and circular economy, followed by references for further reading section.

1.1 Digital Transformation

The cyberspace is endlessly expanding, which offers huge opportunities to digitization in the digital world. In this space, digital data are the raw material. These data are generated, collected, analyzed, processed, and stored. The driver for digital data is the digitization, which offers many advantages to process digital data. Digital data can be manipulated, reproduced, stored, and distributed using data processing systems. These machine-readable digital data make them faster to process and search data. In addition, compression algorithms applied to digital data, which significantly reduce storage requirement, enable the evolution and challenges of the digital transformation. The term digital transformation itself broadly defines the use of computer-based technologies in industrial, public, and private organizations. In industrial production processes, business strategies improve their performance through digital transformation. In this regard, digital transformation is concerned with changes that the digital technologies bring into industry's business models, production, services, or organizational structures. Moreover, this is the most pervasive managerial challenge for incumbent industries of the last and decades coming. In this context, digital transformation is the integration of digital technology in all areas of industrial business models, which fundamentally changes the ways to operate and deliver value to customers. Therefore, digital transformation needs, beside digital technologies, skilled employees and executives, in order to reveal its transformative power due to resultant innovative business models and altered consumer expectations, which has enormous pressure on traditional business models [1–3]. What does it entail in information technology (IT), digitized operations, digital marketing, and new ventures? In IT, existing IT is modernized; in digitized operations, existing business models are optimized; in digital marketing, digital tools for marketing is used; in e-commerce, customer acquisition is introduced, and new business models and intelligent production processes are created, to name a few. In this context, essential criteria with a huge impact on IT are the required capabilities and the Key Performance Indicator (KPI) in the digital transformation era. However, digital transformation is a disruptive process, which open up entirely new business opportunities and rise to completely new customer expectations. Since customer services is a key tenet in industrial survival because the customer experience should be front and center in digital transformation projects. Therefore, digital transformation, on the one hand, results in adjustments of existing business processes and, on the other hand, in fundamentally new and/or different business approaches that provide intelligent services. The trend big data and analytics in digital transformation help to achieve advanced objectives. Nevertheless, the idea of digital transformation is instead of only replicating the existing process into a digital form to transform it into an intelligent and interconnected process, which can be accessible, connected, controlled, and digitally designed [4]. Thus, digital transformation receives an outstanding level of attention [5]. Industrial sectors deal with this narrative, as the digital transformation fundamentally changes their business environment [6]. As a result, the different industrial sectors must transform themselves to survive in their volatile industrial environment [7].

The transformation of industry through technological innovations occurred in several technological waves. These technological waves are the first, second, third, and now the fourth technological wave in industrial transformation. In this context, the fourth industrial transformation optimizes the computerization of the third industrial transformation through a manifold of innovations such as wireless connectivity, wireless infrastructure devices and networks, intelligent algorithms, virtualization, cloud computing and cloud services, and others, which result in the digital transformation paradigm [8, 9]. In this regard, the digital transformation refers to the advent of digital innovations that comprises implanting new combinations of digital and physical components to produce novel products [10]. However, it includes embracing the outcome and the design phases in a broader sense. Therefore, digital innovation is the creation of, and the consequent change in, market offerings, business processes, and/or models, which result from the use of digital technology [11]. Similarly, digital innovation is referred to as a "product, process, or business model perceived as new, requiring some significant changes on the part of adopters, and embodied in enabled by information technology" [12]. Their development aims to transform all resources into services available to customers, partners, and end-users. Services are autonomous and platform-independent entities used to build decoupled applications through the Internet. With the development of flexible service entities and options for adapting to changing requirements, the cloud computing concepts of Everything-as-a-Service (XaaS) facilitate on-demand provision. Furthermore, this integration enables significantly better support in realizing digital business models [13]. Such designed digital ecosystems comprise interconnected and accessible IT resources that act as entities [14]. This digital ecosystem built upon suppliers, customers, trading partners, applications, third-party data service providers, and required essential emerging technologies. As a result, organizations are forced to transform in order to remain competitive in the constantly changing digital ecosystems. However, digital innovations have changed the relevance and scope of the digital transformation process [15] not only to replicate existing processes and procedures in a digital form but use digital and emerging technologies to transform that process into something intelligent and extraordinary. In addition to these technological aspects, studies on strategies [16], organizational change, and business model transformation [17, 18] have created a holistic view of digital transformation.

The confluence of digital and new technologies is fundamentally changing how organizations can operate their business to be competitive to survive in their market. The market is constantly evolving and is leading to groundbreaking digital- and emerging technologies (see Sect. 1.2) such as

- *Additive manufacturing*: Group of technologies that create products by adding material rather subtracting it, based on a digital model. With this model, the manufactured objects made by deposing a constituent material or materials onto a subtractive layer by a minuscale laser. The tools used to layer the material in this procedure are digitally controlled and operated.
- *Artificial intelligence (AI)*: Simulation of human natural intelligence by computing machines. Specific application of artificial intelligence include expert systems, machine vision, natural language processing, and speech recognition.
- *Blockckain*: Record-keeping technology designed to make it impossible to hack the system or forge the data stored on it, making it secure and immutable.
- *Cloud computing and cloud service*: Term for anything that involves delivering hosted services over the Internet. These services divided into four main categories or types of cloud computing: Infrastructure-as-a-Service (IaaS), Platform-as-a-Service (PaaS), Software-as-a-Service (SaaS), and Function-as-a-Service (FaaS).
- *Industrial Internet of Things (IIoT)*: Network of connected intelligent objects to build systems that monitor, collect, exchange, and analyze data.
- *Machine learning*: Creating artificial intelligence (AI), which in turn is one of the primary drivers of machine learning use in industrial, public, and private organization.
- *Virtualization*: Creation of a virtual rather than physical version of an entity, such as network resources, server, storage device, and others.
- Others [11].

These enforce established industrial organizations to transform their business in order to remain competitive [19, 20]. To promote the alignment of resources in the digital transformation implies to create new and/or modify existing business processes, business culture and customer experience of the entire ecosystem, to meet changing business and market requirements. The scope of this development is innovating and/or modifying business processes to meet changing business and market requirements, and to transform all offers into services. These services are available to customers and end-users in the digital ecosystem. One of these services is, for example, predictive maintenance, a forward-looking approach that proactively maintains machines and devices in order to keep downtimes low. For this purpose, predictive maintenance process measures values and data recorded by sensors.

In an ideal case, faults predicted before they have an impact or failure. The actual occurrence of the fault can be predicted proactively initiating maintenance measures at an early stage. In this context, services are autonomous entities to build decoupled applications through the Internet. This results in the integration of Information and Communication Technology (ICT)-supported production capabilities that offer significantly better support in a digital ecosystem model [13]. Furthermore, on-demand services in digital ecosystems have a fundamental impact on business from a vendor-centric view, to survive and thrive where technology is the key economic driver. In this context, a digital ecosystem is a group of interconnected IT resources that act as an entity to the impacts on the future economy like individual services, which result in service aggregation and value chains. This function-based approach corresponds to value networks or individual competitive advantage, which results in the strength of the digital ecosystem.

Data with certain data structures are required to ensure a function-based implementation format. A popular data format is XML that allows structured data handling but any other data format is also suitable, as long as the system can handle it. Moreover, the digital ecosystem model quickly influences changes in various industrial sectors. Moreover, the integration of Business-to-Business (B2B) practices, Business Process Management (BPM), business application, and data within an ecosystem enables an industrial organization to control new and/or old technologies, and build automated processes around them that consistently grow their business. In this regard, BPM is a systemic approach to capture, document, execute, shape, measure, monitor, and steering automatic and nonautomatic processes, to reach coordination and sustainable industrial organizations targets, improving corporate performance, by optimizing and managing business processes of the industrial organization [14]. In this context, the digital transformation and the digital ecosystem interact with each other combining an individual digital transformation with ecological thinking of the digital ecosystem. The strategic goal is to clarify what an industrial organization has to work with: to guarantee they have the right tools supporting their goals and ensure they are being as efficient and effective as possible in achieving those goals. However, digital transformation is not a unified entity; it much more refers to four major characteristic subject topics:

- *Business transformation*: Focus on certain aspects of the industrial organization's business model. Changes in the business model are at fundamental aspects of how to provide value in a particular industrial sector. In essence, businesses are utilizing digital transformation to change traditional business models.

- *Cultural transformation*: Individuals mostly use different approaches. It is challenging getting different users using the same approach and to accept resulting big changes. However, it will pay back if business provides better overall services for customers.
- *Domain transformation*: Occurs when one industrial organization effectively transitions into another.
- *Process transformation*: Means transforming business processes, services, and models by adding emerging technology, which utilize employee's talents, accomplishments, and possibilities.

Against this background, digital transformation transcends traditional roles in development and design, marketing and sales, customer service, and others. In this regard, digital transformation begins and ends with enabling intelligent digital technologies in development and design, to think about and engage with customers from paper to spreadsheets to smart applications, to manage all business processes and digital tasks. Therefore, digital transformation is changing the way business can perform in different shapes to create entirely new business options. In terms of content, digital transformation refers to a holistic organizations change process, driven over time by changes in the organization with regard to value creation and value appropriation [15]. However, from a more general perspective, digital transformation also causes a digital disruption. The term disruption has several connotations. It refers to a specific process that explains how entrants in transformation can successfully compete with the incumbents. The competitive relationship between incumbents and entrants, and the specific means through which the latter enter the market, which are key boundary conditions for digital disruption of digital transformation, rarely exhibited by examples of digital disruption [16, 17].

Digital transformation also requires revisiting everything essential to do, from internal objectives on the shop floor to customer interactions, both vital and physical. This result in the question to answer: How to change business processes to enable better decision-making, game-changing efficiencies, and/or better customer relation with regard to personalization? Hence, collaboration, adaptability, inclusivity, transparency, and connectivity are the important keys to digital transformation efforts. Another question deals with the potential of technological enhancements achieved through digital transformation: What is the technology used capable of, and how can business and processes be adapted to make the most of the technology investments? A possible answer is: It's all about building relationship and listening to people inside and outside IT to create cross-functional alignment around a singular goal with technology at the core [21].

Therefore, advanced competencies must be made accessible, available and aware, the 3A principle, as essential resource and scope in digital transformation. This includes awareness and knowledge in digital and emerging technologies. Key competence in digital transformation also include understanding the potential of emerging technologies, which raise the question: What is the digital- and new technology used capable of, *and* how it can adapt to the required business *and* production processes to make the most out of technology investments *and* transformation?

Against this background, digital transformation efforts are as follows:

- Challenges that are changing products or services to become digital.
- Challenges that are embracing digital to change how industrial organizations engage with their customers, regardless of whether their products are digital or on premises.
- Challenges that are transforming internal infrastructure to change how the industrial organization works.

In this context, digital transformation enables a business to serve better its principal business and stakeholders: customers, business partners, employees, and shareholders. Moreover, digital transformation also leads to an increased agility within IT and engineering teams that help these teams in executing projects and processes in a better and faster way. To achieve these goals, the development stage is essential to create the required digitization for digital transformation. This includes the deviant value logic approach promoted to digitization that causes a digital disruption. It may also call for proactive measures to avoid outmaneuver or counteract obstacles hindering or constraining digitization in the development stage [15]. Furthermore, digital transformation enables computer-based digital and emerging technologies usage to support business to achieve the following features:

- Speed up with development and design of products to minimize time-to-market with new products and services.
- Increase employee productivity in all areas as part of Continuous Improvement Process (CIP).
- Increase sustainable responsiveness to customer requests.
- Gain more insight into individual customer requirements to better anticipate and personalize products and services.
- Improve sustainable customer service, especially in providing more intuitive and more engaging customer experiences.

This sophisticated variety of advanced and intelligent technologies require to understand, develop, and dominate the digital transformation narrative, to transform business processes in the context of digital, innovative, intelligent, efficient and effective processes, as much as possible. Therefore, advanced competences in digital systems and digital network processes as well as deep knowledge in digital and emerging technologies are essential to dominate the digital transformation. Thus, gaining successful usage of digital transformation requires a clear vision and a deep understanding of business models and production processes across the completely industrial organization. The respective key success factors considered are as follows:

- Leadership in digital transformation is set in place.
- Investing in talent and skills development is set in place.
- Ensuring clear communication is achieved.

- Digitizing tools and processes are set in place.
- Cybersecurity awareness is implemented.

These all play key roles in driving success in industrial organizations digital transformation, spurring a growing need for it across industrial sectors. In this context, it may be nearly impossible to know how this innovation will look like at the end of this evolutionary step. It is the process of rapid digitization that started in the late twentieth century and underwent rapid acceleration in the first two decades or the twenty-first century with continuous learning through experience and reiteration along the way, gathering expertise and practice that make the difference in gaining the specific knowledge in digital transformation. As described in [18], "companies and organizations that figure out how to breath big data, how to harness the power of this new resource and extract its value by leveraging the cloud, AI, and Internet of Things (IoT), will be the next to climb out of the data lake and master the new digital land." This includes digital awareness, knowledge in digitization, and emerging technologies that are important to digital transformation.

1.2 Emerging Technologies in Digital Transformation

Digital transformation represents an ambitious and well-founded model for the innovative and technological industrial advancements through emerging technologies [19]. In this context, emerging technologies are technologies that, alone or in combination with other technologies, achieve significant expansions or leaps in possible applications and specific scope of services. In this context, it is characteristic that the industrial sector in the course of rapid technological development undergoes radical innovations and changes [20]. For example, the Internet of Things is an emerging key technology. It creates networks of everyday objects embedded in digital electronics, software, and network connectivity. In this way, users can communicate with objects to control them or receive necessary information [21]. Technologies used in the context of digital transformation are divided into emerging technologies or future technologies and available technologies. Some emerging technologies are shown in Table 1.1.

The aim is to develop emerging technologies in such a way that manufacturability of products becomes more sustainable from an economic, ecological, and social point of view. In addition to competitiveness, the energy efficiency and effectiveness is important in order to ensure the sustainability of the production process, making the economy more resource and energy efficient. In contrast, there is already a variety of technologies available today that serves a wide range of applications. Therefore, attention draws to those emerging technologies that represent the technological foundation of the digital transformation and have a decisive influence on it. These technologies are as follows: Big Data and Analytics, Blockchain, Cloud Computing and Cloud Services, and Internet of Things. The Internet of Things (IoT), used in a wide range of value-added applications, ranging from industrial production

Table 1.1 Examples of emerging technologies

Emerging technologies	
Technology	Capability
Intelligent manufacturing	Through the integral use of big data technologies, AI, machine learning technologies, and others, intelligent manufacturing is a future (emerging) technology manifested in the future project industry 4.0 (I4.0). In this way, critical events predicted or preventive measures for problems expected in the I4.0 production process can be identified and a solution can be provided at an early stage in order to avoid serious effects on the production process [22]. Furthermore, in the future, product development can be interactively adapted to the respective application using intelligent technologies. Genetically intelligent technologies work together with functional machine tool components to create a "sensitive machine." this approach uses measured machine data and simultaneous simulation process data [9]. Research in this area focuses, among other things, on intelligent optimization of production, minimization of resource, and energy consumption by moving toward a circular economy [11].
Material science	Cross-sectional technology of high economic importance that enables new properties or functions or enhances the properties of know materials. Research in material development is indispensable for innovation required technological solutions. These include glass artificial intelligence that uses light to recognize and differentiate icons [23], biomaterials stronger than steel and biodegradable, material capable of absorbing carbon from the atmosphere, silicon X that contains a matrix of silicon nanoparticles and other nanoparticles, material substances for efficient batteries, and others. Another new area of materials research focuses on two-dimensional material, graphene, a layer of crystalline carbon just one atom thick. Research on new materials also focuses on needs such as haptic interaction in communication or intelligent solutions that conduct discharges at low thresholds, or composing material in such a way to minimize the consumption of energy and resources during processing.
Nanotechnology	Enables the creation of objects on the tiny nano- and sub-nano scale. Nanotechnologies are essential prerequisites for novel components and concepts in digital electronics, multicore semiconductor chips, optoelectronics, pervasive computing in ICT, biological agents, Nano scalable sensors and actuators, as well as new technologies for a digital circular economy and others [24]. Beside this, nanotechnology has potential advances and impacts on agriculture, the environment, and human health.

and healthcare to the so-called smart home and other. In a study on IoT, around 300 application scenarios in different environments were examined [25]. In this regard, the most important concerns about IoT networks are the security of the connections of IoT nodes and the Transport Layer. However, new IoT hardware platforms view this absence of security by containing hardware-increased cryptographic capabilities. Despite all pros and cons, the application of emerged technologies are the principal driver of digital transformation, are briefly described in the following sections.

1.2.1 Artificial Intelligence

The term artificial intelligence (AI) was coined in 1956 by John McCarthy and was defined as the science and engineering of making intelligent machines. AI is a branch of computer science concerned with building smart machines capable of performing tasks that typically require human intelligence. In other words, AI is the capability of a machine to imitate intelligent human behavior. An ideal intelligent machine is a flexible rational agent that perceives its environment and takes actions that maximize its chance of success at an arbitrary goal. Furthermore, the term AI is applied when a machine uses cutting-edge techniques to completely perform or mimic cognitive functions that are intuitively associated with human behavior, like learning and problem solving. In a more global sense, AI is

- Academic field of study on how to create machines and software that are capable of intelligent behavior.
- Constituted by machines and/or software.
- Study of design of intelligent agents, whereby an intelligent agent is a system that perceives its environment and takes actions that minimize its chances of success.

In this context, AI splits in the literature into the categories listed in Table 1.2.

AI is also one of the emerging technologies that tries to emulate human reasoning and one of the core technologies in digital transformation. AI helps businesses to scale up through its integration and deployment across a variety of sectors such as national security, industrial security, healthcare, logistics, education, and others. Furthermore, AI leverages computing machines to mimic the problem-solving and decision-making capabilities of the human mind, based on a set of methods to mimic human intelligence. The four types of AI are as follows [27]:

- *Reactive machines*: Capable of only using its intelligence to perceive and react to the world in front of it. A reactive machine cannot store in a memory and as a result cannot rely on experiences to inform decision making in real time.

Table 1.2 Strong versus weak AI, own presentation based on [26]

	Strong AI	Weak AI
Definition	Form that has same intellectual ability as a human or even surpasses him in it Intended "think" on its own, which means being aware of context and cognitive issues to make decisions	Developed or used for specific applications Represents a collection of technologies that relay on algorithms and programmatic responses to simulate intelligence with regard to a specific task
Capabilities and domains	Communication in natural language Decision-making in case of uncertainty Learning Logical thinking Planning Use these abilities to achieve a common goal	Character recognition Expert system Navigation system Suggestions for corrections in searches Voice recognition

- *Limited memory AI*: Has the ability to store previous data and predictions when gathering information and weighing potential decisions. Essentially looking into the past for clues on what may come next. This AI type is more complex and has greater possibilities than reactive machines by three major machine learning models that utilize limited memory AI:

  - *Reinforcement Learning*: Learns to make better predictions through repeated trial-and-error approach (see Chap. 8).
  - *Long short-term memory (LSTM)*: utilizes past data to predict the next item in a sequence. LTSM view more recent data as important when making predictions from further in the past, still utilizing it to form conclusion.
  - *Evolutionary generative adversarial networks (EGAN)*: Evolves over time, growing to explore slightly modified paths based on previous experiences with every new decision. This model is constantly in pursuit of a better path and utilizes simulations and statistics, or chance, to predict outcomes throughout its evolutionary mutational cycle.

- *Theory of mind*: Technological and scientific capabilities necessary to reach this level of AI.
- *Self-awareness*: Relies on human researchers understanding the premise of consciousness and then learning how to replicate it to implement it into machines. It possesses human-level consciousness *and* understands its own existence, as well as the presence *and* emotional state of others. It is able to understand what others may need, based on not just what they communicate to them but how they communicate it.

The most important outcome of AI is machine learning and deep learning, whereby deep learning is one approach out of machine learning techniques. Machine learning as an application of AI provides systems with the ability to automatically learn and act without being explicit programmed. In this regard, machine learning is an approach to data analysis that involves building and adapting models that allows programs to learn through experience. Machine learning also involves the construction of algorithms that adapt their models to improve their ability to make predictions. In this regard, machine learning technique is divided in several types, called predictive or supervised learning, descriptive or unsupervised learning as well as reinforcement learning, as shown in Chap. 8.

1.2.2 Additive Manufacturing

Additive manufacturing (AM) is a new technology in the industry, involving both old and new technologies that constantly evolve, to bring important changes in the manufacturing supply value chain, also known as 3D printing. The term 3D printing covers different technologies and a variety of processes in which material joined or solidified under computer control creating 3D objects, with material added together, typically layer by layer. AM is a transformative approach to industrial production enabling the creation of lightweight, extremely strong and heat-tolerant objects in specialized applications, e.g., aerospace sector, automotive sector, medical sector, and other. AM methods applies in Industry 4.0 (I4.0) and thus in digital transformation. It is widely used to fabricate small batches of customized products that offer construction advantages, making use of new technologies such as stereolithographic, selective laser sintering, 3D printing, inkjet processes, fused deposition modeling, laminated object manufacturing, and other. AM use CAD software or 3D object scans to control the AM hardware components to build precise 3D objects, based on layer-by-layer manufacturing, adding layer-upon-layer of material, whether the material is plastic, metal, concrete or any other. In this context, AM technology represent a promising approach for true 3D micro manufacturing, employed efficiently to fabricate complex 3D micro-components. AM technology has tremendously improved in the last decade and progressed from primarily being used to prototype and fabricate individual components to end used products. However, it will take some time for the AM industry to broaden its scope from specialized use cases to mainstream equipment and consumer products. Nevertheless, advances in use of metal-based AM enable possibilities for innovative fabrication in I4.0. Indeed, developing materials in 3D printing, a priori data sets required that sufficiently specify the physical, chemical, electrical and other characteristics of the material developed. For known material, there are extensive data in data base systems available. Based on existing data from diverse and different materials, these data build the basis for the required steps in material design, to analyze these with regard to their characteristic properties in order to derive property hypotheses on which new materials be composed. Since these are large amounts of data, appropriate data analysis methods are used. For this purpose, data reduction carried out in order to identify the most valuable data. Big data and analytics used to predict which data is most important for big data management in machine learning, achieving transparency, efficiency and effectivity in the development of new materials. The datasets extracted are basis for further profiling of the new material searched, from a materials science perspective. By means of this profiling, feature extraction carried out based on a suitable dimension in the context of machine learning. The attributes of the underlying feature vector are the new material properties taken into account, identified from clusters of the records by machine learning. Depending on the characteristics (dimensions), the learning process can become complex and time consuming. However, the human expert is very important, because he finally interprets the clustered feature sets and evaluates them with regard to the new material characteristics. In order to reduce the effort involved, interactive learning techniques used to identify and highlight the most meaningful features of the data to support intuitive visualization and ultimately plausibility checking.

Beside the foregoing introduced material design approach, a variety of different AM processes is available to manufacture a customized product as per the customers' requirements in lesser time and cost. Due to its flexibility in design and manufacturing this technologies quickly launch new products, using the manifold of available AM technologies [28]. However, additive manufacturing in digital transformation is still evolving and the complete picture is not available right now. The types of AM today well known are:

- *Advanced/composite materials*:

  - Advanced materials enables creation of highly precise blends of materials.
  - Composite materials creating more performance breakthroughs, and reducing material tradeoff decisions precisely varied physical and chemical properties.

- *Automation/robotics*:

  - Automation improve consistency of work across production units.
  - Robotics limiting risk, overhead, and waste for more consistent, faster, and cheaper products.

- *Laser machining/welding*:

  - Laser technology allows for rapid and high-precision processing of parts.
  - Welding.

- *Nanotechnology*:

  - At the forefront of many industries designers are aiming to add more functionality into smaller sizes as possible.

- *Network/IT integration*:

  - Network access to every part of the manufacturing process provides instant, pinpointed notifications on issues and potential repairs, saving time and costs.
  - IT integration provides Internet connectivity between machines and systems.

AM in digital transformation is also used for low volumes of complex products like functional properties [29].

1.2.3   *Augmented Reality*

Augmented reality (AR) is a technology that layers computer-generated enhancements atop an existing reality to make it more meaningful through the ability to interact with it. The three properties of AR are presenting virtual and real objects together in a real environment, allowing interaction with virtual and real objects in real time, aligning virtual objects with real objects. The additional information brought by the augmentation could assist users when performing real world tasks [30]. AR system contains four hardware components: a computer, a display device, a tracking device, and an input device. The computer is responsible for modeling augmentations and controlling the devices connected and adjusting the position of augmentations in the real scene with regard to the position of the user by using information gathered from the tracking device. Typically, a stationary transmitter radiates electromagnetic signals intercepted by a mobile detector attached to the user's head. The display device is required to display the augmentations on top of the user's real vision. The most widely used technologies in AR are the head-mounted display, the hand-held display, the spatial display, and intelligent glasses or lenses, viewing details about images, nearby objects, and others, to take appropriate actions. The used tracking device is for tracking the exact position and orientation of the user, because it is important that augmentations fit properly to the desired position. Finally, the used input device enables the user to interact with the AR system. In this regard, the four hardware components used in advanced AR-systems are: space mice's, data gloves, headphones, and haptic devices. Haptic devices encompass rich sensory information user gleans from holding an object. Data gloves can monitor the status of the user's fingers. An attached separate tracker is to the user's wrist to monitor its position and orientation.

In an industrial maintenance process, traditional the break between the digital and physical world mitigates through AR. The solution integrates AR technology and human-dependent tasks into one digital solution to achieve a higher level of efficiency and effectiveness, reduce possible human errors by providing workers with real-time information to improve work capability and decision-making on demand, and make maintenance data usable in real time, which means

- Leverage manufacturing machine data to accurately predict the best time to perform manufacturing machine maintenance and avoid downtime before it effects production.
- Use AR to guide technicians through their maintenance workflow and to record their proof-of-service and inspection findings digitally, with no risk of human error.
- Digitally document data from the manufacturing machine processes and environments, making it available and reusable to other business units in real time.

In this regard, AR data trapped in 2D transferred into 3D. Thus, AR represents a bridge between reality and digitalization, whereby the cognitive distance significantly is reduced, and tied mental capacities are released. Thus, AR systems capabilities include information overlay, knowledge capture, and training. Information overlay allows the AR system to present information about the equipment, a technician is working on, to help streamline the maintenance and/or repair process. In contrast, knowledge capture enables a technician to capture information, showing the current state of a problem or document a maintenance and/or repair and then store it in the e-book of knowledge. Moreover, technicians can share videos of the repair with other technicians, who encounter the same equipment. Furthermore, AR is also used as computer vision and man–machine interaction environment, transforming volumes of data and analysis into images of animation that are overlaid on the real world scenario. For example, technicians may receive instructions on how to replace a particular part or component of a complex system, displayed directly in the technician's field of sight using units such as display devices. Another application is virtual plant-operator training that uses a realistic, data based 3D environment with AR system display devices, to train plant personnel to deal with regular and/or dangerous or high-risk system states. In this context, operators learn to interact with machines by clicking interactively on a cyber-representation. They also can change parameters and retrieve operational data and maintenance instructions for further decision-making purposes. Besides this, AR also has many options of use cases in enterprise workforce and operator training in conjunction with AI and machine learning. In case of operator training, a realistic 3D environment of the manufacturing shop floor together with AR glasses or lenses is used to enable training the operational manufacturing job floor activities, by clicking on a cyber-representation device. They also can change parameters and retrieve operational data and maintenance instructions. The main benefits of AR in digital transformation shown in Table 1.3.

1.2.4   Autonomous Robots

Traditional industrial robots placed in designated spaces, programmed to perform repeatedly and continuously predefined sequences of action to deal with complex assignments. However, robots are evolving for even greater and smarter applications with regard to embedded abilities of computing communication and control. In the near future, intelligent and thus more Autonomous Acting Robots (AAR) designed to interact with its environment on its own decisions, achieved through intelligent sensor technologies. This allows performing actions accordingly without human intervention. These robots are adaptive and flexible, cooperative and intelligent through the usage of AI methods, interacting with each other, and work safely side-by-side with human workers to learn from them, based on watch-and-learn or see-and-do strategies. Against this background, intelligent and interconnected autonomous robots have a greater range of capabilities than those used for assembly in the manufacturing environment of Industry 3.0. In Industry 4.0 (I4.0), manufacturing environments with intelligent robots learn dynamically and interact collaboratively

Table 1.3 Benefits of AR tasks and capabilities

AR task	Capabilities
Assembly	Involves manipulation and joining of separate parts to build a whole product in manufacturing. For complex products (high number of different and specific parts), assembly can become a rather difficult tasks to manage.
Collaboration	Important in product design and development since both processes consist of many iterative steps, e.g., pieces of a product have to fit with each other geometrically and functionally and be prepared for production and servicing processes, which cannot archived by a single operator person.
Maintenance	Most crucial process in manufacturing involves activities such as alignment, analysis, assembly, installation, rebuilding, removal, repair, or servicing of engineered systems.
Training	Crucial to improve capability, productivity, and performance of technicians. However, using AR for training is more all-encompassing. It includes tapping into AR applications to pull up a schematic map or diagram. It also could also include collaborations with another expert connected remotely.

and autonomously with each other, representing the required crucial elements of the digital transformation in innovative, intelligent and collaborative interconnected manufacturing shop floors. They are allowing easier manufacturing of different products by recognizing the components of each product. This segmentation allows decreasing production costs, reducing production and waiting time in operations. Additionally, autonomous robots are unique in manufacturing systems, especially in optimization of design manufacturing and assembly phases. For example, assigned tasks divided into simpler sub-problems and then constitute a set of modules in order to solve each sub-problem task. At the end of each sub task completion, integration of the modules to reach an optimal solution is essential. One of the sub technologies autonomous robots achieved from co-evolutionary robots that are energetically autonomous are scenario based "thinking" and reaction focused working principles [31, 32]. They enable new effectivity and efficiency and change the way to fabricate products and organize the manufacturing shop floor. These collaborative, autonomous robots are interconnected enabling them to work together and automatically adjust their interactions to fit the next unfinished component or product in line, and being aware about individualization of the products manufactured. In this context, these robots are capable of interpreting (learning) and deciding (decision-making), using their intrinsic ever-improved predictive model of the world to become an ever better interacting collaborative robot. In this regard, autonomous robots are programed machines, based on dynamic AI (see Sect. 1.2.1) to perform tasks with no human intervention or interaction. They can vary significantly in cost, dexterity, functionality, intelligence, mobility, as well as size and can recognize and learn from their surroundings and make decisions independently. However, autonomous robots expected to grow over the next five to ten years within supply chain operations with regard to individualization of customized products, potentially dangerous or high-risk tasks. Thus, autonomous robots will become ubiquitous in the supply chain of the future, and advancements make them operate with more human-like abilities. For example, improvements in haptic sensors will allow autonomous robots to grasp objects ranging from fragile assembly objects to multisurface assembly objects, without changes in programming or autonomous robotic components. This require improvements in haptic sensors and computer vision for safe interaction and object recognition, as well as AI and machine learning (see Sect. 1.2.1 and Chap. 8). Furthermore, autonomous robots enhance problem solving and learning analytics to enable being responsive in real time by integrating developments in AI and machine learning, being responsive to audio, haptic, thermal, and visual capabilities.

Connecting autonomous robots to a central server, a database or a programmable logic controller allows them to take coordinated and automated actions in greater extent than ever before. Hence, material or components of products, transported across the manufacturing shop floor through autonomous mobile robots, avoiding obstacles, coordinate with fleet mates and identifying where pickups and drop-offs needed in real time [33]. Therefore, autonomous intelligent robots primarily drive supply chain innovation and value in the digital transformation by reducing direct and indirect operating costs and increasing revenue. Against this background, autonomous robots enhance primary potential benefits, such as [34]

- Increase efficiency and productivity.
- Reduce error, re-work, and risk rates.
- Improve safety for workers in high-risk work environments.
- Perform lower value mundane tasks so that humans can work collaboratively to focus on more strategic efforts, which can't be automated. This potential benefit is an important issue enhancing the I4.0 paradigm in the direction toward Industry 5.0 [35]. I4.0 represents a solid ambition and a sound guiding principle for the innovation and further technological development of industry in digital transformation. Industry 5.0 complements the existing I4.0 paradigm by highlighting research and innovation as drivers for a transition to a sustainable, human-centric, and resilient industry [36].
- Enhance revenue by improving perfect order fulfillment rates, delivery speed, and, ultimately, customer satisfaction.

There are secondary potential benefits:

- Enhanced workers value through focus on strategic work instead of mundane tasks.
- Focus on workers safety by minimizing work in hazardous areas for workers.
- Boosted corporate brand by signaling leading-edge practices and implementation of innovative technology.
- Exponential learning by collecting and analyzing machine data.

1.2.5   Big Data and Analytics

Data is core element in digital transformation and key to manufacturing companies to take good decisions about products, services, employees, strategy, and more. Using data efficient and effective is therefore essential. The term big data is dazzling and its edges are not only fuzzy but permanently in development, but also used to describe large amounts of unstructured and semistructured data sets from a variety of sources. Big data is an information asset, characterized by high volume, velocity and variety to describe the characteristics of information, achieved by specific technological and analytical methodological requirements. This requires proper use of information for the transformation of information into insights that create economic value to organizations. In this regard, data modeling and analytics are integral part of almost any data-driven decision making in the digital transformation era, where big data technologies are shifting from data collection to data analytics and outcome [37]. However, big data represent data that is just too large to be managed by traditional databases and processing tools. The problem of big data lies in the diverse data structures and that they are difficult to analyze or incorporate in a traditional structural database. Not only collection of data increased like an avalanche but also the possibilities of their evaluation. Hence, it is no wonder that at the same time as the advent of big data rises, the corresponding advances in data processing increase. Nevertheless, analyze such gigantic amount of data no longer be analyzed without advanced computational aids that even the usual methods of digital data processing are no longer sufficient. Decisive for this trend are advances in the fields of computer science and applied mathematics. In digital transformation industrial Original Equipment Manufacturers (OEM), and their tier suppliers need to analyze data from a variety of sources to benefit from the great advantages in the Industrial Internet of Things (see Sect. 1.2.10) and its related domains, such as cloud computing and cloud services (see Sect. 1.2.7), edge- and fog computing (see Sect. 1.2.8), and other.

Beside knowledge like customer trends, operational effectiveness and efficiency, and other, data generated from all sorts of sources, and data become standard to support real-time decision-making, optimizing manufacturing quality, improving services, and others, whereby big data can have the characteristics shown in Table 1.4.

Improved processing power of machine learning in data analytics, enable more out come of all of the data. Another important issue in big data analytics is real-time analytics and reporting. Some interesting reports in this regard can be found in [38–42]. Thus, the first step in big data processing is data reduction. Storing and analyzing any more than hundreds of megabytes per second deemed most valuable become impractical with current technologies. Therefore, big data analytics requires sophisticated and smart predictive informatics tools to determine which data are most significant to manage big data, achieving transparency and productivity within digital transformation applications. However, many traditional manufacturing shop floors are not able managing big data due to the lack of smart analytic tools. As more AI and machine learning based software integrated in the I4.0 industrial production lines, predictive technologies can further intertwine intelligent algorithms with digital technologies and tether-free intelligence. These technologies will be used to predict product performance degradation, and autonomously manage and optimize product fabrication needs, e.g., adaptive learning for machine clustering on the job floor [43]. Thus, big data and analytics help manufacturers getting actionable insights in smarter decisions and better business outcomes [44]. Therefore, big data analytics is an attractive topic for almost every

Table 1.4 Characteristics and capabilities in big data

Characteristic	Capability
Volume	Describes the size of data and hence the ability to analyze large volumes of data. The larger the data pool, the more one can trust its forecasts.
Velocity	Concerned with the speed of data access and processing, which means the speed data comes into the system, and how quickly it requires analyses. Some data require in-flight or in-memory analysis; other data may be stored and later analyzed.
Variety	Describes the degree of complexity and heterogeneity in data.
Veracity	Problems with big data appear beyond collecting and storing vast amounts of data and analyze the data stored, using the three V's and consider is the data actually true. Thus, veracity represents the inherent trustworthiness of data.
Value	Big data application designed to increase enterprise value. Investments are made where added value arises.

manufacturer in the digital transformation era. However, in addition to the storage of large volumes of data in highly distributed systems, the variety of structures is another difficulty that be addressed to a limited extent with classic data sources. In relational databases, binary data formats be saved using the data type binary large objects, but this form of data management is mostly unsuitable for larger data files. Managing large numbers of distributed actions, other database concepts are required, such as the NoSQL database technologies. NoSQL enable efficient storage and analysis of data, but technologies are required to better analyze large and multistructural data. These technologies scale, e.g., horizontally, and distribute the data to the machines involved within the cluster. Distributing the data in this way, analyzes can be carried out in parallel on sub-sets. First, this results in processing the data partition on a local node in clusters. A second step summarizes the partial results. The advantages of a distributed calculation are the parallelization and the data locality.

Furthermore, batch processing is used in big data analytics, collecting data over a certain period-of-time and then processed together. However, this method is not suitable for time critical, such as real-time detection of anomalies, e.g., in sensitive business data. Anomalies are rare observation events that raise suspicions by differing significantly from the majority of the data. Thus, methods are required in which data streams continuously analyzed, which enables near real-time results while data is still in motion. There are different algorithms for anomaly detection available shown in Table 1.5.

Table 1.5 Algorithm for anomaly detection

Algorithm for anomaly detection	Characteristics
Welford algorithm	Single-pass method for computing the running variance or running standard variation. If used in anomaly detection a simple modification required by introducing a lower and an upper limit expressed as X of a standard deviation. The higher X, the more false negatives assumed; the lower X, the more false positives obtain. However, the choice of X is not trivial. A simple implementation of the Welford algorithm is given in [45].
Z-score method	Assuming value of anomaly detection data is a Gaussian distribution with some skewness and kurtosis. Skewness is the extent to which data distribution is symmetrical. If distribution for data stretches toward the left or right tail of the distribution, then the distribution refers to skewed. Kurtosis is a measure of whether the distribution is too peaked, meaning a very narrow distribution with most of the responses in the center [46]. Hence, anomalies are the data points far away from the mean of the population.
Random first algorithm (RFA)	Autonomous machine learning algorithm for anomaly detection that assigns an anomaly value to each data record. Anomaly values greater than three standard deviations from the mean standard deviation are usually considered unusual or anomalous. As its name implies, it consists of a large number of individual decision trees that operate as an ensemble. Each individual tree in the RFA spits out a class prediction, and the class with the most votes becomes our model's prediction.

1.2.6 Blockchain

A blockchain is a distributed, public database used to manage transactions. The term chain in blockchain refers to the chronological order, in which transactions added or executed. Blockchain is an electronic ledger of digital records, events and transactions, cryptographically hashed authenticated and maintained through a distributed or shared network of participants, using a group consensus protocol. Instead of being stored centrally, blockchain is a virtual database that is stored in a network, where each user within the network has a local copy of the database. The decentralized structure and the use of cryptographic procedures ensure that a user can secretly manipulate no information in the database. In addition, each transaction be sent as a cryptographically protected block to the entire network and be verified (proof-of-work method). With blockchain it is possible to implement fully automated transactions in different business areas. This process is reliable and secure in a common sense, and extremely fast. Therefore, employment of blockchain technologies and the possibility of applying them in different situation enable applications through increased efficiency and security; enhanced traceability and transparency and reduced costs [47].

1.2.7 Cloud Computing and Services

Cloud computing is a huge and highly scalable deployment of compute and storage resources, accessible from anywhere, which are used by multiple industrial sectors. Cloud computing and centralized data processing be today's predominant architectural paradigms. The providers of cloud resources incorporate assortment of pre-packaged services, which also be used for, e.g., Internet of Things (IoT) operations, making the cloud also a platform for IoT deployments. Furthermore, cloud computing offers highly scalable resources and services to tackle complex analytics. In this regard, cloud computing helps industrial application and environments to process the growing volumes of data and analyze them successfully to automate and optimize processes. Moreover, cloud computing includes technologies and business models of innovative IT resources. Instead of implementing own businesses IT resources in their organizations data centers, industrial organizations using cloud computing rescues in different cloud abstraction layers. For the public cloud, differences, tradeoffs and use cases for each level of Cloud-as-a-Service (CaaS) model exist. This model is available through the Internet, whereby the business model represents how an industrial application and/or environment can generate value for its users, and ensure a return on investment (ROI) for the industrial organization. Thus, industrial organizations can reduce their long-term IT capital expenditures by deploying cloud computing as their IT resource. IT resources of various types flexibly deployed in a service-based manner referred to as Everything-as-a-Service (XaaS) model. This model approach has several classes of cloud services [19], as shown in Table 1.6.

Table 1.6 Characteristics and abbreviations of cloud services

Cloud service	Abbreviation	Characteristics
Everything-as-a-service	XaaS	Describes a category of services related to cloud computing and remote access in a digital ecosystem. XaaS is available over the internet.
Software-as-a-service	SaaS	Represents the top layer in the cloud model, where provider provides its own application for the users.
Platform-as-a-service	PaaS	Provides a programming model and developer tools to create and execute cloud-based applications, available over the internet.
Infrastructure-as-a-service	IaaS	Provides basic IT resources such as computing power, storage or network capacities, available over the internet.
Business process-as-a-service	BPaaS	Delivers business process outsourcing services sourced from the cloud and constructed for multitenancy, available over the internet.
Function-as-a-service	FaaS	Application packages developed, managed and executed by developers. No separate infrastructure required for the management, provided by service provider. Functions deployed in isolation in so-called containers.

The highest level of abstraction is the SaaS model. Within this model, the cloud provider offers established complete customizable software applications to the users. Users establish accounts with the provider to gain access to these applications through a network, sharing hardware and platform IT resources, but without noticing or interacting with each other. The usage charge of SaaS is on a monthly fee, based on the number of users and application features. Many business sectors, e.g., Customer Relationship Management (CRM) or Enterprise Resource Planning (ERP), are supported by cloud services. Common commercial examples of SaaS include Salesforce, Dropbox, SAP Concur, Zoom, and Microsoft Teams.

A lower level of abstraction in the public cloud is the PaaS model. Different applications and tools are available, which the user can access across the Internet through a web browser. Thus, users establish accounts with the PaaS provider, and the user pays a recurring monthly fee similar to the SaaS model. While SaaS holds created content on the provider's side, PaaS leaves content, for example, developed applications and data, on the user's side. Let's assume a PaaS development create an application. That application remains the property of the user and the user can sell, deploy and use it if desired, even without the underlying platform. Common examples of PaaS are software development frameworks and toolsets such as Google PaaS App Engine, Heroku, Microsoft Power Apps, or Salesforce, as well as orchestration services such as Amazon-Web-Services-Platform, Elastic Beanstalk or Red Hat OpenShift.

The low level of abstraction in the public cloud is the IaaS model, which works as a virtual data center in the cloud, hosting applications and data. Industrial organizations IT teams use IaaS to assemble a virtual infrastructure of cloud resources and services, capable of operating an application and available to employees, business partners and users. The primary benefit of IaaS is to shed costly local data center infrastructure in favor of flexible cloud resources that are available and paid for only as needed. Public cloud providers offer a comprehensive array of infrastructure services including Amazon-Web-Services-Platform, Microsoft Azure or Google Cloud. Other IaaS providers include Rackspace and DigitalOcean.

Finally, the lowest level of abstraction in the public cloud is the BPaaS model. BPaaS allow customers to outsource all of their processes to a cloud provider and implement them through business process technologies. Therefore, the provider offers all of the IT resources but not the IT-based services customers' needs to support their business processes. Thus, BPaas abstract more from IT resources and focusses on the customers' business processes. Common applications of BPaaS include primarily the area of personal management and controlling as well as employee pay slips to simplify document management, procuring and contracting, as well as advertising and marketing.

Industrial organization are already using cloud-based services for some organizations and analytics applications, but within the digital transformation in the context of I4.0 that are more production-related, whereby undertakings will require increased data sharing across sites and industrial organizations boundaries. At the same time, the performance of cloud technologies will improve, achieving near real-time reaction times. As a result, machine data and functionality are increasingly deployed to the cloud, enabling more data-driven services for manufacturing shop floors in the digital transformation era. However, from a more general perspective, data is rarely static and often moves from where users are collecting and using it to the cloud or to a central data center for analysis, processing, and storage. Nevertheless, data centers and clouds are often far from where the data are collected in the digital transformation industrial environment, whereby transmission of data gathered takes time and inserts latency and inefficiencies into data processing. Therefore, a new computing form is required, which act in near real-time. This new approach is edge computing. Edge computing is the way to move computing and storage closer to where data generated and used. Since data generated increasingly at the edge of industrial organizations networks, it is more efficient to process data at the edge of the network, or downstream data on behalf of cloud services and upstream data on behalf of IoT services (see Sect. 1.2.9).

1.2.8   Edge- and Fog Computing

Edge computing is the deployment of computing and storage resources at the location where data generated. That puts compute and storage at the same data source point at the network edge. Edge-as-a-Service (EaaS) provides a solution for processing data at the edge, which enables users to affordably implement, manage, and scale their edge computing devices without big support of technical staff, and delivers available storage for data protection. Edge computing applies to anything that involves placing service provisioning, data and intelligence closer to users and devices. Thus, the edge approach emphasis reducing latency and providing more processing of data close to the source. Therefore, edge computing used in a spectrum of computing and communication applications and technologies to place computing workloads neither on-premises nor in the cloud, but at the edge.

In practice, an application with several servers and some storage devices installed within a manufacturing shop floor, to collect and process data produced by sensors within the shop floor, is an important edge computing use case. The produced sensor data often feed into some sort of local gateway, which use that data to take an action that needs to happen quickly, such as deciding to continue or to stop the production due to a mismatch of parts assembled at the machinery station. This process is also considered, as Industrial Internet of Things (see Sect. 1.2.10), with its connected devices. The result of any such processing then is sent back for review, archiving, and to be merged with other data results for broader analysis.

Beside edge computing, the term fog computing exist that originally was coined by Cisco. Cisco was a hardware complentator to leading companies to the rapidly growing category of cloud platforms. The rapid growth in IoT applications raises the need for decentralized data processing closer to devices. Hardware companies saw an opportunity to lead this fundamental shift in the computing paradigm. Cisco created a decentralized technology, named fog computing, to manage data via distributed mini-clouds, located near the physical devices, while remaining a viable complementator in the cloud ecosystem [48]. The perspective of Cisco was fog computing is considered as an extension of the cloud computing paradigm from the core of a network to the edge of a network computing infrastructure in which data work. More specific, fog computing describes a highly virtualized decentralized computing infrastructure in which data, applications, computing, storage, and networked services are located between where data originates, the end devices and traditional cloud computing data centers. Its goal is to improve clouds' intelligence, processing, compute and storage capabilities closer to the data for faster analysis and processing. Like edge computing, fog computing eliminates inefficiencies that come with data transmission and solves privacy and security issues inherent in data transmission.

In practice, fog-computing environments can produce huge and distributed amounts of sensor or IoT data, generated across expansive physical areas that are just too large to define an edge. Thus, fog computing can be used in smart utility grids, where data can be used to track, analyze, and optimize the system utilities and services, and guide long-term planning. A single edge deployment is not effective and efficient enough to handle such a load. Fog computing can operate a series of fog node deployments within the scope of the environment to collect, process and analyze data. In this context, edge computing is any computing and network resources along the path between data sources and cloud data centers [49, 50].

However, edge and fog computing differ to the location of intelligence and computing power. With fog, intelligence resides on the local area network (LAN), data moves from endpoints to a fog gateway and then routed for processing. With edge computing, intelligence and computing are on an endpoint or a gateway. Edge devices determine whether to store data locally or send it elsewhere for additional analysis. Fog computing is more suitable and has a broader, more holistic view of the network.

1.2.9 Internet of Things

The Internet, created in the second half of the twentieth century, introduces a new paradigm of communication caused by using digital technologies. The Internet is a global system of interconnected computer networks that use the standard transmission control protocol/internet protocol (TCP/IP) to serve billions of worldwide users daily. It is a network of private, public, academic, business, and government networks, from local to global in scope. Originating from the Advanced Research Projects Agency Network around 1970, the Internet became available in the 1980s. By 1990, it had grown from the initial communication framework into the most popular network in use and had its latest achievements, by the creation of the social media, also called social networks, such as Facebook, Twitter, and LinkedIn, to name a few.

The Internet of Things (IoT) is an information network of Internet-connected physical devices (things, objects, entities) that collect and exchange data using their embedded components, which characterizes the unprecedented growth of the Internet. This allows interaction and cooperation of devices (things, objects, and entities) to reach common goals, covering the real and the virtual worlds. A device in the IoT is any possible item in the real world that joins the communication chain. Each device is uniquely identifiable through its embedded unique identification code and is able to interoperate within the existing Internet infrastructure. Therefore, the initial main objective of the IoT was to combine communication capabilities characterized by data transmission. However, there are different ways of creating value by IoT applications, mostly shaped through industry specific dynamics. In this regard, the IoT appears one step further on the path to ubiquitous computing. This is possible with the introduction of radio frequency identification (RFID) tags, wireless sensor networks (WSNs), but also other technologies such as robotics, nanotechnology, and others, to enable IoT services in industrial organizations. RFID uses electromagnetic field to transfer data for automated identification and tracking of tags attached to IoT devices. RFID tags attached at IoT devices contain relevant and specific information about the tagged component(s). In the manufacturing industry, RFID is adopted in supply chain management, production scheduling, and many others. WSNs represent spatial-distributed autonomous nodes that can sense the manufacturing environment. However, WSN can not only monitor the condition of machines or devices on a manufacturing shop floor but also support multihop wireless communication.

Another concern of IoT is how to support the connectivity of heterogeneous objects, things or entities when a huge number of them is connected. These technologies make the IoT services an interdisciplinary field where most of the human senses are somewhat reproduces and replaced in the virtual world [51]. Furthermore, the IoT embedding computing anywhere and programming it to react automatically. In this regard the Internet Protocol version 6 (IPv6) uses a 128-bit address which theoretically allows access to 2^{128} IP addresses for identification and location [9, 52]. Thus, IPv6 allows larger address space than the earlier IPv4. IP addresses identify nodes in the Internet and serve as locations for routing. In the IoT a large address space is needed to cover the identification of the tremendous number of connected objects/things.

1.2.10   Industrial Internet of Things

The Industrial Internet of Things (IIoT) is the industrial form of the IoT, introduced in late 2012. Things and machines in the IIoT are networked and independently and exchange and process information with each other. In contrast to the IoT, the IIoT does not represent the consumer-oriented concepts, because it concentrates on the application of the IoT in the industrial environments. In this context, IIoT semantically describes a technological movement. When it comes to IIoT, I4.0 and machine-to-machine communication used, whereby IIoT simplifies processes and increases their efficiency in order to achieve cost reductions and faster production. Thus, IIoT describes the industrial transformation in the context of connected machines and devices in industries, which allows interaction and cooperation of these connected machines and devices to reach a common goal. Perceive the IIoT in intelligent manufacturing environments as a large connected industrial system of inventory, logistics, machines, material, parts, tools, and other, which relay on data and communication with each other. Moreover, the advantages result from the consistent implementation of the concepts and technologies of the IIoT such as production and transportation of parts and products with greater operational efficiency. In addition, upcoming new business areas and production processes be automated on the basis of collected and processed data, which results in better real-time flexibility to changing requirements during operation. Furthermore, the machines used to recognize independently when maintenance is required. Under certain circumstances, maintenance performed by the machines themselves. The number of faults and production interruptions decreases while throughput and production capacity increase at the same time. In some areas, the IIoT offers the opportunity to blur the boundaries between the physical and the digital world. Physical products and objects developed, designed and produced based on purely digital requirements. Individual products or small series produced at prizes as realized in large series in traditional production industry. Another advantage be the improved safety of workers and machines in the industrial environment. Therefore, the IIoT based industrial ecosystem combines intelligent, connected, and autonomous machines, advanced predictive analytics and machine-to-machine collaboration to improve productivity, efficiency and reliability. In this context, IIoT is a network of systems, objects, platforms, and applications that communicate and share intelligence with a huge potential to the application in the I4.0 manufacturing environment. This allows field devices to communicate and interact with one another and with more centralized controllers. It also decentralizes data analytics and decision-making, enabling real-time responses [53]. Moreover, components or objects identified and tracked used for manufacturing, e.g., by RFID tags, whereby the RFID enables objects wirelessly promote their identity. RFID tracking allow getting the right objects to the right place at the right time. In this regard, RFID on the one hand support the intelligent manufacturing cell in knowing which manufacturing steps be performed next for a specific product and can adapt to perform the specific manufacturing operation. On the other hand tracking analysis across the manufacturing job floor allows analyzing the overall manufacturing line process at a glance and therefore real-time feedback of the actual job floor situation, e.g., object/part congestion in front of one of the most critical manufacturing cells.

The intrinsic connectivity of IIoT drives the convergence of Operational Technologies (OT), e.g., robots, conveyer belts, and others, as well as information and communication technology (ICT). Furthermore, IIoT unlock the potential of differentiated tracking of detailed processes and overall effects at a global scale. In this regard, a profusion of devices working collaboratively is powerful and transformative. IIoT also enable involving closer cooperation between business partners, like OEM, their Tier suppliers, and customers, as well as employees, providing new opportunities of mutual benefit. In addition, research areas in the context of future trends in IIoT are manifold, like cyber-manufacturing, intelligent communication and networking, intelligent cyber-physical manufacturing systems, and others, as published in [53].

In addition, IIoT cloud computing infrastructure solutions offer the possibility of further developing existing business models or successfully new ones or creating methods increasing value(s). Many IIoT solutions designed as some kind of a product with the focus on IIoT-as-a-Service (IIoTaaS) model. This model build on a cloud-based infrastructure, which create opportunities for new business model offers and value delivery, which include predictable and recurring revenue streams and customer lock-in for a certain time. A sensor-based based IIoT systems that detects machine wear and proactively alerts the operator in maintenance needs before critical parts wear out stated as IIoTaaS approach. Combining emerging and new technologies with available data in the IIoTaaS paradigm, allows optimizing industrial plant operations and processes, to increase productivity, profitability and reliability. An example is the digital twin model, which is a replicate of a physical process or service in industrial production, using computer-aided engineering, integrated with IIoTaaS and machine learning (see Sect. 1.2.12 and Chap. 8) and big data analytics (see Sect. 1.2.5), in which real-time data from its physical industrial production process is merged on an interactive interface. IIoT, also known as Industrial Internet, is a term coined by General Electric, which refers in particular to industrial environments.

1.2.11   Fiber Optics

The rapid development in communication technology and devices causes an exponential growth in data. The swift data growth overtakes the capability of the prevailing communication. The communication process with the help of fiber optics replaced the electronic components. Fiber optics connect everything and achieving massive data transmission through improved bandwidth, stability, latency and reliability. This enables fiber optics ultra-HD media communication, emerging new VR and AR applications, advanced IoT services and other services that relay on fast Internet, which allows expecting reliably great transmission speed and quality. As new inventions become commonplace, unique need for fiber optics will continue to grow, because fiber optic technology is future proof. In this context, fiber optical networks providing the platform and the foundation that enable transformation solutions by minimizing latency and maximizing bandwidth and scalability, so industrial organizations can deliver real-time processing and insights, essential for digital transformation activities.

1.2.12   Machine Learning

Machine learning refers to three different forms with their respective sub-types, as shown in Fig. 1.1.

Supervised learning is the first type of machine learning. In this form machine learning means to learn a mapping from inputs x to output y, given a labelled set of input-output D as a training set, and N as number of training samples (see Chap. 8). In the simplest setting each training input x is a D-dimensional vector of numbers, called features, attributes or covariates, which often is stored in a $N = D$ design matrix. However, x_i could also be a complex structured object, such as image, sentence, time series, graph, and others. Output y in principle is anything but machine learning methods assume that y_i is a categorical or nominal variable from some finite set y_i or a real value scalar. In case y_i is categorical, the machine learning method called classification or pattern recognition, and if y_i is real-valued called regression [54].

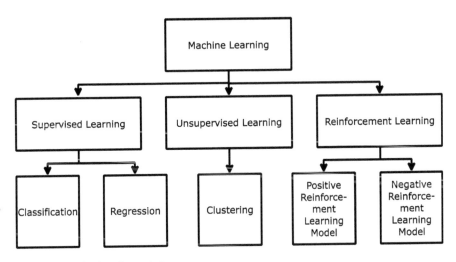

Fig. 1.1 Machine learning techniques

The second type of machine learning called descriptive or unsupervised learning. The goal is finding interesting patterns in data, also called knowledge discovery. Unsupervised machine learning algorithms using datasets containing features to learn useful properties of the structure dataset.

The third type of machine learning is reinforcement learning. Reinforcement learning algorithms interact with an environment, so there is a feedback loop between machine learning and human experiences. This method is useful for learning how to act or behave for given occasional reward or punishment information.

The different forms of machine learning partially perform complex processing tasks, based on different kinds of algorithms. Each of them has advantages and disadvantages to the application domain. In this context, definition of machine learning categories, problem solving types, and best suited for issues shown in Table 1.7.

A comparison for supervised versus unsupervised and reinforcement machine learning based on the same evaluation criteria, to identify the respective pros and cons of machine learning models, is given in Chap. 8.

Machine learning is essentially a form of applied statistics with increased emphasis on the usage of computers to estimate statistically complicated functions and less on proving confidence intervals around these functions. It is used to identify objects in images, transcribe speech into text, match new items, selecting relevant results of search, and many others. Machine learning tasks describe how to process to solve a problem. A problem is a collection of features quantitatively measured from the object under test that the machine learning method should process. Typically, a problem can represent a vector

$$x \in \mathfrak{R}^n,$$

where each entry x_i of the vector is another feature. For example, the features of an image are usually the values of the pixels in the image. Therefore, machine learning models can be assumed as set of n input values $x_1, \ldots, x_n$ and associate them with an

Table 1.7 Capabilities of supervised, unsupervised, and reinforcement learning

Machine learning techniques			
Features	Supervised learning	Unsupervised learning	Reinforcement learning
Definition	Learning model learns from labeled data sets with guidance		Agent interacts with its environment performing actions and learns from errors or rewards
Output types	Regression and classification	Association and clustering	Reward based
Best suited for...	Interactive software system or application	Automation and classification	Supports and work in AI, human interaction

output y. These models learn a set of weights $w_1, \ldots, w_n$, and compute their output, such as

$$f(x, w) = x_1 w_1 + \ldots + x_n w_n.$$

In this context, it can be stated that the continuous technological development drives individual industrial organizations to enhance and adjust their traditional business and process operations, adapt to new information systems and technologies, and keep their existing systems up to date. These new information systems and technologies have a huge impact on business processes and organizational strategies. Machine learning techniques help business organizations to find efficient solutions extracting meaningful information and knowledge, which are useful for decision-making using the different types of machine learning (see Chap. 8). However, new concepts and technologies, such as machine learning, require massive adaptions from the industrial organizations building a strategy to guide them up to date with the pace of technology changes. Furthermore, machine learning methods applied to defend threat event attacks to achieve cybersecurity by quickly scanning large amounts of data and analyzing them, using statistical methods. Therefore, machine leaning is a powerful method used to cyber secure components, systems and network in the digital transformation against cyberattack malware.

1.2.13 Machine-to-Machine Communication

Machine-to-machine (M2M) communication describes any technology that enables networked devices to exchange data and perform actions without manual assistance of humans. AI and machine learning facilitate the communication between machines and systems, enabling them to make their own autonomous choices by interpreting data and make decisions based on which preprogrammed automated actions are triggered and executed.

M2M communication was first adopted in manufacturing and industrial settings, where other technologies, such as Supervisory Control and Data Acquisition (SCADA), a category of software application program for process control, gathering data in real time from remote locations, helping to manage and control equipment and conditions [55]. A common question when accelerating industrial security for SCADA systems is how to align Information Technology (IT) and Operational Technology (OT) efforts for maximum impact [56].

The main components of an M2M system include the following:

- Sensors: A sensor is a device that measures physical input from its environment and converts it into data that can be interpreted by either a human or machine.
- Radio Frequency Identification (RFID) Tags: Are automatic ID systems that consist of two basic components: a tag and a reader.

- Wireless-Fidelity (Wi-Fi): Refers to the Institute of Electrical and Electronics Engineers (IEEE) 802.11 wireless local area network (LAN) standards.
- Cellular Communication Links: Cellar communication channels that connect two or more devices for the purpose of data transmission.
- Autonomic Computing Software: Programmed to support a network device, to interpret data and make decisions.

These M2M features translate the data, which can trigger execution of preprogrammed, automated actions.

Beyond remotely monitor manufacturing equipment and systems, the essential benefits of M2M include:

- Boost revenue by revealing new business models for servicing products in the field.
- Improve customer service by proactively monitoring and servicing manufacturing equipment before it fails or only when needed.
- Reduce costs by minimizing manufacturing equipment maintenance and downtime.

However, M2M systems face a number of security issues, from unauthorized access of cyber-physical systems to wireless intrusion to critical device hacking. Therefore, physical security, fraud, and the exposure of mission critical applications are considered. Segmenting M2M devices onto their network and managing device identity, data confidentiality, and device availability can help in defending M2M security risks.

1.2.14 Network-as-a-Service

Network-as-a-Service (NaaS) is a business model delivering organization-wide area network services virtually on a subscription basis. It provides networking hardware, software, and operational and maintenance services as an operational expense instead of the traditional upfront expense. In this regard, NaaS enables users to easily operate the network and achieve the outcomes they expect from it, without owning, building, or maintaining their own infrastructure. Like other cloud services, the service provider delivers for a fixed fee to manage NaaS. Thus, users can scale up and down as demand changes, rapidly deploy services, and eliminate hardware costs, which mean that NaaS simplifies how hardware and software technologies are used and managed. Moreover, the software-defined wide area network (SD-WAN) architecture deployed as a value-added service with NaaS to enhance application experience, performance, redundancy, and security. Moreover, the SD-WAN enable leveraging any combination of transport services to connect users securely to applications.

1.2.15 Network Virtualization

Network virtualization (NV) is a method combining the available resources in a network to consolidate multiple physical networks, divide a network into segments, or create software networks between virtual machines (VM). Network virtualization manages the environment as a single software-based network, optimizing network speed, reliability, flexibility, scalability, and security. It also improves productivity and efficiency of network administrators by performing many of those tasks automatically, thereby brilliant disguise the complexity of the network [57].

1.2.16 Network Function Virtualization

Network Function Virtualization (NFV) improves the flexibility of network services provisioning and reduces the time taken to market new services effectively. By leveraging virtualization technologies and commercial off-the-shelf programmable hardware such as general-purpose servers, storages, and switches, NFV decouples the software implementation of network functions from the underlying hardware. Thus, software takes over the network functions previously used by dedicated hardware. The combination of server and software can be a variety of network devices, from switches and routers to firewalls and Virtual Private Networks (VPN) gateways. These new software devices can run on physical servers, in VM, controlled by hypervisors, or a combination of both. The technology leverages developments in virtualization technology and hardware optimization included in the current generation of processes and network interfaces, integrated to the need for traditional, dedicated network devices. As an emerging technology, NFV brings several challenges to network operators, such as the guarantee of network performance for virtual appliances, their dynamic instantiation and migration, and their efficient placement [58].

1.2.17 Simulation and the Digital Twin

Simulation analysis is a powerful problem-solving approach whose origins lie in statistical sampling theory and analysis of complex probabilistic physical systems, as well as in linear continuous systems, system dynamic techniques, time-varying and time-invariant system approaches, lumped and distributed parameter systems, and others. There are many benefits to simulation and analysis. Applications of simulation analysis have been in the product development lifecycle in the manufacturing sector, but increasingly, fruitful models are developed for the aerospace sector, the automotive sector, and all kinds of real world systems. The value that results from the simulation analysis is remarkable, because simulating objects in a digital format

Table 1.8 Model types with characteristics

Model type	Characteristics
Closed loop or open loop	Closed loop models used if the output is fed back and compared to some desired level or goal to alter the system such that it maintains, or approximates, the desired value Open loop models have no output that is fed back as input to modify subsequent output
Continuous or discrete	Continuous models used for real systems continuous in time Discrete models used for real systems discrete in time
Deterministic or probabilistic	Deterministic models used if variables, whose values may be stated with certainty, can sufficiently describe the system behavior Probabilistic models used if any random variable is present
Descriptive or prescriptive	Descriptive models used if the system behavior and its optimization leaves totally in the hand of an analystPrescriptive models used to formulate and optimize a given problem providing the one best solution
Dynamic or static	Dynamic models used if the model variables change over time Static models used if the model variables don't change over time

reap advantages in physical space. Therefore, use simulation early in the design process support good design decisions and enable to verify performance and determine whether all requirements are satisfied, through a lifecycle analysis. In this regard, simulation can use different X-in-the loop methods to verify the logic of the code developed and their interactions with the actuated hardware and software components. There are different types of models used in simulation as shown in Table 1.8.

The power of simulation lies in executing the simulation model. In this context, developing a mathematical or logical model of a real system or process and then conducting computer-based simulation scenario analysis with the developed model is used to describe, explain, and predict the behavior of the real system or process through the developed simulation model. In this context, simulation is the computer-aided replication in which not analyses at the real system or process is carried out, rather the developed model of the real system or process with its intrinsic behavior. Simulation as reproduction of the static and/or dynamic behavior of a real system or process refers to material or immaterial real boundary conditions through the developed model, in order to be able to draw conclusions from the simulation results to the properties of the real system or process. Against this background, simulation is based on the three R paradigms, which means.

3R : Reductionism, Repeatability, and Refutation.

Reductionism refers to decomposing any system or process into a set of components that follow the laws of physics. Repeatability is a validation criterion to correctness in simulation, which means that diversity reduction of the real system or process within the chosen model fits well to repeatability. With repeatability, one can make intellectual progress by refutation of hypothesis. In this regard, simulation is the imitation of the operation of real systems, technical processes, business processes and other, which support to visualize the design and identify the problems that might occur in a much earlier design stage [59].

In digital transformation, Industry 4.0 (i4.0) refers to an intelligent manufacturing environment. Simulation application in I4.0 requires real-time data to mirror the physical world in a virtual model, which include manufacturing machines, manufactured processes, manufacturing products, and other. Hence, simulation allows plant operators to test and optimize, e.g., the machine settings of the manufacturing job floor processes for the product next in line in a (virtual) process model, before its physical implementation. Furthermore, it is possible driving down machine setup times on the one hand and increasing manufacturing quality on the other hand through simulation aided production planning. For this purpose, the data of the complex product and its fabrication made available for the simulation. Simulation itself enables comprehensive analyzes of the complex I4.0 manufacturing processes in current projects and creates transparency in the underlying manufacturing processes. This also leads to better decision-making principles and finally it significantly increases planning flexibility and security.

Through digital transformation a new generation of information technologies in I4.0 based manufacturing systems simulation is coming up, the concept of the digital twin, an integrated multiphysics, multiscale, probabilistic simulation of complex systems or processes like I4.0. The digital twin represents a digital plant simulation model, which shows the same behavior and communication as the real physical plant, used for virtual commissioning, optimize the manufacturing process, and other, to promote better production efficiency and worker safety. Thus, the digital twin is a realist image of the real behavior of an individual system, product or production process that accompanies its real counterpart to enable high availability, minimized maintenance times with appropriate downtimes as well as a reliable machine or equipment over-all functionality. This requires knowledge of the actual conditions of the machines and the manufacturing environment. In the physical manufacturing environment, embedded sensors capture relevant data, which are basis for a condition assessment. However, not all of these data provide the important properties for the rating. A solution for this problem is the digital twin that provide the required information through software based virtual sensors, used for a continuously updated calculation through simulation and analysis. Such advanced knowledge of the real-time machine condition is critical to condition-based maintenance. Therefore, the digital plant model is ultra-realistic and considers one (or more) essential manufacturing machinery systems, the respective maintenance history data, and other for the necessary real-time plant simulation and analysis. Moreover, manufacturing anomalies, that may affect the fabricated product, can also explicitly considered, evaluated and monitored through the digital twin. Thus, a digital twin is an evolving digital profile of the historical and current behavior of a physical system or process that support optimize operations in work cells and production lines to improve manufacturing business performance. As introduced in [60], digital twins are of two types:

- *Digital Twin Prototype*: Describes a prototypic physical artifact. It contains the information necessary to describe and generate a physical version that duplicates or twins the virtual version.
- *Digital Twin Instance*: Describes a specific corresponding physical product that an individual digital twin remains linked to throughout the life of that physical product, operated within a Digital Twin Environment (DTE), an integrated multidomain physics application space for operating digital twins for a variety of purposes.

Moreover, based on the digital twin paradigm, the Digital Twin Shop Floor (DTSF) is proposed as new paradigm for product fabrication entity. The DTSF is composed of the physical manufacturing shop floor, the virtual shop floor, the shop floor service systems, and the shop floor digital twin data. Since the concept of the digital twin is proposed and applied in many industrial fields, it has demonstrated its great potential. With regard to the characteristics of a digital twin, especially synchronous linkage and ultrahigh fidelity between physical product and corresponding virtual product, the digital twin has high potential to solve problems existing in Product Lifecycle Management (PLM) as mentioned in [61].

Development of a digital twin starts with the design process, which raises questions such as: What are the processes and integration issues for which the twin will be developed? Standard process design techniques used which includes the respective business processes, business applications, data, and the essential physical interaction. The application specific process flow requires real-time data and other information, to develop the digital twin. The design process is augmented by attributes such as cost, time, asset efficiency and other for improvement. These are the basic assumptions based on which the digital twin development should start.

Hence, the digital twin is introduced as virtual representation of a physical manufacturing system across its life cycle. Real-time data used to enable learning, and dynamically recalibration, for improved decision making, which require the respective level of detail in creating the digital twin model.

While a simplistic model may not yield the value a digital twin promises, a too broad approach can result in getting lost in the complexity of thousands of sensors, several thousands of signals the sensors generate, and the massive amount of technology to make sense of the model. Fig. 1.2 shows a possible approach representing an optimistic concept of the process flow in the digital twin [62].

1.2.18 Software Defined Network

Software Defined Network (SDN) is a networking architecture that improvise conventional network in terms of scalability, security, and availability. It separates that network in separate layers for controlling the network configuration, the control plane, the data plane and the data transport in the network. Thereby, it is possible to use the analysis and control plane to virtualize, administrate and configure via a

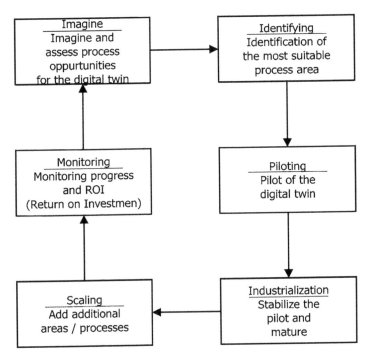

Fig. 1.2 Process flow in a digital twin

central console to control. Today, the SDN principle is used in cloud computing applications, because cloud providers using SDN to shift more workloads to cloud environments. As networks become divers and new workloads increasingly migrating at the edge of the network, SDN are part in these new complex environments. However, at the same time SDN is vulnerable to security threat incidents that requires solutions against cybersecurity risks [63].

1.2.19 Wireless Sensor Networks

Wireless Sensor Networks (WSN) represent a collection of nodes, organized into a cooperative network. Each node consists of processing capability like one or more microcontrollers, one or more types of memory, a radio frequency transceiver, a power source and accommodates various sensors and actuators. The nodes communicate wireless and are often self-organized after being deployed in an ad hoc fashion. WSN is deployed at an acceleration pace in communication infrastructures to monitor and record conditions at various locations. The integrated microprocessor processes and stores the output of the sensor, and the receiver receives commands from a computer and transmits the data. WSN technology has an unlimited potential for numerous application areas, Its various challenges from the theoretical, practical and algorithm perspective is elucidated in [64, 65].

1.2.20   5G and 6G

5G refers to the fifth generation of wireless broadband cellular technology for mobile communication, developed to increase the speed and responsiveness of wireless networks, to satisfy the high demand on real-time traffic. 5G is based on the IEEE 802.11, a standard with potential peak speeds of 20 gigabits per seconds. (GB/s). This speed exceed wireless network speed and offer latency of below 5 milliseconds or lower, which is useful for applications that require real-time feedback. 5G enable an increase in data transmitted over wireless systems due to available bandwidth and advanced antenna technology. Thus, users have a smooth connectivity to the network. 5G also promotes integration of industrial Internet applications, based on the development of 5G-based industrial Local Area Network (LAN) in manufacturing environments, which allow stationary and mobile robots working together to complete tasks in perfect harmony, where a difference of a few milliseconds of latency is problematic. 5G networks also expand applications of IT, such as AI, IoT, big data and analytics, and others.

6G is the successor to 5G. 6G offers better connections, improve precision and efficiency of using big data to carry out supra-regional and interdisciplinary mega science projects for the public and private organizations. Speed of 6G is assumed exceeding 125 GB/s, enabling zero-latency for machine-to-machine communication in intelligent digital manufacturing and global high-speed Internet coverage over extensive satellite networks. However, the rapid transmission of huge amounts of data could also raise concerns about costs and data security that could result in leaks or security risks not noticed. Working in conjunction with AI, 6G computational infrastructure will be able to identify the best place for computing, which includes decisions about data storage, processing, and sharing [66].

1.3   Challenges in Digital Transformation

Digital transformation impact results in changing traditional isolated processes into fully integrated and connected data flow driven processes. This requires advanced, intelligent and connected entities from the cyber and the physical world, designed by security measures to avoid threat event attacks. Besides this, digital transformation is a process of change that goes hand in hand with high-speed innovation cycles, achieved by the inherent dynamic development of digital technologies. At the same time, this paves the way for further technological innovations. In addition to short digital technological innovation cycles, such those seen in the development cycles of smartphones, the driving force behind is the change in customer require-ments, which can be served through innovative and emerging technologies that maintain the digital transformation process. However, this require align employees on the digital transformation plan before implementing new digital technologies, and choosing the right technologies to avoid frictions.

To adapt the digital transformation economy, the capacity for sensing challenges and opportunities, as well as for fast adapting processes and models in all sectors is essential. However, this requires answering the question How the new developed model is aware about securing data and business assets? The reason is that cyber-security is the most critical and crucial issue in digital transformation security, to avoid threat event attacks by cybercriminals. Threat event attack is a term used to any ongoing threat event incident on computers, systems or networks. Threats (see Chap. 2) intruded from careless user behavior of internal computer systems against threat event attacks or careless behavior using passwords, or external (outsider) who finds vulnerability in a web application and exploits it. Threat event attacks are primary concerns for businesses, loosing ransom money, facing brand damage and litigation costs, should a threat event attack successfully breach into an organizations system to steal data or manipulate data and/or a system for unauthorized shutdown. This requires appropriately practical and technological knowledge and competences, essential to lead the digital transformation at the respective cybersecurity level. All in all, this is a very essential high level, because none of the previous technological waves has had a truly disruptive potential like digital transformation. Largely implemented it will show disruptive steps, in contrast to evolutionary steps, which not only substitute solutions but also create new markets and business models that will change the social life of societies, as the Internet has changed it. Therefore, digital transformation with its clear need for continuous innovation in digital tech-nologies and cybersecurity awareness is a continuous change in progress. This will be achieved connecting anything with everything and allow accessing and control-ling anything, whereby everything will be recordable and programmable. However, this require a clear roadmap showing what should be done first, what are the accelerators, what are the barriers and what are the challenges. The answer can have a positive or negative effect on both, the industrial organization and the customer site. Indeed, the ultimate impact that industrial organizations want to leverage through digital transformation is value creation. Therefore, digital transfor-mation is essential to business success, but industrial organizations may fail planning on digital transformation in their business model. Common reasons for the failure are:

- Budgetary constraints.
- Data privacy and security concerns.
- Immature digital culture.
- Inadequate management support.
- Lack of employee engagement.
- Lack of employee skills.

- Limited in-house technological skills.
- Limited innovation skills.
- Poor or nonexistent cross-functional collaboration.
- Poor regulatory and legislative issue awareness.

This requires strategical criteria for a successful planning for the digital transformation, as shown in Table 1.9.

Return on investment (ROI) is a widely used financial metric for measuring the profitability of gaining a return from an investment. It is also a performance measure comparing the efficiency of different investments, e.g., pushing the digital transformation process. Therefore, ROI directly measures the amount on return on a particular investment, in relation to the investment's cost. Thus, ROI evaluating investments judges how well a particular investment has performed to a special investment, by comparing the gain or loss from an investment relative to its cost. In business analysis, ROI and other cash flow measures, such as internal rate of return and net present value, are key metrics evaluating and ranking the attractiveness of investment for a given project and for an alternative. Overall, an industrial organization ROI is a way to grade how well managing the industrial organization is. ROI typically calculates the benefit (or return) of an investment, divided by the cost or the

Table 1.9 Strategic criteria for successful digital transformation planning

Strategic criteria	Capability
Data	Key enabler of digital transformation success. Collecting data and perform analytics for better insights. This requires converting existing volumes of data into high valuable, actionable assets, to achieve digital transformation success.
Digital technologies	Requires multiple ongoing initiatives that involve investments in emerging technologies, new hard- and soft skills, updated workplace culture, and organizational restructuring.
Digital technology risk	Requires risk awareness and risk management while integrating emerged technologies to avoid unsuccessful embedment or unprecedented failures due to a manifold of cybersecurity risks, using emerging technologies.
Digital transformation project	Demonstration of a compelling return on investment (ROI) as vital for digital transformation to support digital initiatives. ROI measuring is important to improve revenue over time.
Talent	Ability to compete in digital transformation depends on having the right talent, at the right time, in the right place. This requires prioritizing using an elaborate sourcing ecosystems for talent(s) needed, and establish a robust resource recruiting and management program to support planning for future skill requirements.
Technology	To achieve ongoing success requires embracing a platform-based operating model for digital transformation, taking into account all users, employees, customers, and partners in the context of technological enablers, driving customer experience, decisions, process efficiency, scalability, and performance.

investment. The result is expressed as a percentage rather than a ratio. ROI can be calculated using two different methods:

$$ROI = \frac{Net\ Return\ on\ Investment}{Cost\ of\ Investment} \times 100\%$$

$$ROI = \frac{Final\ Value\ of\ Investment - Initial\ Value\ of\ Investment}{Cost\ of\ Investment} \times 100\%$$

ROI calculation includes the net return in the numerator, because returns from an investment can be either positive or negative. When ROI calculations yield a positive figure, this means that net returns exceed total costs. Alternatively, when ROI calculations yield a negative figure, this means that net returns produces a loss. To calculate the ROI with the highest degree of accuracy, it is important to consider the total returns and total costs. To measure the ROI delivered by digital transformation, the following aspects are important:

- Identify objectives or goals of a particular measure that supports the industrial organizations digital transformation process or strategy.
- Determine the components and associated costs delivered as constraints to the measures.
- Define the metrics that will determine whether, and how well, the identifies objectives and goals identified were achieved, using those compared against costs to determine ROI at various points along the industrial organizations digital transformation process or strategy.
- Ability to measure the success of digital transformation, which will be critical as industrial organizations are expected to invest staggering amounts in their digital transformation programs in the upcoming years.
- Digital transformation challenge for industrial organizations that were not digital refer to legacy systems and applications, which are older technologies that do not support digital activities, but can't easily be replaced. Refusing to invest to replace old technologies or fails to garner the necessary executive level support for investing in new technologies, digital transformation is unlikely to happen.
- Obsolete systems and applications cannot be maintained and utilized forever. At some point, an industrial organization has to upgrade hardware, coding language, Operational Technology (OT) that monitors and manages industrial process assets and manufacturing/industrial equipment or the specific application. In this regard, modernization and migration involve refactoring, repurposing, or consolidating legacy software programming to realign with current industrial organizations business needs. The goal of a modernization project of a legacy application is to create new business value from existing applications.
- For industrial organizations that are strapped for resources, replacing obsolete systems and applications, one at a time instead of migrating the entire systems and applications to digital transformation standards and technologies, seems to be a

good practice. However, this migration principle is not as practical and cost-effective as it looks like at the first view.

- The better practice is boosting the efficiency by reducing the Total Cost of Ownership (TCO) and gaining actionable insights from data that come along with migration in the context of digital transformation, and then, further modernizing systems and applications.
- Furthermore, modernizing the legacy workloads too increases their value and business return assets.

Against this background, the digital transformation challenge refers to emerging themes and technological principles such as

- Connected (smart) buildings.
- Digital billing.
- Digital commerce.
- Preventive maintenance.
- Smart grid.
- Smart home.
- Smart cities.
- Smart health.
- And others.

Technological methods used are (see Sect. 1.2):

- Big data and analytics.
- Cloud computing and services.
- Connectivity.
- Cybersecurity.
- Interactive devices and objects.

In this context, an emerging theme and frontier in digital transformation is the digital sovereignty. Digital sovereignty is a term under the pressure to digitize. However, the debate of digital sovereignty currently is not sufficient-based on evidence, but more used to act in political debates on the national level in the Internet age. From the technological perspective, this means that parties must have sovereignty over their own digital data within the essential resources in the digital transformation ecosystem, which involves considerations how data and digital assets are treated. A definition of digital sovereignty is given in [67]: "Digital sovereignty is the ability of an entity to personally decide the future of identified dependencies in digitalization and to possess the necessary powers." However, this entails prerequisites in order to decide personally. These prerequisites take into account concrete dependencies to be recognized, analyzed and evaluated, and own abilities and possibilities identified. This raise the question: "What competencies are necessary for this and if they are lacking, is it possible to create the realistically and over what time horizon?" Another important question is: "How does one handle problems in the meantime?" [67]. To pave the way towards further digital transformation success, the European Commission has set in motion several initiatives, the Digital

Markets Act and the Digital Services Act [68] and the GAJA-X project, to develop common requirements for a data infrastructure, supported by representatives of business, science and administration. The GAJA-X project is an open digital ecosystem in which data made available, merged, confidentially and securely shared, to meet the next generation of a networked data infrastructure that meets the demands of digital sovereignty and promotes innovation. The GAJA-A project provides a homogenous user-friendly system for the networking of decentralized infrastructure services, especially cloud and edge instances. The resulting networked formed of other data infrastructure strengthens, both the digital sovereignty of cloud service users, and the scalability and competitive position of European cloud providers. The solution approach of the GAJA-X project is driven by the developments in Industry 4.0 (I4.0), cloud- and edge computing, and others. It shows the importance of multicloud strategies and, thus, rapidly increasing interoperability. Users expect flexibility, functionality, user-friendliness, worldwide availability, cross-company services, specialization of services, distributed data processing, and data management. Nevertheless, easy migration to other cloud or edge providers must be possible too. Users demand interoperability quality while finding it efficiently using of services through new data service intermediaries. This also includes transparency about the services offered. Users also want to be able to distribute data processing flexible across many providers and at the same time, which rely on robust processes to fall back on [69]. Thus, the next wave of digital transformation has to capture this challenge.

1.4 Applications in Digital Transformation

An outcome of the recently published report on Digital Transformation of the Boston Consulting Group (BCG) [70] states, "Only 30% of industrial organizations navigate a digital transformation successfully. And navigating it in the midst of uncertainty is especially difficult because new behaviors and expectations take shape and evolve at warp speed." Although digital transformation is not a new imperative for industrial organizations leaders, yet in many cases, organizations have a long way to go. So many organization have yet to apply digital technologies and ways of working at scale, others create a culture that embraces change or are planning end-to-end transformation. Nevertheless, many industrial organizations are successfully in navigating the business in digital transformation for improving the way of running business as illustrated in [71]. Three enablers of digital transformation in industrial organizations are [72], besides others, are shown in Table 1.10.

One big application domain of digital transformation is Industry 4.0 (I4.0). I4.0 is a term coined for the innovations of traditional manufacturing with intelligent and connected industrial practices introduced by the digital transformation. In the past, manufacturing of equipment has been embedded in a closed, hard-wired network environment. The reason was that industrial sensors, controllers, and networks are expensive and upgrading projects in existing facilities are not easy to achieve.

Table 1.10 Digital transformation enablers

Digital transformation enablers	Target applications and themes
Product design	Advanced simulation and analysis: Support today's complex system design by a model-based engineering approach Digital twin: Virtual representation of a product or plant that exists before anything is physically built Creating, managing, and disseminating documentation: Using an e-book of knowledge Managing extant definitions of digital transformation: Conceptual clarity challenges
Manufacturing and production	Digital thread: Digital form of individual business processes—Activities, tasks, and decisions, also supporting automation, traceability, and standardization efforts Ability to capture and stream sensor data Which is cleaned, standardized, and normalized, providing the most accurate information about entire streaming Building a true single source of truth: Finding the documentation needed, stops searching
Service and maintenance	Streaming data for products Delivery software updates Delivering operational data

However, the growth of utilization of the Internet of Things (IoT) at the consumer side has driven cost reductions in sensors, controllers, and communication devices through high volume production, e.g., smart components and/or smart devices. Furthermore, industrial standard equipment is constrained by a huge installed base of legacy equipment and standards. That is why capital, energy, human resources, information, and raw material acquired, transported, and consumed in transforming the raw material into value-added components and/or products, in order to increase benefit. However, in recent years, the manufacturing industries have been faced substantial challenges due to the necessity in coordination and connection concepts. To accomplish this goal the manufacturing industry is working worldwide to achieve the next level of innovation in industrial production by digitization, the continuing convergence of the real and the virtual worlds through technological innovations as main drivers of the manufacturing process of digital transformative, which results in the I4.0 paradigm. The technological innovations, with huge influence on the digital transformative event of I4.0) including advanced analytics, cloud computing, cyber-physical systems, IoT, machine learning, human–machine interfaces and other, and are part of the mix of driving innovative and intelligent technologies. These technologies are already making a difference throughout today's industry in general but they make specific sense within the manufacturing sector. These advancements cause an extension of the developments in manufacturing systems and Information and Communication Technology (ICT) enabling dealing with the exponentially growing amount of data generated by the different affordable interconnected technologies, which came along with the digital transformation in I4.0. In this regard, Cyber-Physical Systems (CPSs) that are physical and engineered systems whose operations are monitored, controlled, coordinated, and integrated by a computing and communicating core, the IoT, and the IIoT play a vital role in mastering today's digital transformation processes and securing a competitive position for the leading global manufacturing industry. In Germany, this effort is promoted under the heading "Industry 4.0 (I4.0)," a term that was first used in 2011 at the Hannover Fair in Germany. I4.0 was introduced as a vision on how to maintain and expand the competitiveness of the industry. Originally, I4.0 was promoted as a future project of the high-tech strategy of the German Federal Government, to motivate research in the intensive efforts for the I4.0 project. It has now become widely established in the public debate in Germany. This result in the technical integration of CPS in manufacturing plant environments and logistics, as well as the application of the IoT and the Internet of Data and Services in industrial processes, including value creation, new business models, downstream services and organizations business services.

Beside Germany, in the United States the Smart Manufacturing Leadership Coalition (SMLC) is founded, which drives the infusion of intelligence that transforms the way Industries conceptualize, design, and operate the manufacturing enterprise and Industrial Internet, to drive and facilitate the broad adoption of manufacturing intelligence. China has initiated a similar project called Made in China 2025. Other countries have initiated similar projects supported by their governments to support industrial competiveness in the world of digital transformation.

Other applications of digital transformation are [73] summarized in Table 1.11.

1.5 Leadership in Digital Transformation

Digitalization is a top priority for industrial organizations, and other, through the digital transformation with its torrent of data generated and its impact on the transformation on all business processes. Furthermore, the dynamics of the current development in the digital transformation make industrial business development a continuous task. This requires that industrial organizations prepare for significant disruption to their routine business activities and processes, typically for a longer time, because it involves large-scale migration to data-driven technology-based business processes, value acquisition for stakeholders, customers and effective streamline processes. Therefore, industrial organizations and other have to follow the digital and technological advancements, prepared for the digital transformation, gaining knowledge on emerging trends in cybersecurity such as AI and machine learning solutions, intelligent authentication, intelligent access management, data security solutions, social engineering threat protection and other. Despite being up-to-date in the digital age, industrial organizations need to understand how to protect business assets and to prevent them against threat event attacks, avoiding a dystopian future. This requires on the one hand a leadership in detection and prevention of threat event attacks and on the other hand master digital advantages turning digital

Table 1.11 Examples in digital transformation

Digital transformation in	Characteristics
Banking	The financial services revolution involves much more than the shift to mobile banking and cash apps. Consumers now use cash airdrops, card less payments through services like apple pay on phones or watches, and completely resort to bank-less banking.
Healthcare	Healthcare often lags in digitization on the patient side. Now digital innovation initiatives come up like e-portals for patients, real-time health monitoring, and virtual health visits. The innovation is improving the patient-institution relationship and leading to better health outcomes and engagement in healthcare decisions.
Insurance	Insurance is an industry with a lot of tradition. However, now starting to see innovation in the insure market like never before. This includes significant technological changes related to the customer experience.
Manufacturing	Many manufacturing companies undergoing a digital transformation are doing so for greater efficiency. They use predictive analytics to reduce supply chain costs and maintenance needs, as well as decrease energy and water consumption, and other as a B2B industry, the main attraction for their clients is the ability to reduce costs in all manners, and even to increase yield.
Retail business	Retail businesses are fighting to meet customer expectations due to the massive push towards e-commerce in the market. Even Walmart has undergone a transformation to keep up with Amazon and other grocery stores turning to things like online ordering and delivery services.
Transportation	Tesla pushed many car manufacturers into a new type of competition. With the rise in concerns over global warming and the world's overreliance on oil, the automotive sector has recently embraced a move towards electrification and automated driving.
And others	./.

technologies into technological and business transformation that will pay back. Therefore, leadership in digital transformation is required, which means to understand and master the digital upheavals and innovations to develop future-proof business strategies, design structures, marketing concepts, data- and process security concepts for the industrial organization. This requires significant human and financial capital and technological skills, aligned to facilitate successful results to employ leaders, capable directing this change to influence this process through their knowledge and experience [74]. However, beside knowledge in the latest technological innovations, an appropriate leadership style is essential, which align with the organizations needs to make the transition somewhat easier and to improve managerial engagement and enthusiasm regarding the change. In this context, transformation leaders can provide an example for others to follow, inspire other to be successful, develop a shared vision and empower creativity [75]. Therefore, leadership in digital transformation has to combine digital, technological, financial and specific leadership capabilities to achieve a performance greater either dimension can deliver on its own. Thereby, digital- and technological capabilities make new initiatives easier and less risky for the leadership, providing revenue advantage to transform the digital business [76]. In this regard, leadership in digital transformation relies on data-driven insights to decide how to proceed to drive efficiency and effectiveness. This requires building a governance model with its respective committees that suits organization's needs. However, the steps require deciding how to develop digital leadership capabilities of the organization, the key to innovation and sustained value creation. Thus, an innovative blueprint of industrial organizations business model embrace the wave of digital transformation. Therefore, a digital transformation strategy necessitated the collaboration of leaders, resources and teas to ensure that an implementation was successful and did not significantly disrupt operations. However, this does not appear aligned to one specific type of leadership style, found out in the research study of [77].

1.6 Cybersecurity

Cybersecurity, also known as IT security, is the protection of connected systems and applications from cyberattacks by combatting them, regardless of whether these cyberattacks come from inside or outside of an attacked organization. The term is used in a variety of contexts, from business application domains to mobile computing, and can be grouped into a number of general categories: application security, computer security, data security, information security, network security, operational security, disaster recovery and business continuity, and others. Hence, cybersecurity is the practice to protect against unauthorized access to sensitive data, business assets, computerized systems, and other.

1.6.1 Introduction to Cybersecurity

Analyzing the impact of digitization and advancements in technologies in digital transformation, the focus is on the topics: i) knowledge about the interaction of digitization and advanced technologies, ii) knowledge about resulting intrinsic cybersecurity risks through the rapid advancements in technologies, shifting the threat event landscape with their direct impact on cybersecurity risk(s). This potentially has an adverse impact to organizations and their business objectives with its imperative for organizations to manage these cybersecurity risks effectively. Therefore, cybersecurity risk assessment becomes an integral part of organizations cybersecurity risk management. Conducting a cybersecurity risk assessment organizations must be able to ensure the following:

- Identify malicious incidents that often result of threat event attacks that could lead to undesired business operational consequences.
- Determine cybersecurity risk levels organizations are exposed.
- Create a culture of cybersecurity risk awareness within the organization.

There are several definitions of cybersecurity risks, such as:

- Cybersecurity risk is the likelihood of a threat event attack on vulnerabilities of organizations assets occurs.
- Impact of the probable occurrence of the threat event attack that result in cybersecurity risk(s).

In this context, threat event attacks against organizations assets based on identified vulnerabilities probability of cybersecurity risk(s), meaning that a threat event attack is able to exploit an identified cyberattack through vulnerabilities. Let's assume that a cybersecurity flaw exist and threat event actors exploiting this identified vulnerability to compromise the targeted computer system, network, infrastructure resource, and other. In case the latest software patches that fix such a problem are applied, then the vulnerability can't be exploited, and the threat event attack has no chance being successfully executed. In this regard, impact measures how much disruption an organization will face if the threat event attack occurs. Combining likelihood and impact, also known as consequence, result in a residual risk, rating the risk tolerance score between low, medium, medium high, high or very high, as described in Sect. 2.1.3. Against this background, cybersecurity risk assessment is turning into an important narrative of organizations risk management process (see Sect. 2.1.4). To perform a cybersecurity risk assessment needs the following three steps:

- *Risk identification*:

  - Identify and prioritize organizations assets: Asset is an object of value that has been identified, such as servers, client contact information, sensitive or regulated data, documents, information or sensitive partner documents, trade secrets, and others.
  - Identify threat events: A threat event is anything that could cause harm (see Sect. 2.1), such as hardware failures, malicious code, malicious behavior, and others.
  - Identify vulnerabilities: A weakness that enable a threat event actor to harm an organizations asset(s) by exploiting the identified vulnerabilities through threat event(s).
  - Analyze controls: Either in place or in the planning stage, supporting minimizing or eliminating the probability for a threat event to exploit a vulnerability.

- *Risk analysis*:

  - Determine the likelihood of a threat event: Likelihood is the probability that a threat event exploited actually, taking into account the type of threat event, the capability and motivation of the threat event source and the effectiveness of controls.
  - Determine the impact a threat event could have: Compromise the confidentiality, integrity, or availability of assets (see Sect. 1.6.2).

- *Risk evaluation*:

  - Determine, prioritize, and understand the significance of the risk level that compromises.

- Determine and prioritize risk by making use of an m-by-n matrix for determining the risk level.
- Recommend control by using the risk levels as basis to determine the action (s) required to mitigate the risk.
- Documentation of results by developing a risk assessment report to support management to make appropriate decisions in the context of existing measures, current risks, treatment plan, progress status, response to risk, budget, policies, procedures, and others.

In this regard, cybersecurity is the practice and application of technologies, processes, and controls to protect computer systems, networks, infrastructure resources, devices, imperative data, and others from unauthorized intervention through threat event attacks. Therefore, the major concern of cybersecurity is to reduce the cybersecurity risk of threat event attacks and vulnerabilities, and protecting computer systems, networks, and other against unauthorized exploitation. In general the term security categorize into cybersecurity and physical security, which are put in place to prevent any form of unauthorized access to computerized systems or database systems. In this context, the fundamental role of security and cybersecurity is protecting the confidentiality, integrity, and accessibility of data, commonly known as CIA triad (see Sect. 1.6.2), also referred to as pillars of cybersecurity that have to be followed stringently. Therefore, cybersecurity remains a top priority for industrial, public and private organizations business, as high-profile hacking incidents have highlighted the risks for threat event attacks, data breaches, identity thefts and other in today's highly connected world. However, cybersecurity is more than installing an antivirus program. Cybersecurity is the application of controls, practices, policies, processes, services, strategies, and technologies, designed to protect computer systems, critical and crucial infrastructure, networks and other from a wide range of threat event attacks. The cybersecurity topic itself is a computing-based discipline, which deal with the presence of adversaries and their cyberattack repertoire through the Internet. Within computer science, cybersecurity spans many areas, including (but not limited to) data security, cryptography, hard- and software security, network and systems security, as well as cybercrime and cyber-forensic research. Thus, cybersecurity is fundamental to both, protecting and defending computer systems, micro-electronic devices and systems, mobile devices, networks, servers and other from malicious threat events. Against this background, cybersecurity breaches through threat event attacks become reality and industrial, public and private organizations face a constant danger of cybersecurity risk(s), because hackers routinely try to convince authorized users to hand over their credentials. These threat event attacks occur through phishing mails and legitimate-sounding phone calls among other tactics. Therefore, end users in industrial, public and private organizations must learn to be suspicious on email attachments, rouge websites and unauthorized applications. This helps to avoid that threat event attacks executed successfully and damage or manipulate an internal process through an unauthorized access. Table 1.12 shows the various cybersecurity domains.

Table 1.12 Security domains with their capabilities

Security domain	Capability
Application security (APPSEC)	Security measures at the application level to make them more secure by finding, fixing, and enhancing the security of applications, preventing data or code within the application from being compromised, corrupted, stolen or hijacked. Much threat event attacks happen during the development phase, but it include tools and methods to secure apps once they deployed. This is becoming more important as hackers increasingly target applications.
Computer security (COMPSEC)	Protecting computer system or network from harm, theft, unauthorized use, disruption or misdirection of services they provide. Becoming important due to increased reliance on computer systems and growth of smart devices. Computer security refers to four major threat events: 　Theft of data 　Vandalism or destruction of data by a computer virus 　Fraud 　Invasion of privacy like accessing protected data
Data security (DATASEC)	Set of standards and technologies to prevent unauthorized access to computer systems, databases, websites and other to protect data from intentional or accidental destruction, corruption, modification or disclosure. Data security is an essential IT security topic supporting use of techniques and technologies, including administrative controls, physical security, logical controls, organizational standards, and other safeguarding techniques to minimize or avoid access to unauthorized or malicious users or processes.
Disaster recovery (DR) and business continuity planning (BC)	Practices with the goal to limit threat events risk and get organizations running their business tasks as close as possible to normal after an unexpected cyber threat incident. As threat events increase and the tolerance for downtime decreases, DR and BC gain importance. Therefore, there is a continuing trend to combine DR and BC.
Information security (INFSEC)	Strategy designed to protect the confidentiality, integrity and availability (CIA triad of information security) of computer systems, data or networks from malicious intensions. Information security includes a set of strategies for managing the processes, tools and policies necessary to prevent, detect, document, and counter threat event attacks to digital and nondigital information. Thus, information security is the practice preventing unauthorized access, destruction, disclosure, disruption, inspection, modifications, or use of unauthorized information and others.
Network security (NETSEC)	Term that covers a multitude of technologies, devices and processes to secure networks. The term refers to a set of rules and configurations designed to protect the confidentiality, integrity, and accessibility of computer

(continued)

Table 1.12 (continued)

Security domain	Capability
	systems or networks as well as data using both software and hardware technologies. Today's network architecture is complex and faced with a threat environment that is always changing and threat event attacks that are always trying to find and exploit vulnerabilities. These vulnerabilities can exist in a broad number of areas, including devices, data, applications, users, and locations. Thus, network security is the practice preventing and protecting against intrusion incidents into corporate public and private networks.
Operational technology security (OTSEC)	Term that includes the industrial operational processes and decisions for handling and protecting data assets. Thus, OTSEC is a security and risk process that prevents sensitive information from getting into wrong hands. OTSEC is a risk management approach, discovering potential threat event attacks and vulnerabilities in sensitive organizations processes, meaning the way they operate their production shop floor machines, hard- and software. It also controls the permissions users have when accessing organizations sensitive OT-based systems or networks and the procedures that determine how and where data may be used, stored or shared, taking appropriate countermeasures to keep their most sensitive OT data secure.
End-user-education (EUE)	One of the most vital aspects of organizations is their security posture because threat event attacks have shifted their attention to humans to break in to organizations information systems. It makes sense that most secure users will have a fundamental understanding of common cybersecurity risks. Unfortunately, many users do not have a strong understanding of basic threat event attacks, and they too often rely on others to take care of security for them. This requires education of end users to security/cybersecurity awareness with an objective to reduce security/cybersecurity risks that occur due to human related vulnerabilities, which is still paramount. However, this kind of education usually enlists to willing participants.

Nevertheless, one of the most problematic elements of cybersecurity is the fast and constantly evolving nature of cybersecurity breaches increased over time, which requires the fast and constantly evolving nature and expansion of cybersecurity methods. The traditional approach in IT security focus on resources on the most critical assets and processes and protects them against the biggest known threat event attacks, which leaves some less important assets undefended and exposed to some threat event attack risks. Such an approach is insufficient for the current advancement in digitization and new technologies in industrial public and private organizations computer systems, networks, and infrastructure resources. Against this background, cybersecurity professionals assume that the traditional approaches to IT security can become unmanageable because the progress in threat event attacks has become impossibly complex. However, manual and semi-automated cybersecurity monitoring checks and interventions can't keep up with the constantly evolving threat event landscape. Hence, IT-security identified as vulnerable to cybercrime because of their omnipresent accessibility and connectivity, which makes them vulnerable to threat event attacks. In this context, cybersecurity teams worldwide are trying to analyze vulnerabilities in order to gain deeper knowledge about them, to build up upon efficient and effective cybersecurity strategies, such as the National Institute of Standards and Technology Cybersecurity Framework (NIST CSF) with its core five functions: Identity (ID), Protect (PR), Detect (DE), Respond (RE) and Recover (RC) (see Chap. 5). Besides IT security Operational Technology (OT) security is an important threat event attack issue, because it claims industrial organizations to closely assess the cybersecurity risks they face. Operational Technology Security (OTSEC) enables cybersecurity teams to fine-tune traditional, digitized and new technology-based OT processes, reducing the potential of threat event risk(s) and safeguarding them against malware-based threat event attacks. This requires an effective and efficient OT security strategy to prevent the inadvertent or unintended exposure of critical and sensitive data, and infrastructure resources. This enables industrial organizations to prevent their OT-based industrial process and production activities, capabilities, and future intensions from threat event attacks. Leaking such information, threat event attacks cause major damage like identity fraud or theft if employees reuse their login credential across multiple online services [78]. A best practice OTSEC analysis contains a five-step approach, as shown in Table 1.13.

Digital transformation requires that OT systems work together with IT systems. OT network components and industrial networks are connected to IT network components such as processors, data storage, and systems management. With an OT-IT integration, the data collected from physical devices and IIoT devices can also be used to detect security problems or increase systems efficiency. Computer systems perfectly integrate computation with real processes, and provide abstractions, modeling, analysis, and design techniques for their overall advanced digital technology based conceptualization. Their integrated computational and physical capabilities interacts Over-the-Air (OTA) by wireless devices, connecting the cyber with the real physical systems, and processes through new modalities in the era of digital transformation dedicated tasks. However, the consolidation of cyber and real physical components within the digital transformation enable new categories of vulnerabilities, with regard to interception, replacement, or removal of data from the communication channels, which result in malicious attempts by threat event attacks to capture, disrupt, defect, or fail the computer system operations and other. This requires trace the reason for this new vulnerability to understand the way in which the cyber and the real physical components of digital systems are integrated. In this vulnerable space, the cyber-component provides computing capability, processing, control software and sensory support to facilitate the analysis of data received from various sources of the organizations digital systems. However, this

Table 1.13 OTSEC actions with meanings

OTSEC task	Meaning of OTSEC task
Identify sensitive assets	Identification, classification, and periodization of assets of OT systems and devices. This includes detailed shop floor information, customer information, intellectual properties of production, and product research information. It is vital for organizations to focus their resources on protecting this critical data.
Identify possible threat event attacks	Analysis of data traffic to determine potential threat event attacks to OT-based systems. This include continuous behavioral analysis in OT networks to provide security teams with actual and accurate data about what is happening in the network in the context of what, where, when, who, and how. Data from known and unknown threat event attacks continuously is collected for this analysis (see Sect. 2.2.1).
Analyze vulnerabilities	Analyze potential vulnerabilities in the OT-based systems to prevent threat events materialize. This involves assessing the processes and technological solutions that safeguard data and identifying loopholes or weaknesses that attackers could potentially exploit.
Analyze threat event attack level	Each identified vulnerability has a threat event attack level attributed to it. The vulnerabilities ranked on the likelihood of attackers targeting them, the level of damage caused if exploited and the amount of time and work required mitigating and/or repair the damage. The more damage is inflicted and the higher the chance of a threat event attack to occur, the more resources organizations should place in mitigation the risk(s).
Plan to mitigate threat event attacks	Provides organizations with everything they need to plan to mitigate the threat event attack identified. The final step in OTSEC is putting measures in place, to eliminate threat event attacks and mitigate cybersecurity risk(s). Typically this include updating e.g. operational hardware, creating policies around safeguarding sensitive data and providing employee training on security best practices and corporate data policies.

accessibility provides an entrance for launching threat event attacks that can result in:

- Defective operation if the attack affects the digital systems.
- Denial of service, which is common in the cyber domain.
- Destruction and exfiltration.
- Information/data corruption.

These threat event attacks not only have tremendous impact on the cyber entity but cover the overall cyber and real physical system entities. Hence, the defense against threat event attacks is an essential must. In this context, cybersecurity is one of the crosscutting issues in digital transformation today. Cybersecurity is fundamental that authorized messages are delivered at any time, at any place, at the right time in real time, without any disturbance and without malicious incidents. Thus, cybersecurity reduces the risk of threat event attacks and protects industrial organizations and individuals from the unauthorized exploitation of cyber and physical entities.

Unfortunately, nowadays, online users getting access to data or information on the Internet for which they not authorized, which refers to users called hackers (see Sect. 1.6.2). In this regard, it is important to note that the ability to hack is a skill, involving the manipulation of technological assets in some specific way or form. Those, who can successfully attack, called cyber attackers, suggesting that they have the capability to hack. To differentiate between benign and malicious hackers, the hacker community introduced the term cracker, to characterize people who engage in cybercriminal or unethical acts using hacking technologies (see Sect. 1.6.2). This term suggests that a cracker is different from a hacker and has be treated accordingly. Thus, information security or more in general cybersecurity are essential needs that focus on measures to protect authorized messages from malicious cracker attacks.

The fundamental objective in information security refers to protecting information and information systems from unauthorized access, destruction, disclosure, disruption, modification or usage. Therefore, the three fundamental principles in information security are confidentiality, integrity and availability, commonly referrd to as CIA Triad. This is a model to guide policies for security within the industrial public and private organizations, which also form the main objectives of any basic security program. Sometimes, the model refers to as AIC Triad, which means availability, integrity and confidentiality, to avoid confusion with the CIA term that refer to the US Central Intelligence Agency.

1.6.2   CIA Triad

The CIA Triad is a widely accepted model in information security. Information security is a compliance that requires industrial, public and private organizations to use policies, processes and procedures protecting and acting against cybersecurity risk. However, it is not a singular approach and there was also no single author, rather model has developed over time. In this context, concepts pulled from different documents, a 1976 paper for the US Air Force and a paper written in the 1980s about the difference between commercial and military computer systems [78]. As a result, the CIA Triad information security model consists of three core components to guide the security procedures and policies in industrial, public, and private organizations, as shown in Table 1.14.

The CIA Triad, shown in Fig. 1.3, is a well-known model for the development of security policies and procedures, to identify cybersecurity problem areas along with required solutions in the area of computer systems, data, information, infrastructure, network security, and others. It contains policies and procedures that are basis for a successful solution approach to protect information, and adapts to changing threat event possibilities, to withstand and recover rapidly from cybercriminal attacks. However, an effective information security program requires a strategic approach,

Table 1.14 CIA Triad core elements and capabilities

Core elements	Characteristics of core elements
Confidentiality	Vital security characteristic in the era of digital transformation that means protecting information from unauthorized access and misuse, e.g., by a set of rules that limits access to information. Measures undertaken to ensure confidentiality to prevent sensitive information from reaching the wrong people, making sure that the right people get it. Federal Code 44 United States Code, Section 3542, defines confidentiality as *"preserving authorized restrictions on access and disclosure, including means for protecting personal privacy and proprietary information."* This requires a number of access controls and protection as ongoing monitoring, testing and training. Data encryption is a common method of ensuring confidentiality. Thus, user IDs and passwords constitute a standard procedure. Other options include biometric verification, as well as security token, which is a small hardware device that the owner carries to authorize access to a network service, and key fobs, which are small, programmable hardware device that provides access to a physical object, or soft token, a software-based security token, that generates a single-use login PIN. However, to satisfy the desired security requirements the solution should include a holistic consideration.
Integrity	Involves maintaining the consistency, accuracy, and trustworthiness of information and data over its entire life cycle. This covers the topics data integrity and system integrity. Data integrity is the requirement that information, data, and programs changed only in a specified and authorized manner, while system integrity refers to the requirement that a system performs its intended function in an unimpaired manner, free from deliberate or inadvertent unauthorized manipulation. Against this background, a deficiency in integrity can allow for modification of information, data, and programs stored on the digital systems memory, which can affect the crucial and critical operational functions of the digital systems, without ad hoc detection.
Availability	Information, data, and programs are accessible by authorized users when needed is an essential requirement in the era of digital transformation. This ensure rigorous maintaining all system hardware, immediately performing hardware repairs when needed, and maintaining correct functioning operating system environment, free of software conflicts. If crucial and critical operational systems cannot access the needed data when required, the information, data, and programs of operational systems are not secure. Thus, availability is a fundamental feature of a successful deployment of the digital systems in the era of digital transformation. To prevent data loss, a backup copy may be stored in a geographically isolated location, perhaps even in a digital safeguard. Extra security equipment or software such as firewalls or proxy servers can guard against downtime and unreachable data, information, and programs in case of malicious activities such as denial of service (DoS) attacks, and network intrusions.

because it provides a holistic plan on how to achieve and sustain a desired level of cybersecurity maturity (see Chap. 7).

In Table 1.15 attack intentions to the CIA Triad classified in the context of external and internal threat event attackers with regard to the specific skills gained, working in industrial, public and private organizations, as well as contractors or business associates.

Fig. 1.3 CIA Triad

Table 1.15 CIA Triad and respective cyber attacker's skill background

Confidentiality		Integrity		Availability
External	Internal	External	Internal	Possibilities
Cracker:	**Hacker:**	**Cracker:**	**Hacker:**	**Cracker/hacker:**
Malicious: Stealing internal information	*Insider/employee:* Downloading/ exporting internal information	Malicious: Modifying, creating, deleting, secret information	*Insider/employee:* Maliciously modifying creating, deleting, internal information	*Ransomware:* Attack rendering data unusable until backups are accessed or encryption key obtained
	Hacker:	**Hacker:**	**Hacker:**	**Cracker:**
	Insider/employee: Accessing internal information as unauthorized user	*Vendor:* Modifying secret internal information	*Insider/employee:* Maliciously modifying internal information	*Malicious (DoS):* Degrading network performance and affecting organizations operations.
	Employee:			**Cracker/hacker:**
	Losing: Unencrypted thumb drive with internal secret organization information			*Server failure:* At organizations or vendor site.
				Cracker:
				Remote access: Crackers sneak remotely in the attacked network (s) setting up phishing scams, duping users downloading malware-ridden files, which are executed to commence cyber-threat attack like ransomware or others.

In the context of digital transformation, industrial, public, and private organizations try their best to improve their cybersecurity programs (see Chap. 2). However, managing third-party risks becomes overwhelming, especially as they need to incorporate more cloud-based vendors, to support streamline business operations. This requires organizations to pivot their cybersecurity programs to become more proactive, which requires third-party risk management policies and procedures, to act as foundational guidelines for creating a strategy for cybersecurity risk management for vendors, which reduces the potential negative impact on the respective organizations critical computer systems, networks and infrastructure resources. The reason for this strategy is that third parties may pose a variety of cybersecurity risks [79], as shown in Table 1.16.

1.6.3 Cybersecurity Is Still Paramount

Cybersecurity risk refers to threat event attacks to computer systems, networks, and infrastructure resources in industrial, public, and private organizations. However, the requirements for cybersecurity differ in a number of ways to enable a safe and reliable operation. Within today's, expanded protection sphere an enhanced cybersecurity level is required; focusing on central objectives of the CIA Triad (see Sect. 1.6.2), meaning that cybersecurity is still paramount. Tools are therefore necessary to monitor computer systems, networks, infrastructures resources and other supporting detecting threat event intrusion, to respond actively to the treat event attack(s). Threat intelligence is an evidence based actionable knowledge approach

Table 1.16 Third-party cybersecurity risks, characteristics and capabilities

Third partycybersecurity risk	Characteristics and capabilities
Compliance risk	Industry standards and regulations often incorporate third-party vendor risk as a compliance requirement, ensuring that the own organizations apply the organizations risk tolerance to their third-party business partners. If a primary control within the own organization is to update security patches every thirty days, then they should hold third-parties accountable to that standard and monitor, to verify their controls' effectiveness.
Operational risk	Potential risks by third parties can achieve through a technology integrated to provide continued business operations. If the third party experiences a threat event attack that shuts down the service, the own organization may experience business interruption.
Reputational risk	While operational risk applies to the own organizations businesses ability to provide customers a service or product, reputational risk applies to the fact how customers view their partner organization. If the own organization experience a third-party data breach, then the own organization may experience decreased customer trust or loyalty in the aftermath (reputation loss).

and an essential method supporting cybersecurity (see Chap. 2). Furthermore, the method of intrusion detection and prevention (see Chap. 3) of threat intrusions requires algorithms that identifies and analyze data, generated by the various threat intrusions on critical computer systems, networks; infrastructure resources, and others.

Once a sophisticated suspicious or anomalous activity has been identified, a cybersecurity risk analysis must be carried out to determine the likelihood of impact, and the possibly consequence of this risk (see Sect. 2.1.3). This approach relates to risk quantification that evaluates the identified cybersecurity risk(s) to respond actively to the identified threat event attack(s) in real-time to defend is. In this regard, cybersecurity is the probability that a threat event attack provide serious harm under specific conditions. In this regard, a cybersecurity risk is a combination of two factors:

- Probability that a threat event attack occurs.
- Consequences of the threat event.

  Various statistical techniques are

- Program Evaluation and Review Technique (PERT).
- Critical Chain Analysis.
- Decision Tree Analysis.
- Monte Carlo Analysis.
- Sensitivity Analysis.
- Statistical Sampling.

These statistical techniques are used to evaluate and quantify cybersecurity risks. Furthermore, the objective of quantification is to establish a way ranking cybersecurity risks in the order of likelihood and consequences. In the context of likelihood and probable consequence of risk levels qualitative risk analysis applied. Qualitative cybersecurity risk analysis is appropriately early activated in a cybersecurity strategy project and is effective in categorizing which potential cybersecurity risks should be combatted or not and what action must immediately be applied. Qualitative analysis techniques based on classification of acceptance criteria of acceptable risk levels, in conjunction with the security criteria referred to in the CIA Triad (see Sect. 1.6.2). An acceptable risk refer to the likelihood of a threat event attack, whose occurrence probability is low and whose consequences are so slight, that the responsible cybersecurity manager decide to monitor the threat event attacks behavior before combatting it. However, the acceptance criteria have to assure the security requirements for accessing a crucial digital device. Some acceptance criteria examples in relation to the CIA Triad are shown in Table 1.17.

As shown in Table 1.17, the qualitative risk analysis does not give precise values about the risk, but is effective when only little time is available to evaluate risks before they actually happen. However, understanding the impact and potential consequences of risks requires a solid understanding of risks and vulnerability types to decide about likelihood and impact (see Sect. 2.1.3). Nevertheless, many risk models suffer from vague, nonqualified outputs based on partial information or

Table 1.17 Security criteria and likelihood

Security criteria	Likelihood
Confidentiality	If likelihood is higher than low, e.g., an unauthorized individual gets access to sensitive data of a crucial system or network, which causes cybersecurity risk(s), then immediately response is required defending the unauthorized access in real time.
Integrity	If likelihood is higher than low, e.g., an unknown anomaly in system or network behavior is identified, which causes a cybersecurity risk, then immediately response is required to reveal in real time before the identified incident gets executed.

Availability	If likelihood is higher than low means that a potential threat event attack can cause a
	    Denial of services (DoS) attack to a crucial system or network
	    Malfunction of data of a crucial system or network
	If likelihood is higher than moderate, e.g., a service is unavailable for a period of time that happens not more than for example once for every month, then the events behavior should be monitored for further decision.

on unfounded assumptions. Therefore, threat intelligence (see Chap. 2) provides an approach that enable cybersecurity risk models to define risk measurements and be more transparent about their assumptions, variables, and outcomes, helping answering questions such as

- Which threat event actors using this specific threat event(s) and do they target the business or industrial operational production processes?
- Is the threat event(s) observed found more often?
- Is the trend of this threat event(s) up or down?
- Which vulnerabilities does this threat event(s) exploit and are those vulnerabilities present in business or industrial operational production systems?
- What kind of malicious-damage, technical- and/or financial-wise has the threat event(s) caused in industrial organizations?

Based on this background, a simple classification scheme is derived for cybersecurity risk levels, in accordance with the acceptance instance in the CIA Triad. These distinct cybersecurity risk levels based on scores such as extreme, high, moderate, medium and low, as shown in Sect. 2.1.3. A methodological approach that use qualitative values for likelihood levels, described as frequency values, exploits threat event(s) (see Sect. 2.1.3). However, threat event attacks in organizations have different vulnerability types, like:

- Internal, external and remote threat event attack(s).
- Intrinsic vulnerability of digital system security.
- Magnitude of hazards when digital system security is compromised.
- Threat event actors potential motivation.

Vulnerabilities are weaknesses enabling a threat event actor to denial, disturb or disrupt transmission of information required for undisturbed system operation.

Therefore, identifying threat event attack(s) is an essential issue knowing the actor (s) behind the threat event attack(s), because the options on threat event actor's Tactics, Techniques, and Procedures (TTP) are constantly evolving. However, the source of threat events remain at the same time. It's always someone with an intension, which is the real source of threat event attack(s). Thus, preventing and defending threat event attack(s) require threat intelligence (see Chap. 2), a knowledge enabling to prevent or mitigate threat event attack(s). To enhance knowledge of the potential impact of threat event attack(s) requires a cyberattack model and scenario analysis (see Chap. 4).

Today, cybersecurity is still paramount. In the context of the European Union (EU) Directive on the Resilience of Critical Entities (RCE), cyber-resilience of critical infrastructure operation is required. Therefore, the EU initiate a Security of Network and Information Security (NIS) directive, to regulate cybersecurity objectives, to achieve a high common level of cybersecurity across the EU-Member States. The background of NIS are the growing cybersecurity demands in the era of digital transformation with its manifold of digital-based infrastructure resources. The further development of NIS result in NIS2, a new cybersecurity-strategy with the focus on the directive of Critical Entities Resilience (CER), a network of Security Operations Centers (SOCs) and new measures to strengthen the EU Cyber-Diplomacy Toolbox, because the threat events landscape has changed considerably. Against this background, the NIS2 Cybersecurity Directive refers to the following:

- Reinforced rules for a high common level of cybersecurity across the EU.
- Supervision for medium and large organizations across more than a dozen key sectors.
- Establishment of a higher level of mandatory, reviewable, and sanctionable cybersecurity measures for risk management, security governance, incident reporting/recovery, resilience and network, system and application security.
- Risk reviews of security practices for major connected third party services providers.

NIS2 correlates with the CER directive and both seem to address the hybrid cyber-physical nature of OT. *The CER Directive is* transposited in parallel to the NIS2 Directive. Together they address current and future online and offline risks, from cyberattacks (NIS) to physical attacks and natural disasters (CER). CER focuses on physical rather than digital resilience measures for critical entities. National authorities perform reviews, and assess the effectiveness and accountability of the risk management process, and provide significant support to entities. The CER directive is, like NIS2, currently in the final publication state.

Currently, the NIS2 directive is prepared for official publishing by the EU parliament. Thereafter, adoption by national legislative bodies across the EU Member states is done so that NIS2 can come into effect no later than 2024.

In 2020, the enisa NIS investment report [80] presents the findings of a survey, examining approaches to cybersecurity of some EU Member States. Compared to their US counterparts, data shows that the EU organizations allocate on average 41% less to cybersecurity than their US counterparts [81]. This is an alarming situation given the availability of ever more sophisticated spyware. Spyware is software with malicious behavior that aims to gather information about a person or organization and send it to another entity in a way that harms the targeted user.

Pegasus is such a spyware developed by the Israeli cyber-arms company NSO Group, named after Pegasus a winged horse of the Greek mythology. Pegasus is a Trojan horse computer virus. As of 2022 Pegasus was capable of reading text messages, tracking calls, collecting passwords, location tracking, accessing the target devices (iPhones) microphone and camera, and harvesting information from apps [82–85]. Pegasus is being widely used against high-profile targets. News of the Pegasus spyware caused significant media coverage that called the most sophisticated smartphone attack ever. It was the first time that a malicious remote exploit used iOS jailbreaking to gain unrestricted access to an iPhone [82]. Jailbraking is the use of a privilege escalation exploit to remove software restrictions imposed by the software manufacturer.

Another commercial spyware capable of infiltrating mobile phones and stealing everything digital inside of them like videos, pictures, text messages, search history, passwords, call logs and other is Predator. Predator spyware typically sold to high-paying government clients by a company called Cytrox, a secretive surveillance firm based in North Macedonia. Cytrox is owned by an Israeli parent company called Intellexa [86]. The New York Times obtained a copy of a nine-page Intellexa pitch for Predator to an Ukrainian intelligence agency in 2021, the first full such commercial spyware proposal to be made public [87].

In 2014, the Executive Office of the president of the United States has published the Cybersecurity Enhancement Act (Public Law 113-274) that requires the National Science and Technology Council and the Networking and Information Technology Research and Development Program to develop, maintain, and update every 4 years a Cybersecurity Research and Development (R&D) Strategic Plan. This plan also addresses priorities established by the 2018 National Cyber Strategy of the United States or America, including both its domestic and foreign priorities, and by the Administrations FY 2021 Research and Development Budget Priorities Memorandum. The Plan identifies the following goals for cybersecurity R&D [88]:

- Understand human aspects of cybersecurity.
- Provide effective and efficient risk management.
- Develop effective and efficient methods for deterring and countering malicious cyber-activities.
- Develop integrated safety-security framework and methodologies.
- Improve systems development and operation for sustainable security.

To realize the goal of a secure cyberspace, the Plan carries forward the essential concepts from the 2016 Federal Cybersecurity Research and Development Strategic Plan [89], including the framework of four interdependent defensive capabilities [88]:

- *Deter*: Ability to discourage malicious cyber-activities by increasing the cost to, diminishing the spoils of, and increasing the risks and uncertainty of potential cyber attackers.

- *Protect*: Ability of components, systems, users, and critical Infrastructure to efficiently resist malicious cyber-activities, and to ensure confidentiality, integrity, availability (see Sect. 1.6.2) and accountability.
- *Detect*: Ability to efficiently detect, and even anticipate, cyber attackers decisions and activities, given that perfect cybersecurity is not possible and that computer systems, networks, infrastructure resources and other should be assumed to be vulnerable to malicious cyber-activities.
- *Respond*: Ability of defenders, defenses, and infrastructure to react dynamically to malicious cyber-activities by efficiently adapting to disruption, countering the malicious activities, recovering from damage, maintaining operations while completing restoration, and adjusting to similar future activities.

These four elements are similar but not identical to the five core functions in the National Institute of Standards and Technology's (NIST) Cybersecurity Framework (CSF) for improving critical infrastructure cybersecurity.

To advance the priorities and objectives of the 2018 National Cyber Strategy of the United States of America and the Administration's FY 2021 Research and Development Budget Priorities Memorandum, the Plan outlines research objectives in the following priority areas [90]:

- Artificial intelligence (AI).
- Quantum Information Science.
- Trustworthy Distributed Digital Infrastructure.
- Privacy.
- Secure Hardware and Software.
- Education and Workforce Development.

Advancements in the defensive capabilities and priority areas critically depend on progress in human aspects, research infrastructure, risk management, scientific foundations, and transition to practice [88].

Thus, enhancing the trustworthiness of computer systems, networks, infrastructure resources, and others is still paramount, which requires enhancing creation and deployment practices of cybersecurity. Promising research involves AI and machine learning to detect errors in programs, identify security vulnerabilities, and make it easier for software engineers to design security into their systems.

1.7 Digital Transformation and Circular Economy

Circular economy is a term describing an industrial system in which the potential use of materials and energy is optimized and created object's return to the system at the end of their viable life cycle. The circular economy contains development, production, selling, consumption and waste management to minimize the generation of waste. Against this background, the circular economy is in direct opposition to today's consumerist culture, considered as linear structure, sometimes expressed as take, make and dispose. In the consumerist culture waste is an integral element of consumerism because its emphasis is on promoting the purchase of new goods, which often translates to disposing of older but still viable products, materials and other. In this context, circular economy requires the manufacturing- and production industry to rethink more than just their resource consumption footprints and energy efficiency. Circular economy based on preserving the value of products, materials and other resources within the economy for as long as possible, and generating as little waste as possible, towards a sustainable, low carbon resource-efficient and competitive economy. In this context, circular economy aims to keep products, components and materials at their highest utility and value at all times. With regard to the technical cycle, materials remain in a closed-loop system of reuse, remanufacture refurbishment or recycle activities [91]. As reported in [92] three global changes support the transition towards circular economy as indicated in Table 1.18.

Against this background, IIoT enables intelligent machines in intelligent manufacturing and production environments. They gather, analyze and exchange data, sense changes in the manufacturing and production job floor and process the results obtained for further decisions. In this context, IIoT represents an intelligent information network of physical entities such as sensors, actuators, machines, devices and other that allows interaction and cooperation between them. This enables to reach innovative common goals through the comprehensive connection of industrial entities, supply chain sustainability and supply value chains in manufacturing and production job shop floors. Supply chain sustainability is a holistic view of supply chain processes, logistics and technologies that affect the environment, social, economic and legal aspects of supply chain components. Sustainability also means identifying the source of raw materials, ensuring good conditions for workers and reducing the carbon footprint. Against this background,

Table 1.18 Context and characteristics in circular economy

Context	Characteristics
Resource	Limitation of raw natural material resources leads to constant increase in prices and is becoming more volatile that in the past, making the recovery of raw materials from products at the end-of-life more attractive.
Emerging technologies	Big data and analytics Cloud computing Internet of things (IoT) Industrial internet of things (IIoT) 3D printing Machine learning And others Enabling creation of new business model that enhance Products utilization Reuse Remanufacturing Refurbishment Recycling
Numbers of green-oriented consumers	Rapidly growing clientele that increasingly occur among customers behavior show instead of ownership attitude an increasingly share behavior

supply chain sustainability and Product Life Cycle Sustainability (PLCS) are integral to the circular economy. Materials and products have a closed-loop life cycle and all elements that used creating a product are reused, recycled or remanufactured rather than disposed. Moreover, IIoT promotes new business models and activities, such as reshaping industrial productions and services in the context of digital transformation, to achieve high-quality and low carbon footprint in the industrial ecosystem. In this regard, IIoT is a new digital paradigm in industry that shifts the focus from individual digital technologies to an integrated and systematic approach [35].

Depending on their market volume, industrial production proceeds differently selecting raw material for the design, produce, sell and dispose of their products. Sustainability in circular economy requires along the manufacturing value chain downstream recycling processes. Therefore, a circular economy requires a starting point as general perspective, the Proactive Design for Manufacturability (PDM) approach with the scope optimizing all manufacturing functions towards a sustainable, low-carbon, resource efficient and effective competitive economy [93]. PDM is a systemic approach that covers "the full range of policies, techniques, practices and attitudes that cause to be designed for the minimum manufacturing cost, the optimum achievement of manufacturing quality and the optimum achievement of life-cycle support, serviceability, reliability and recyclability" [94]. Therefore, everyone in the product development team has to be aware how products manufactured to achieve the circular economy goal. This requires early and proactive participation from

- Marketing (and even customers).
- Manufacturing and engineering (and even quality assurance).
- Factory workers (and even early training in digital transformation technologies).
- Specialists (bring in new know how at early stage).
- Quality assurance (continuously).
- Service/repair/warranty (know how database and digital field reporting).
- Finance.
- Purchasing (100% know how about parts/components).
- Vendors (trustable, same Cybersecurity level standard).
- Regulation Compliance.
- Lawyers.

Based on the proactive PDM approach, digital transformation concepts such as digital tagging and tracking technologies be integrated in materials and products to generate core data for a sustainable material product database. This approach enable a sustainable product life cycle that include elements such as reuse, recycling and disposal of valuable materials. The sustainable material product data-base requires a strategic decision developing a national (or global) data base to foster the circular economy approach with a domain specific ontology about raw materials, manufacturing, assembly, consumers, services, reuse, recycle, disposal and environmental impact. In this context, the ontology organizes essential domain knowledge, properties and relationships, which provide convenient formalism for modeling and for implementing solutions to application tasks.

The PDM model of circular economy is based on the sustainable material product database approach with which product life cycle, which contains valuable raw material, can be recorded and evaluated in real-time [95]. This enables

- Monitoring of material(s), (resource(s)), product flow(s), and process(es).
- Early market detection for end consumer environment-friendly products.
- Cross-references between exported waste and sales to end consumers.

In addition to the conceptual approach of the sustainable material product database, its implementation and integration into a circular economy concept, it requires proactive design to:

- Cost (Target Costing & ROI).
- Customization (Satisfying Customers).
- Delivery (On Time).
- Ease of Assembly (Seamless Assembly).
- Ease of Recycling (CO_2 Footprint).
- Ease of Service (Self-Awareness & On Demand Spare Parts – 3D Printing).
- Environmental Compliance (Minimizing Greenhouse Footprint).
- Expandability (Software-as-a-Service; Re-Configurability).
- Quality (Quality Management).
- Regulatory Compliance (National and International Laws).
- Reliability (If-Then Clauses).
- Testability (Cloud-as-a-Monitor).
- Safety and Security (Against Misuse, Checking for Anomalies).

Depending on the industrial sectors and markets, production companies proceed differently to select their raw materials, design, production, sales, and dispose of their products. In this context, digital transformation offers a sustainable solution for circular economy along the value chain through innovative digital technologies. Furthermore, it create new and innovative solutions to be used in circular economy, e.g., based on digital tagging and tracking to create core data records for a sustainable product and material database. This approach support disposal and recycling of valuable materials. However, this come up with the problem of cybersecurity. The new and innovative digital technologies usable in the circular economy and the sustainable material product database have to be cyber-secure, which requires further efforts. In addition, the conception of the sustainable material product database also requires legislative and technical means, necessary for the structure and updating of entries.

The necessary optimization for the extraction of secondary raw materials be qualitatively and quantitatively analyzed and the strategic acceptance and political decision-making can be supported to

- Mobilization of processing and production residues.
- Cycle of old products in the context of an end-of-life perspective.
- Assessment of sustainability issues in the context of innovative and environment-friendly development.

Employees in contact with older products not specially trained to correctly classify and treat disposal, which require specific training on technical solutions. Remedy achieved through technical detection and recognition solutions through sensors. Using the sustainable material product database approach requires an adequate technical solution to place sensors in easy-to-use portable devices or integrate them into return machines. Recorded products promptly registered in sustainable material product database demonstrate proper disposal of valuable materials and products. This provides recycling companies with information required about type, origin, manufacturability, raw materials and other, contained in a product. Hence, in circular economy, digital tagging and tracking of objects at the very early beginning at the design and development process is mandatory. Thereafter, digital tagging and tracking during orchestration within the overall manufacturing shop for product development and manufacturing in the circular economy environment is mandatory, generating core data records, essential to feed the sustainable material product database with real product and material data.

At the recycling site, (semi) automatic intelligent classification tools for materials and end-of-life-cycle products must be available with assess to sustainable material product database, to identify and assign the material and/or product as e-Waste or other type of valuable waste for the circular re-cycling processes.

Developing recycling systems that autonomously can recycle eWaste or other type of valuable waste at a high-level of efficiency is an important issue. Against this background, a (semi) automated mapping is essential that can be used by recycling companies to extract products containing raw materials from informal landfills. For this approach, it is necessary to provide appropriate access to a sustainable material product database. Nevertheless, the aforementioned technical solutions also require an integrated cybersecurity solution, to prevent possible misuse of the systems and to keep whereabouts of affected resources up to date (see Sect. 1.6 and Chap. 2).

1.8   Exercises

1.8.1   Digital Transformation

What is meant by the term *Digital Transformation*?
Describe the characteristics and capabilities of the Digital Transformation.
What is meant by the term *Fourth Industrial Revolution*?
Describe the characteristics and capabilities of the Fourth Industrial Revolution.
What is meant by the term *Industry 4.0*?
Describe the efforts of Industry 4.0.
What is meant by the term *Business Transformation*?
Describe the characteristics and capabilities of the Business Transformation.
What is meant by the term *Cultural Transformation*?
Describe the characteristics and capabilities of the Cultural Transformation.
What is meant by the term *Process Transformation*?

Describe the characteristics and capabilities of the Process Transformation.
What is meant by the term *Emerging Technologies*?
Describe some emerging technologies used in Digital Transformation.
What is meant by the term *Artificial Intelligence*?
Describe the characteristics and capabilities of Artificial Intelligence.
What is meant by the term *Autonomous Robot*?
Describe the characteristics and capabilities of the Autonomous Robot.
What is meant by the term *Big Data and Analytics*?
Describe the characteristics and capabilities of Big Data and Analytics.
What is meant by the term *Cloud Computing?*
Describe the characteristics and capabilities of Cloud Computing.
What is meant by the term *Cloud-as-a-Service*?
Describe the main four cloud services in detail.
What is meant by the term *Edge Computing*?
Describe the advantages and disadvantages of Edge Computing.
What is meant by the term *Internet of Thing*?
Describe the characteristics and capabilities of the Internet of Things.
What is meant by the term *Industrial Internet of Things*?
Describe the characteristics and capabilities of the Industrial Internet of Things.
What is meant by the term *Simulation*?
Describe the characteristics and capabilities of Simulation.
What is meant by the term *Digital Transformation Challenges?*
Describe the strategic planning process.
What is meant by the term *Return on Investment*?
Describe the meaning of Return on Investment in the Digital Transformation.
What is meant by the term *Leadership in Digital Transformation?*
Describe the advantages and disadvantages of it.

1.8.2 Cybersecurity

What is meant by the term *Cybersecurity*?
Describe the characteristics and capabilities of Cybersecurity.
What is meant by the term *Risk Analysis*?
Describe the characteristics and capabilities of Risk Analysis.
What is meant by the term *Application Security*?
Describe the characteristics and capabilities of Application Security.
What is meant by the term *Computer Security?*
Describe the characteristics and capabilities of Computer Security.
What is meant by the term *Data Security*?
Describe the characteristics and capabilities of Data Security.
What is meant by the term *Information Security*?
Describe the characteristics and capabilities Information Security.
What is meant by the term *Operational Security*?

Describe the advantages and Disadvantages of Operational Security.
What is meant by the term *CIA*?
Describe the characteristics and capabilities of the C, I, and A.
What is meant by the term *Vulnerability*?
Describe the characteristics and capabilities of the Vulnerability.
What is meant by the term *Compliance Risk*?
Describe the advantages and disadvantages of the Compliance Risk.
What is meant by the term *Operational Risk?*
Describe the advantages and disadvantages of the Operational Risk.
What is meant by the term *Cybersecurity is still Paramount*?
Describe the advantages and disadvantages of it.

1.8.3   Circular Economy

What is meant by the term *Circular Economy*?
Describe the advantages and disadvantages of the Circular Economy.
What is meant by the term *Supply Chain Sustainability*?
Describe the characteristics and capabilities of a Supply Chain Sustainability.
What is meant by the term *Product Life-Cycle Sustainability*?
Describe the characteristics and capabilities of the Product Life-Cycle Sustainability.
What is meant by the term *Design for Manufacturability*?
Describe the advantages and disadvantages of a Design for Manufacturability.
What is meant by the term *Product and Material Data Base*?
Describe characteristics and capabilities of a Product and Material Data Base.
What is meant by the term *Cybersecurity in Circular Economy*?
Describe the characteristics and capabilities of Cybersecurity in Circular Economy.

References

1. Hess, T., Benlian, C., Wiesbck, F.: Options for Formulating a Digital Transformation Strategy. In: MIS Q. Exec., Vol. 15, No. 2, pp. 123139, 2016
2. Nadkarni, S. Prügl, R.: Digital Transformation: A Review, Synthesis and Opportunities for future Research. In: Management Review Quarterly, Vol. 71, pp. 233–341, 2021
3. Verhof, P.C., Broekhuizena, T., Barth, Y., Battachayaa, A., Donga, J.Q., Fabiana, N., Haenleine, M.: Digital Transformation: A Multidisciplinary Reflection and Research Agenda. In: Journal of Business Research, Vol. 122, pp. 889–901. 2021
4. Möller, D.P.F.: Cybersecurity in Digital Transformation: Scope and Applications. Springer Nature 2020
5. Vial, G.: Understanding Digital Transformation: A Review and Research Agenda. In: The Journal of Strategic Information Systems, Vol. 28, No 2, pp. 118–144. 2019
6. Matt, C., Hess, T., Benlian, A.: Digital Transformation Strategies. In: Bus. Inf. Syst. Eng. Vol. 57, pp. 339–349, 2015

7. Yoo, Y., Boland, R.J., Lyytinen, K., Maichrzak, A.: Organizing for Innovation in the Digitized World. In: Organization Sciencs, Vol. 23, No 5, pp. 1398–1408, 2012
8. Lasi, H., Fettke, P., Kemper, H.-G., Feld, T., Hoffmann,, M.: Industry 4.0. In: Business and Information Systems Engineering, Vol. 6, No. 4, pp. 239–242. 2014
9. Möller, D.P.F.: Guide to Computing Fundamentals in Cyber-Physical Systems: Concepts, Design Methods, and Applications. Springer International Publisher, 2016
10. Hund, A., Wagner, H.-T., Beimborn, D., Weitzel, T.: Digital Innovation: Review and Novel Perspective. In: Journal of Strategic Information Systems. Vol. 30, No. 4, 101695, 2021
11. Nambisan, S., Wright, M., Feldman, M.: The Digital Transformation of Innovation and Entrepreneurship: Progress, Challenges and Key Themes. In: MIS Quarterly, Vol. 41, No.1, pp. 223–238, 2017
12. Fichman, R.G., Santos, B.I., Zheng, Z.E.: Digital Innovations as a Fundamental and Powerful Concept in the Interformation System Curriculum. MIS Quaterly, Vol. 38, No 2, pp. 329–343, 2014
13. Li, L., Su, F., Wei, H., Zheng, W., Mao, J.Y.: Digital Transformation by SME Entrepreneurs: A Capability Perspective. In: Special Issue Paper, Wiley and Sons Publ., 2017
14. Lazzazzara, A., Ricardi, F., Za, S.: Exploiting Digital Ecosystems: Organizational and Human Challenges, Springer Nature, 2020
15. Chanias, S., Hess, T.: Understanding Digital Transformation Strategy: Insights from Europe's Automotive Industry. In: PACIS Proceedings, 2016. https://aisel.aisnet.org/pacs2016/29 (Accessed 12.2022)
16. Bharadwaj, N., Noble, C.: Finding Innovations in Data Ricj Environments. In: Journal in Production Innovation Management, Vol. 34, No.5, pp. 560–564, 2017
17. Grassmann, O., Sutter, P.: Digital Transformation (in German). Hanser Publ. 2013
18. Remane, G., Hildebrandt, B., Hanelt, A., Kolbe, L.M.: Discovering New Digital Business Model Types: A Study of Technology Start Ups from the Mobility Sector. In: PACIS 2016. Proceedings provides by AIS Electronic Library 2016
19. Möller, D.P.F., Haas, R.E.: Guide to Automotive Connectivity and Cybersecurity – Trends, Technologies, Innovations, Applications. Springer Nature, 2019
20. https://en.wikipedia.org/wiki/Enabling_technology. (Accessed 12.2022)
21. Grabovia, M., Peezer, D., Popic, S., Knezevic, V.: Providing Security Measures of Enabling Technologies in Industrial Internet of Things – A Survey. In: ZINC Proceed., pp. 28–31, 2016
22. Lou, D.: AI made from a sheet of Glas can recognize numbers just by looking. In: New Sientist Techology, July, 8. 2019
23. Cartwright, J.: Glass that will store your info forever. In: New Scientist Technology, October, 8. 2014
24. Adee, S.: Heat scavengers promise heat bonanza. In: New Scientist Technology, October,8, 2014
25. Manyika, J., Chui, M., Bisson, P., Woetzel, J., Dobbs, R., Bughin, J., Aharon D.: The Internet of Things: Mapping the Values beyond the Hype.IcKinsey Global Institute Report, 2015
26. Paschek, D., Fraghici, C.T., Draghici, A.: Automated Business Process Management: In Time of Digital Transformation using Machine Learning or Artificial Intelligence. In: MATEC Web of Conferences, 121, 04007, 2017. https://doi.org/10.1051/matecconf/20171210. (Accessed 12.2022)
27. https://bulitin.com/artificial-intelligence,2020. (Accessed 12.2022)
28. Haleem, A., Javaid, M.: Additive Manufacturing Applications in Industry 4.0: A review. In: Journal of Industrial Integration and Management, Vol.4, No.4, World Scientific Computing, 2019
29. Beyca, O.F., Hancerliogullari, G., Yazicio, I.: Additive Manufacturing and Applications. In: Ustundag, A., Cevikcan, E. (Eds.): Industry 4.0: Managing the Digital Transformation, pp. 216–234, 2018

30. Esengün, M., Ince, G.: The Role aof Augmented Reality in the Age of Industry 4.0. In: Ustundag, A., Cevikcan, E. (Eds.): Industry 4.0: Managing the Digital Transformation, pp. 210–215, 2018
31. Salkin, C., Oner, M., Ustundag, A., Cevekcan, E.: A conceptual Framework for Industry 4.0. In: Ustundag, A., Cevikcan, E. (Eds.): Industry 4.0: Managing the Digital Transformation, pp. 3–23, 2018
32. Wang, S., Wan, J., Zhang, D., Li, D., Zhang, C.: Towards Smart Factory for Industry 4.0: A Self-Organized Multi-Agent System with Big Data based Feedback and Coordination. In: Computer Networks, pp. 1–11, 2016
33. https://aethon.com/mobile-robots-and-industry4-0. (Accessed 12.2022)
34. Fitzgerald, J., Quasney, E.: Using Autonomous Robots in Drive Supply Chain Innovation: Future Trends in Supply Chain. Deloitte Research Report, 2021
35. Möller, D.P.F., Vakilzadian, H., Haas, R.E.: From Industry 4.0 Towards Industry 5.0. In: Proceedings IEEE-EIT International Conference on Electro-Information Technology (EIT), pp. 61–68, 2022
36. Carayannis, E.G., Dezi, L., Gregort, G., Calo, E.: Smart Environment and Techno-Centric and Human Centric Innovation for Industry and Society 5.0: A Quintuple Helix Innovation System View Towards Smart Sustainable and Inclusive Solutions. In: Journal of the Knowledge Economy, Vol. 13, No. 2, pp. 926–955, 2022
37. Loshin, D.: Big Data Analytics: From Strategic Planning to Enterprise Integration with Tools, Techniques, NoSQL and Graph. Morgan Kaufman Publ. 2013
38. Househ, M.S., Kushnirak, A.D., Borycki, E.M. (Eds.): Big Dat, Big Challenges: A Healthcare Perspective. Springer International Publ. 2019
39. Morabito, A.: Big Data and Analytics: Strategic and Organizational Impact. Springer International Publ. 2016
40. D'Onofrio, S., Meier, A. (Eds.): Big Data Analytics: Basics, Case Studies and Potential Uses. Springer International Publ. 2021
41. Raj, P., Raman, A., Nagaraj, D., Duggirata, S.: High-Performance Big Data Analytics. Springer International Publ. 2015
42. Sedkaoui, S.: Data Analytics and Big Data Analytics: Information systems, Web and Pervasive Computing. Wiley & Sons Publ. 2018
43. Lee, J., Kao, H.-A., Yang, S.: Service Innovation and Smart Analytics for Industry 4.0 and Big Data Environment. In: Proceedings 6th CIRP Conference on Industrial Product Service Systems, pp. 3–8. 2014
44. Jain, K.H.: Big Data andH adoop. Khanna Book Publ. 2017
45. https://gist.github.com/alexalemi/2151722 (Accessed 12.2022)
46. Hair, J.F., Hult, G.T.M., Ringle, C.M., Sarstedt, M.A.: A Primer on Partial Least Squares Structural Equations Modeling, 2nd Edition. Sage Publ. 2017
47. Al-Jacobi, J., Mohamed N.: Industrial Applications of Blockchain. In: 9th IEEE Annual Computing and Communication Workshop and Conference (CCWC), pp. 550–555, 2019
48. Khanaghal, S., Ansari, S., Paaroutis, S., Oviedo, L.: Mutualism and the Dynamics of New Platform Creation: A Study of Cisco and Fog Computing. In: Strategic Management Journal, Vol. 43, pp. 476–506, 2021
49. Shi, W., Cao, J., Zhang, Q., Li, Y., Xu, L.: Edge Computing: Vision and Challenges. In: IEEE Internet of Things Journal, Vol. 3, No. 5, pp. 637–646, 2016
50. Shi, W., Dustdar, S.: The Promise of Edge Computing. In: Computer Vol. 49, No.5, pp. 78–81, 2016
51. Chaouki, H. (Ed.): The internet of Things: Connecting Objects to the Web.
52. Geng, H.: Internet of Things and Data Analytics Handbook. Wiley & Sons Publ. 2016
53. Jeschke, S., Breecher, C., Song, H., Rawat, D.B. (Eds.): Industrial Internat of Things: Cybermanufacturing Systems. Springer International Publ. 2017
54. Murphy, K.P.: Machine Learning: A Probabilistic Perspective, MIT Press, 2012

55. Manoj, K.S.: Industrial Automation with SCADA: Concepts, Communication and Security. Notion Press, 2019
56. https://www.verveindustrial.com/resources/ot-systems-management-whitepaper-thank-you-page/?submissionGuid=8cf40fb0-d10f-4d70-be9b-f191a5525826 (Accessed 12.2022)
57. Rouskas, G.N.: Network Virtualization: A Tutorial. North Carolina State University, 2012
58. Han, B., Gopalkrishnan, V., Ji, L., Lee, S.: Network Function Virtualization: Challenges and Opportunities for Innovations. In: IEEE Coomunication Magazine, pp. 90–97, 2015
59. Moeller, D.P.F.: Mathematical and Computational Modeling and Simulation: Fundamentals and Case Studies. Springer Publ., 2004
60. Grieves, M., Vickers, J.: Digital Twin: Mitigating Unpredictable, Undesirable Emergent Behavior in Complex Systems. 2015. https://research.fit.edu/media/sitespecific/researchfitedu/camid/documents/Origin-and-Types-of-the-Digital-Twin.pdf (Accessed 12.2022)
61. Tao, F., Cheng, J., Qi, Q., Zhang, M., Zhang, H., Sui, F. : Digital twin-driven product design, manufacturing and service with big data. In: Int. J. Adv. Manuf. Technol. Springer Publ. 2017
62. Parrot, A., Warshaw, L.: Industry 4.0 and the Digital Twin – Manufacturing meets its Match. In: Deloitte University Press 2017. https://www2.deloitte.com/insights/us/en/focus/industry-4-0/digital-twin-technology-smart-factory.html (Accessed 12.2022)
63. Aziz, N.A., Mantora, T., Khaiduran, M.A., Mrushid, A.F.: Software Defined Network (SDN) and its Security Issues. In: Proceedings IEEE International Conference on Computing Engineering and design, pp. 40–45, 2018
64. Dargie, W., Poellabauer, C.: Fundamentals of Wireless Sensor Networks: Theory and Prctice. Wiley & Sons, 2010
65. Wiczanowski, M., Boche, H., Stanczak, S.: Fubndamentals of Resouce Allocation in Wireless Networks: Theory and Algorithm. Springer International Publ. 2009
66. https://www.twéchtarget.com/searchnetworking/definition/6G (Accessed 12.2022)
67. Steiner, F., Grzymek, V.: Digital Souvereignty in the EI. In. Vision Europe. Bertelsmann Foundation, 2020
68. https://ec.europe.eu/info/sites/default/files/proposal-regulation-single-market-digital-sercice-digital-services-act_en.pdf (Accessed 12.2022)
69. https://www.bmwi.de/Redaktion/DE/Publikationen/Digitale-Welt/das-project-gaia-x.html
70. https://www.bcg.com/capabilities/digital-technology-data/digital-transformation/overview?utm_source=search&utm_medium=cpc&utm_campaign=digital&utm_description=none&utm_topic=digital_transformation&utm_geo=global&utm_content=digital_transformation_consulting&gclid=CjwKCAjwyaWZBhBGEiwACslQo05Hr3pj0crkCMN2t8vw-zX86js_q6s_KcW-aM8VYlC7E3lsDNWtdxoC6v8QAvD_BwE (Accessed 12.2022)
71. Kumar, V., Rezaei, J., Akhberdina, V., Kuzmin, E. (Eds.): Digital Transformation in Industry – Trends, Management, Strategies.Springer Internat. Publ. 2021
72. https://static.sw.cdn.siemens.com/siemens-disw-assets/public/cTs1Jfi7pjGr2y2E53ge2/en-US/LifecycleInsights_Report-ROIofDigitalTransformation.pdf (Accessed 12.2022)
73. https://terakeet.com/blog/digital-transformation-examples/ (Accessed 12.2022)
74. Dahlström, P., Desmet, D., Singer, M.: The Seven Decisons that Matter in a Digital Transformation: A CEO Guide to RE-Invention. In: Digital McKinsey Report, 2017
75. Holten, A.L., Brenner, S.O.: Leadership Style and the Process of Organizational Change. In Journal of Leadership and Organization Development, Vol. 36, No. 1, pp. 2–16, 2015
76. Schellinger, J., Tokarski, K.O., Kissling-Näf, I.: In: From Digital Transformation to Digital Corporate Managemen (Iín German). In: Digital Transformation and Corporate Management, Schellinger J., Tokarski, K.O., Kissling-Näf, I. (Eds.), pp. 1–10, Springer-Gabler Publ. 2020
77. Sow, M., Aborbie, S.: Impact of Leadership on Digital Transformation. In. Business and Economic Research, Vol. 8, No. 3, pp. 139–148, 2018
78. https://www.fortinet.com/resources/cyberglossary/operational-security (Accessed 12.2022)
79. https://securityscorecard.com/blog/how-ti-write-rthird-party-risk-management-policies-tprm-and-procedures (Accessed 12.2022)
80. Negreiro, M.: ENISA and a new cybersecurity act, ERPS, European Parliament, 2019

81. Negreiro, M. The NIS2 Directive – A high common level of cybersecurity in the EU. Briefing EU Legislation in Progress, 2022
82. https://translate.google.com/?hl=de&sl=de&tl=en&text=Dies%20%20ist%20ein%20 bedenklicher%20Zustand%20vor%20dem%20Hintergrund%20der%20verf%C3% BCgbarekeit%20immer%20ausgekl%C3%BCgelter%20Spionage%20softwrae&op=translate (Accessed 12.2022)
83. Marczak, B., Scott-Railton, J., Razzak, B.A., Al-Jizawi, N., Anstis, S., Berdan, K., Deibert, R.: Pegasus vs. Predator: Dissident's Doubly-Infected iPohnes Reveals Cytrox Mercenary Spyware. Citizen Lab Research Report No. 147, University of Toronto, December 2021.
84. Scott-Railton, J., Marczak, B., Razzak, B.A., Anstis, S., Herrero, P.N.-, Deibert, R.: New Pegasus Spyware Abuses Identified in Mexico. Citizen Lab Report No. 159, University of Toronto, October 2022.
85. Parsons, C.: Cybersecurity Will Not Thrive in Darkness: A Critical Analysis of Proposed Amendments in Bill C-26 to the Telecommunications Act, Citizen Lab Report No. 160, University of Toronto, October 2022
86. https://gizmodo.com/thanasis-koukakis-sues-intellexa-over-predator-spyware-1849625793 (Accessed 12.2022)
87. https://www.nytimes.com/interactive/2022/12/08/us/politics/intellexa-commercial-proposal. html (Accessed 12.2022)
88. Federal Cybersecurity Research and Development Strategic Plan. Executive Office of the President of the United States, 2019
89. Federal Cybersecurity Research and Development Strategic Plan. Executive Office of the President of the United States, 2016
90. Federal Cybersecurity R&D Strategic Plan Implementation Roadmap. Executive Office of the President of the United States, 2020
91. Baumgart, M., McDonough, W., Bollinger, A.: Cradle-to-Cradle Design: Creating Healthy Emissions: A Strategy for Eco-effective Product and System Design. Journal Clean Production, Vol. 15, pp. 1337–1348, 2007
92. Bressanelli, G., Perona, M., Saccani, N.: Reshaping the Washing Machine Industry through Circular Economy and Product Service Business Models. In: Procedia CIRP, Vol. 64, pp. 43–48, 2017
93. Möller, D.P.F.: Design for Manufacturability. Lecture at HUAWEI International Expert Conference, Shenzhen, China, 2019
94. Stoll, H.W.: Design for Manufacturability. In: Simultaneous Engineering. Allen C.W. (Ed.), pp. 23–29, SME Press 1990
95. Möller, D.P.F.: Enhancement in Intelligent Manufacturing through Circular Economy. In. Proceedings IEEE International Conference on Electro-Information Technology (EIT), pp. 87–92, 2020

Chapter 2
Threats and Threat Intelligence

Abstract Digital technologies used in digital transformation are essential for every industrial, public, and private organization. In industry, the automation with its connectedness has revolutionized the economic situation of work through the transition of the fourth technological wave, termed Industry 4.0. However, this also enables various types of threat event attacks. Therefore, this chapter introduces us to the virtual world of Threats and Threat Intelligence. The intention of threat event attacks is to inflict harm, intruding viruses, worms, malicious code, and others, to get unauthorized access to computer systems, networks, infrastructure resources, and others, misusing or manipulating operational tasks. Threat event attacks also tries to shut down targeted computer systems, networks, infrastructure resources, and others, making it inaccessible to regular operation tasks or users, which can be achieved by a Denial of Service attack or others, through flooding the targeted object with traffic, or sending it information that triggers a crash. Sometimes, targeted organizations incorporate threat data feeds as simple indicator of artifacts in their systems and/or networks that present a stream of information, e.g., on anomalies in their data flows but not knowing what to do with this additional data. For some reason, they potentially put an additional burden on analysts to decide what to consider dangerous and what to ignore. However, an important prerequisite is that the analysts have the appropriate tools in order to be able to make such decisions at all, which is a reason using Threat Intelligence. Threat Intelligence is evidence-based knowledge, including context, mechanisms, indicators, implications, and action-oriented advice about an existing or emerging menace or hazard to assets. This intelligence can be used to inform decisions regarding the subject's response to that menace or hazard" [1]. In this regard, Threat Intelligence is knowledge that allows preventing or mitigating threat event attacks rooting in data, like who is attacking and what is their motivation and capabilities to get better information for decision-making about the potential cybersecurity risks. In this context, threat perception describes an essential capability and estimated intention to vulnerability and opportunity to really executing the threat event attack(s). Therefore, a solid understanding of the impact and potential consequences of threat event attacks is required, to cyber secure mission critical computer systems, networks, infrastructure resources, and others. This requires a detailed analysis of well-known and documented threat event attacks, which may cause a loss of confidentiality, integrity, and availability, as

D. P. F. Möller, *Guide to Cybersecurity in Digital Transformation*, Advances in Information Security 103, https://doi.org/10.1007/978-3-031-26845-8_2

described in the CIA Triad (see Sect. 1.6.2) of computer systems and data it stores or processes that finally reveal identifiable interactions or dependency patterns. Such recognizable interactions or patterns require further study to highlight their specifications, their severity, and impact and, if possible, to develop a method to reveal them before executed. In this context, Threat Intelligence addresses these issues making use of machine learning (see Chap. 8) to automate data collection and processing unstructured data from disparate sources and connect them by providing context on Indicators of Compromise (IoC) and Tactics, Techniques and Procedures (TTP) of threat event actors. Therefore, Chap. 2 introduces to Threats, Threat Events and –Intensions, Threat Event Types and their Cybersecurity Risk Level, the Likelihood and Consequence Level, and Risk Management and Risk Analysis in Sect. 2.1. Section 2.2 refers to Threat Intelligence, taking into account the problem of *Known-Knowns, Known-Unknowns*, and *Unknown-Unknowns*, Digital Forensic and Threat Intelligence platforms. Furthermore, Sect. 2.2 introduces, besides Threat Intelligence in Threat Event Attack Profiling, Threat Event Lifecycle and Threat Intelligence Sharing and Management Platforms. Section 2.3 contains comprehensive questions from the topics Threats and Threat Intelligence, followed by "References" with references for further reading.

2.1 Threats

The term Threat refers to any event during which a threat event actor acts against an industrial, public, or private organizations asset(s), data and others with the intention to inflict harm, pain, injury, damages, danger, or hostile action on someone in retribution for something done or not done. More in general a threat event attack has the potential for causing undesirable consequences or impacts. Therefore, any threat event attack issued by a cybercriminal actor with the potential impact on organizations normal (regular) operations through unauthorized access, modifying information, Denial of Service (DoS), and others, is termed a Threat or a threat event attack. A threat event actor is either a single actor or a group of actors that take part in an action to cause harm to organizations computer systems, networks, infrastructure resources, and others. Against this background, perception of threat event attacks is the conscious or unconscious estimation that something or someone has the capability and/or intension of malicious actions against targeted individuals or organizations. Therefore, threat event attacks are a cybersecurity risk issue, with the potential to harm organizations' critical and crucial technical systems and networks. Some prominent cybersecurity risks are shown in an alphabetic ranking in Table 2.1 [2].

With regard to Table 2.1, cybersecurity tries to defend threat event attacks to targeted organizations systems, networks, infrastructure devices, and others. Thus, threat event attacks result in a security breach that requires prevention and defense. In this context, threat intelligence is the information industrial, public, and private organizations have to deal with to understand the intension and impact threat event attacks executed by threat event actors. Therefore, gathering information is essential

Table 2.1 Cybersecurity attack risk(s), characteristics, and prevention

Cybersecurity risk	Characteristics	Prevention
Cross-site attack (XSS Attack)	Third party targets a vulnerable website, loads dangerous code onto the site. Once a regular user accesses website, code is delivered either to user system or browser, causing unwanted behavior	Encryption is usually required on the host's side. Users can also install script-blocker add-ons to their browser if they prefer additional browsing control
Crypto-jacking aattack	Attack to install malware, which forces infected system perform crypto-mining, a popular form of gaining crypto-currency	Keep all security apps/software updated and make sure firmware on smart devices is also using the latest version
Distributed Denial of Service (DDoS)	Attack in which malicious parties target servers and overload them with user traffic	Stopping a DDoS requires identifying malicious traffic and halting access
Dridex attack	Financial Trojan that infects computers via phishing emails or existing malware. Able to steal passwords, bank credentials, and personal information, used in fraudulent transactions	Safety patches must be installed regularly
Drive-by-download attack	Malicious code is delivered to a system or device. No action is needed on the user end, where typically, they need to click a link or download an executable	Avoid suspicious websites. Normally, compromised websites are flagged by search engines and anti-malware programs
Eavesdropping attack (traffic interception)	Occurs when third-party listens to info sent between user and host. Information stolen varies based on traffic but is often used to take log-ins or valuable data	Proactive defense avoid compromised websites. Encrypting network traffic – through VPN – is another preventive method
Malware attack	Unwanted piece of programming or software installs itself on a target system, causing unusual behavior: deny access to programs, delete files, steal information, spread by itself to others systems	Combination of caution and antivirus to thwart most malware concerns
Man-in-the-middle attack (MitM)	Third party hijacks session between client and host. Hacker generally cloaks itself with a spoofed IP address, disconnects the client, request information from the client	Encryption and use of HTML5 are recommended
Password theft attack	Password changed and details lost. Unwanted third party managed to steal or guess passwords.	Two-factor authentication is a robust protection method, as it requires an additional device to complete the login
Phishing attack	User typically receives message or email that requests sensitive data. Sometimes, phishing message appears official, using legitimate appearing addresses to click on links that give away sensitive information	Official emails from organizations do not request personal data; this is a giveaway there is malicious intent

(continued)

Table 2.1 (continued)

Cybersecurity risk	Characteristics	Prevention
Ransomware attack	Installs itself on a user system or network. Once installed, it prevents access to functionalities until a ransom is paid to third parties	Keep all security apps/software updated and make sure firmware on smart devices is also using the latest version
Social engineering attack	Method for attempting to deceive users into giving away sensitive details. Occur on any platform. Malicious parties often accomplish their goals, such as utilizing social-media info	Remaining skeptical of suspicious messages, friend requests, emails, or attempts to collect user info from unknown third-parties
SQL injection attack	Data manipulation, to access information that isn't available. Essentially, malicious third parties manipulate SQL queries to retrieve sensitive information	Implementation of smart firewalls is one prevention method; application firewalls detect and filter out unwanted requests
Trojan virus attack	Attempts to deliver its payload by disguising itself as legitimate software. One technique used was an "alert" a user's system was compromised by malware, recommending a scan, whereby the scan actually delivered the malware	Avoid downloading programs or emails from unrecognized senders or those that attempt to alarm the user to a serious problem
Water-Hole Attack	Group infects websites an organization frequently uses. Goal is to load a malicious payload from the infected sites	Antivirus passively identifies dangerous scripts. Keep website scripts off as a default if organization suspects infection
Zero-day exploit attack	Attack takes advantage of an overlooked security problem, looking to cause unusual behavior, damage data, and steal information	Zero-day vulnerability can exist for an extended period before discovered. Users must maintain good safety habits until a fix is released

to identify and to prevent intrusion of malicious code in valuable organizations Information Technology (IT) and Operational Technology (OT) resources. However, it is important to be aware about the difference between IT and OT systems. OT controls Industrial Automation and Control Systems (IACS) and Information Technology (IT) control the data. IT is all about ensuring the confidentiality, integrity, and availability of the CIA Triad (see Sect. 1.6.2) of systems and data. Furthermore, Table 2.1 shows that threat event attacks include different attack types and techniques. Typical common threat events actors are [3, 4]:

- Careless employees
- Cybercriminals
- Disgruntled insiders
- Hacktivists
- Nation States
- Cyber-Terrorists

In this regard, cybersecurity awareness should be the number one priority for all organizations to respond to the numerous challenges due to increasingly persistent and devious threat event attacks. At the same time, there is a need to educate industrial, public, and private organization employees in cybersecurity awareness and cybersecurity risks [5] about motivational factors for cyber-activism, cybercriminal intent, cyber-espionage, cyber-terrorism, and finally cyber-war.

In traditional cybersecurity risk approaches, the focus is primarily on understanding and addressing risks, vulnerabilities, and configurations, which is not sufficient as it should be, because mobile technology and the so-called social media offers entirely new vistas for modern threat event attacks like cyber-swarming [4]. Therefore, an effective prevention and defense against current and future threat events requires a focus on outward, understanding adversary's behavior, capability, and intent. In this regard, threat intelligence provide a form gathering information about *who* is attacking, *what* motivation and capabilities are available at the threat event actor(s)' side, and *what* kind of Indicators of Compromise (IoC) can be identified to support decision-making for developing a successful strategy against cybersecurity risks. In this context, knowledge about threat event actors' motivation and the threat event attack defending team experience is essential, which enables understanding enough about the nature of threat event attacks, organizations face to come up with intelligent Threat defense decisions. Alongside commoditized threat event attacks, advanced capabilities that were rare in the past are now commonplace. The reason for this is that today's cyber-world is a global and multidimensional information and communication technology (ICT) network, through which threat event actors can connect via fixed, remote or mobile data nodes, and virtually move within. Nevertheless, threat event actors' behavior solely is not focused on widespread, disruptive activities, but rather it often involves more targeted, lower-profile multi-stage threat event attacks, an intertwined entirety in the cyber-world that aims specific tactical objectives and establishs a persistent foothold in organizations. Therefore, threat intelligence solutions today make also use of artificial intelligence (see Sect. 1.2.1), big data and analytics (see Sect. 1.2.5), and other emerging technologies (see Sect. 1.2).

The term artificial intelligence (AI) coined in a study of AI submitted by John McCarthy (Dartmouth College), Marvin Minsky (Harvard University), Nathaniel Rochester (IBM), and Claude Shannon (Bell Telephone Laboratories). The workshop took place in 1956, generally recognized as the official birthdate of this new field. AI is the science and engineering approach of making intelligent machines. In this context, AI is the study of how to train machines doing things humans doing better. In computer science, an ideal intelligent machine is a flexible rational agent that perceives its environment and takes actions that maximize its chance to succeed at an arbitrary goal. Meanwhile AI and its tools have attracted attention from the literature to application in business organizations; especially by the advances in machine learning-based techniques. Today, AI technologies are used for a manifold of practical problems, such as

- AI-powered assistants
- Automated vehicles
- Creating smart content
- Fraud prevention
- Personalized learning

Machine learning (see Sect. 1.2.12 and Chap. 8) completely performs or mimic cognitive functions that are intuitively associated with human intelligent behavior, such as learning, understanding, and problem-solving capability. These methods allow automating data collection and processing them. Furthermore, the data used can be unstructured as well as from different disparate sources and used for Indicators of Compromise (IoC) and Tactics, Techniques, and Procedures (TTP) of threat event actors. IoC is the evidence that a threat event attack has taken place. IoC gives valuable information about what has happened but can also be used to prepare for the future and prevent against similar threat event attacks. TTP describes an approach of analyzing an Advanced Persistent Threat (APT) operation, a threat event scenario, showing a series of steps, ending with the threat event actor having an established foothold in the intruded organizations computer systems, networks, infrastructure resources, and others. APT incidents are typically assumed nation state related. However, the same behavior is exhibited by APT actors engaged in conducting cybercrime, financial threats, industrial espionage, cyber-hacktivism and cyberterrorism. Against this background cybercrime covers traditional fraud or forgery, attacks against information systems, DoS or hacking attacks [4, 5], as shown in Table 2.2. In this regard, threat intelligence seeks to understand and to characterize narratives in the context of what sort of threat event attack actions have occurred or are likely to occur, which requires asking questions like [6]:

- How to detect and recognize this action?
- How they mitigated?
- Who are the relevant threat event actors, and what do them intent?
- What are their capabilities with regard to TTP they have leveraged over time and are likely to leverage in the future?
- What sort of vulnerability, misconfigurations, or weaknesses are likely to target?
- What actions they have taken in the past?
- And others.

2.1.1   Threat Events and Threat Intensions

Threat events are information security threats that are malicious actions to corrupt or steal data or disrupt organizations computer systems, networks or infrastructure resources, and others. Thus, a threat event in information security involves an attempt to obtain, alter, destroy, remove, implement, or reveal information without authorized access or permission and thus compromise the CIA Triad (see Sect. 1. 6.2). Such threat event attack cause a loss of confidentiality, integrity, and

Table 2.2 Major characteristics and capabilities of APT

Advanced Persistent Threats (APT)	
Threat form	Characteristics and capabilities
Cybercrime	Primary activity is financial in order to generate profit for the cybercriminal. Include different types of profit-driven cybercriminal activities like ransomware attacks, email and internet fraud, attempts to steal financial accounts, credit card or other payment card information, and others
	Carried out against computer systems or infrastructure resources to damage or disable them. Other activities use computer systems or networks to spread out malware, illegal information, or others. However, some cybercrime activities do both (i) target computer systems to infect them with a computer virus, (ii) which afterward is then spread to other devices, entities, or an entire organizations network. There are many types of cybercrime:
	Credit card fraud: Fraud committed using a payment fraud
	Crypto jacking: Cybercrime where a criminal secretly uses a victims computing power to generate cryptocurrency
	Cyber extortion: Involving a threat event attack coupled with a demand for money or some other response in return for stopping or remediating the threat event attack incident
	Cyber-eSpionage: Silent threat event attack over IT infrastructure
	Identity theft: Cybercrime to obtain personal or financial information of another person to use their identity to commit fraud
	Jailbreaking: Use privilege escalation exploit to remove software restrictions. Jailbroken device permits root access within operating system and provides right to install software
Cyber-hacktivism	Hacking a secure computer system, network, website, or infrastructure resource in an effort to convey a social or political message, The cyber-actor who carries out the cyber act of hacktivism is a hacktivist. Hacktivist use the same tools, methods, and software as hackers. Used attack forms are: malware, viruses, Trojans, computer worms, phishing, and other malicious software, in addition to DDoS attacks, brute-force attacks, and similar attacks
Cyberterrorism	Threat event actors act often premediated, sometimes politically motivated against mission critical systems, infrastructures, programs, and data that threatens violence, damaging, or disrupting them. Sometimes the term including any threat event attack, which intimidates or generates fear to the targeted entities
Cyber-Militia	New term that describes the call in a nation state for hackers to form a cyber-militia, with the aim just destroying, e.g., essential infrastructure, and weakening opponents. However, officially, no one knows exactly who belongs to this but there are also exceptions. Ukrainian Deputy Prime Minister Mykhailo Fedorov officially declared on Twitter: We are creating an IT army. We need digital talents. All operational tasks will be given here: t.me/itarmyofurraine (https://t.me/itarmyofurraine)

availability of the computer system and the data stored or processed. In this context, a threat event attack refers to an occurrence through which organizations computer systems, networks, infrastructure resources, and others is exposed. Hence, a threat event attack is an incident that results in a breach-in into organizations assets as well as their critical and crucial systems or devices. Cybersecurity threat event attacks show major impacts on the industrial, public, and private organizations. There are different possible types of threat event attacks that affect cybersecurity, introduced as knowledge base about threat event attacks, including but not limited to (see Sect. 2.1.2):

- *Active threat events*: Attempt to alter targeted system resources or effect their operations. They involve modifications of the data stream or create false statements. Active attack types are as follows:

  - *Masquerade attacks*: Takes place when someone pretends to be a different entity
  - *Modification of messages*: Means that a portion of a message is altered or delayed or recorded to create an unauthorized effect
  - *Repudiation*: Done by either sender or receiver denying later sending or receiving a message
  - *Replay*: Involves the passive capture of a message and its subsequent transmission to create an authorized effect
  - *Denial of Service (DoS)*: Specific target like network disruption of an entire network by disabling the network by overloading it by messages so as to degrade performance

- *Passive threat events*: Attempts to make use of information form a system but does not affect targeted system resources. They are a more an eavesdropping or a monitoring of transmission. Their goal is to obtain transmitted information being. Passive attack types are as follows:

  - *Release of message content*: Gain knowledge from the content of transmission

- *Traffic analysis*: Assuming an encrypted message. The threat event is unable to extract any information from the message. However, the threat event actor can determine location and identity of communicating host and observe frequency and length of exchanged messages. Then this information might be helpful in guessing the nature of the communication that takes place.
- *Botnet threat event*: The term botnet comes from the words robot and network. In this regard, a bot is an entity infected by malicious code, which then becomes part of a computer system, network, or net of infected devices, all controlled by a single threat event actor or threat event actor group. The infection by malware involves a collection of Internet-connected devices allows threat event actors, also called cyberattackers, to control them remotely, unknown to their user. Malware have different forms used by threat event actors such as Adware, Keylogger, Ransomware, Remote Access, Rootkit, Spyware, Trojan horse, Virus, and Worms. Botnet malware design done to scan automatically computer systems and networks for common vulnerabilities not been patched, in hope infecting as many as possible entities. Thus, cybercriminals use botnets to instigate botnet threat events that include malicious activities such as credential leaks or credential-stuffing threat events, leading to account takeovers, unauthorized access to a device and its connection to a network, data theft to steal data and intrude Distributed Denial of Service (DDoS) threat events that cause unplanned application downtime.

Defending threat event activities, organizations try to minimize risks associated with insider threat events as follows [7]:

- Monitor network performance and activity to detect any irregular network behavior
- Keep the operating system up to date
- Keep operating systems and all software up-to date an install any necessary security patches
- Educate users not to engage in any activity that puts them at risk of bot threat infections or other malware, including opening emails or downloading attachments of messages or clicking links from unfamiliar sources
- Implement anti-botnet tools that find and block bot viruses

However, most firewalls and antivirus software include basis tools to detect, prevent, and remove botnets.

- *Brandjacking threat events*: Brandjacking is the unauthorized use of an organizations brand to hijack a corporate brand's online presence. Typical brandjacking threat events include setting up a bogus social media account, which is affiliated with a particular brand or service, hacking a brand's legitimate social media account or labeling social media posts with a hashtag that is being officially used by a particular brand or service. Against this background, it is difficult for brands to counteract the problems that online brandjacking cause. However, targeted organization may threaten legal action, but censoring brandjacks can backfire, drawing more attention to the situation and further injuring the brand's image. Perhaps the best way for organizations to counteract online brandjacking is to continue to build a social media presence, promote a positive brand image and partners with online brandjacks who are also well-wishers [8].
- *Clickjacking threat event*: Clickjacking events use multiple transparent or opaque layers to trick users clicking on the top of a web page. However, on top of that web page, the threat event actors has loaded an iframe with the targeted email account, and lined up exactly the delete of all messages button directly on top.

As shown in Table 1.15, two major groups of threat event actors are of interest, internal and external actors, whereby internal actors divided into two groups, insiders and employees. Insider attacks occur when individuals or entities, close to an organization have authorized access to computer systems, networks, or infrastructure resources, and intentionally or unintentionally misuse that access to affect negatively the organizations sensitive data, business assets, systems or networks. In this regard, an insider is a wholly or partially trusted individual or entity, which has legitimate access to organizations resources. The negative intention skills of an insider is an additional essential factor, posing threat events by a malicious motivation of intruders. Unfortunately, there may be less cybersecurity against insider threat events, because many industrial, public and private organizations focus on protection against external threat event attacks. An employee of an organization can be an internal threat event actor who performs exploits within the organizations network(s). If authorized he tries to find vulnerabilities in organizations networks to fix them. In case non-authorized employees use threat event flaws for some personal gain, they are called hackers. However, insider data breaches can also occur through accidental means. Insider threat events caused by intentional or malicious threat events from employees as a rouge insider, also called cybercriminal threat actor. With regard to their behavior profile, they may belong to the cracker type, as shown in Table 1.15.

Furthermore, an insider threat event can be a malicious threat to any industrial, public, or private organization, which comes from someone within the organization, such as actual employees, former employees, contractors or business associates, who have inside information concerning the industrial, public or private organizations security practices, data assets, computer systems, networks, and infrastructure resources. The insider-based threat event attack involve frauds, theft of confidential or commercially valuable information, theft of intellectual property or the sabotage of computer systems and networks, whereby this behavior profile belong to the cracker type. Insider threat events including but not limited to be:

- *Compromised actors*: Have access to credential or computing devices, compromised by an outside threat event. These incidents are more challenging to identify, since real threat events coming from outside, and posing a much lower risk of being identified.
- *Emotional actors*: Steal data or destroy industrial, public, or private networks intentionally, such as a former employee who injects malware or logic bombs in corporate computers on his last day at work (see Table 1.15).
- *Malicious insider threat events*: Someone who take advantages of their access to inflict maliciously and intentionally abuses to legitimate credentials, typically to steal information, harm an industrial, public or private organizations, executed from frustrated employees, fired employees, unserious contractors or business associates, and others. This threat event attack is known as turnclock that has an advantage over other threat actors, because they are familiar with the security policies and procedures of an organization, as well as its vulnerabilities. The following best practice steps help to reduce the risk of malicious insiders threat event attacks [9]

  - *Protect critical assets*: Asking questions such as:

    - What critical assets organizations possess?
    - How assets of an organization be prioritized?
    - What does an organization understand about the current state of each asset?

  - *Enforce policies*: Document organizations policies and enforce them to prevent misunderstandings. Everyone in the organization should be familiar with cybersecurity procedures and should understand their rights in relation to intellectual properties (IP) so they do not share privileged content that they have created.
  - *Increase visibility*: Deploy solutions to keep track of employee actions and correlate information from multiple data sources.

- *Promote security cultural changes*: Ensuring security is not only about expertise but also about attitudes and beliefs. To combat negligence and address the
  drivers of malicious behavior, employees should be educated regarding cybersecurity issues and work to improve employee cybersecurity awareness and
  satisfaction.
- *Negligent insider threat events*: Someone who make unintentionally errors or
  disregard policies that place their organization at risk. Also known as careless
  insider.
- *Unintentional actors*: Expose data accidentally, such as an employee who
  accesses organizations data assets through public Wireless-Fidelity (Wi-Fi) without the knowledge that it's unsecured.

Defend activities organizations can take into action to minimize the cybersecurity
risk(s) associated by insider threat events, which include the following activities:

- Limit employees' access to only the resources they need to do their jobs.
- Train employees and contractors on cybersecurity awareness before allowing
  them to access the organizations network.
- Enhance regular cybersecurity training on threat event attack awareness due to
  unintentional and malicious insider threat events.
- Set up contractors and other externals with temporary accounts that expire on
  specific dates, such as the dates their contracts end.
- Implement two-factor authentication that requires each user to provide a second
  piece of identifying information in addition to a password.

External or outsider threat events occur, when an individual threat event actor or a
group of threat event actors seeks to gain protected information by infiltrating and
taking over profile of a trusted user from extern (outside) the organization, hoping to
steal business assets and secrets. External or outside threat event actors, also called
moles, gain confidence of a current employee to get insider access to organizations
computer systems, networks, or infrastructure resources (see Table 1.15), Examples
of external or outsider threat events including, but not limited to be:

- *Cybercriminal groups*: Individuals or entities involved in digital crime. Responsible or more advanced threat events with better organizational background and
  access to greater resources and funding, executing APT, either cybercriminal
  groups or nation states.
- *Distributed Denial of Services (DDoS) attacks*: Threat event attacks intruded
  when a computer system or network targeted become overwhelmed through
  massive traffic and cannot respond to service requests required. These computer
  system or networks infected with viruses, controlled by one over all threat event
  actor.
- *Infiltrator threat events*: External threat event actors obtain legitimate wise access
  credential based on illegal ways, without legal authorization with which they can
  implement malicious software into organizations systems; also known as mole.

- *Session hijacking threat events*: Is a man-in-the-middle threat event attack type, wherein a hijacked session between a network server and a client happen. In hijacking sessions, threat event actors replace it's IP address for the client's and the server continues the session. During this threat event the server believes it's still communicating with the trusted client.
- *Drive-by-download threat events*: Refer to malicious scripts spread malware around the web. The threat event actor looks for insecure websites and plant scripts in the code on one of the pages. Thus, the malicious scripts install malware on the computer of a web page visitor. In other cases, the threat event actors redirect the visitor to a website that the threat event actor own, where they may be hacked. Drive-by-download attacks happen most commonly on web pages, pop-ups and emails.
- *Password threat events*: Passwords mostly used to protect data on the web. However, they are a main area of threat events for attackers and hackers. Having a person's password can open up all sorts of additional hacks. Hackers obtain passwords by sniffing the connection to a network gaining access to the passwords. Hackers also obtain passwords by using social engineering tactics and physically looking around desks and offices.
- *Phishing and spear phishing threat events*: Type of social engineering threat event, used to steal user data, including login credentials, credit card numbers, or to force the victim to do something special. Used to gain foothold in organizations networks as part of a larger incident, such as an APT incident. In this scenario, employees compromised to bypass security perimeters, distribute malware inside a closed organizations environment, or gain privileged access to secured data assets. Likewise, spear phishing is the practice of targeting a specific person or organization in a specific attempt to obtain valuable information or exploit a person or organization.

Thus, the question of how to prevent both groups requires answers about:

- Changing passwords regularly and immediately after an employee leaves
- Installing employee-monitoring software to support reducing the cybersecurity risk of data breaches and the theft of intellectual property by identifying careless, disgruntled or malicious insiders
- Keeping track of employee access levels and changing them accordingly and frequently
- Enhancing training with the goal:
  - Do not share passwords.
  - Do not reuse passwords.
  - Ensure that passwords meet at least medium security level requirements like strong passwords that have at least eight characters, at least one uppercase letter and one under case letter, at least one special character.

However, external data breaches are mostly malicious with the intension stealing internal information, maliciously modifying or deleting secret information, encrypting information to blackmail money, degrading network performance and affecting organizations operations, as well as providing computer systems and server failures, intruded by external threat event actor's (also termed cyberattacker). These types of threat event attacks belongs to the group of unknown threat events which an organization did not expect and do not know when such threat events will happen (see Table 2.10). This is a dangerous situation for organizations, because in such threat event attacks valuable time passed by before finding a solution and take action to implement it, to protect the system against such kind of threat event attack.

Internal threat event intensions by insiders and employees have a broad range of threat event possibilities. They are among the most difficult ones to be detected and prevented threat event intensions. Insiders and employees have gained computer systems and network access as well as computer systems and network knowledge. They have physical access to critical and crucial areas in the organizations, and are able to take access with less effort. Their motivation to hack can result from revenge or entitlement, while employment terminates. Often they take customer or industrial organizations data assets and information with them, when moving to an industrial competitor organization. Another way threat event insiders may crack the organizations system to cause damage are often network planning logic bombs which damage after the employees leave. Thus, an Intrusion Detection System (IDS) and Intrusion Prevention System (IPS) is essential to support detecting and preventing threat events attack intensions (see Chap. 3), like:

- Block access and mirror data if insider or employee terminates job in organization
- Least privilege
- Monitor logs
- Safe and dynamic authentication

However, the question is how to identify a potential insider or employee intrusion as threat event actor. One way doing so is using insider or employee data, based on account identification, which can help tracking back by comparing insider or employee account data with attack data, to identify identical signatures. In the context of confidentiality and integrity, the following cybersecurity risk types happen as a best practice roadmap:

- Access accounts and applications normally not used for their daily jobs
- Conduct furtive instant-messaging chats
- Create network accounts for themselves
- Network access during office hours
- Network access outside office hours
- Perform large downloads and file copying
- Successful login of account for former employee
- Search for web sites that cater of disgruntled employees

Therefore, installed employee-monitoring software enable to reduce the risk of data breaches and the theft of intellectual property by identifying careless, disgruntled, or malicious insiders.

Another important threat event type is password guessing. This is one of the most common threat event types by insiders or employees. The threat event actor knows a login and then attempts to guess the password for it by:

- Common word search
- Defaults
- Exhaustively searching all possible passwords
- Short passwords
- User info like variations of names, birthday, phone number, interests, and others and thereafter run a
- Security breach by successful login

Finally, the question is raised, "What personal traits do these insiders or employees may have?" After analyzing a pool of cracking cases provided by computer crime investigators, for instance prosecutors and security specialists, the researchers conclude that insider or employee computer criminals tend to be [10]:

- Introverted individuals who admit to being more comfortable solving cognitive problems than interacting with others in the workplace
- More dependent on online interactions than on face-to-face interactions
- Ethically flexible individuals who can easily justify ethical violations
- Of the opinion that they are somehow special and thus deserving of special privileges
- Lacking in empathy and thus seeming not to reflect on the impact, their behaviors have on others or on the organization

2.1.2 Threat Event Types

Risky end-users behavior affects all industrial, public and private organizations around the world. These results in implications that can be immediate, like a ransomware attack, or become a threat event attack that lies in wait, like an incident of compromised credentials. This happens, when malicious actor's has unknowingly obtained a legitimate user's credentials and use them to access the organizations assets. Thus, organizations have to be aware about users' personal cybersecurity habits during worktime, and that the cybersecurity team may overestimate end- users' understanding of fundamental cybersecurity awareness in general knowledge about possible cybersecurity risks, as mentioned in [11]:

- Understanding cybersecurity fundamentals like phishing, ransomware, Wi-Fi security, and others
- Password management and attention to physical security measures
- Use of data protection like Virtual Private Network (VPN) and file backups
- Application of best practices related to activities like social media sharing and use of workplace based devices
- And others

Therefore, a much deeper knowledge is required to identify threat events, and prevent and defend them. How critical threat events impact society's infrastructure shows the following examples for the year 2022:

- Hackers used a DDoS attack to shut down the National Telecommunication Authority of the Marshall Islands, disrupting the Internet services.
- A major Israeli telecommunication provider targeted by a DDoS attack taking multiple Israeli governmental websites offline.
- Several oil terminals in some Europe's biggest ports across Belgium and Germany fell victim to a ransomware cyberattack, rendering them unable to process incoming barges.
- Threat event research identified campaigns by two North Korean government-backed groups targeting employees across numerous media, fintech and software companies, using phishing emails advertising fake job opportunities.

Thus, it is essential to identify threat events and to understand the intensions of threat event actors, based on the CIA Triad (see Sect. 1.6.2). Based on that knowledge it's possible to determine the likelihood versus the impact of the threat event attack (see Tables 2.7 and 2.8) to take further action on cybersecurity. In this regard, threat events are typically composed of a single or a combination of different types of options. Most common threat events are as follows:

- *Advanced Persistent Threat (APT)*: Prolonged and targeted threat event attack in which unauthorized actor's access to networks and remain undetected for an extended period. To gain access, APT groups often use advanced attack methods, including exploits of zero-day vulnerabilities, highly targeted spear phishing, and other social engineering techniques. The primary intent of an APT is to steal data, disrupt business operation, and damage infrastructures. Some APTs are very complex that they require full-time administration to maintain the compromised systems and software in the targeted network. APT attackers coordinate their activities with the security measures of their targeted organization and often attack them several times. APT groups often receive instruction and support from governments or government agencies;
- *Botnet*: Sets of computers that are under the control of a malicious controller without the knowledge of their owners, to send files (including spam and viruses) over the Internet to other computers. Every element of this system is a bot. Most of the affected systems are private computers. Bot computers can work in distant directions without the owner's knowledge. However, there are a few clues that can indicate a possible botnet attack: slow computing speed, high CPU usage as well as unnecessary sudden pop-ups. For example, Emotet is a malicious botnet software.
- *Brute-force attack*: A password cracking technique. Trial-and-error method to obtain information such as user password or personal identification number (PIN). Based on software that autonomously generates a large number of consecutive guesses to the value of the desired data. However, this attack need time to run to provide anything usable. This threat event used by cyber-criminal

attackers to crack encrypted data, but also by security analysts to test an organizations network security. Using brute-force attacks is an exhaustive effort rather that employing intellectual strategies. A cybercriminal might break in a safe password by trying many possible combinations, which means a brute-force attack tries all possible combinations of legal characters in a sequence. Thus, brute-force can break weak passwords in short time like minutes or seconds. Strong passwords can typically take hours or days. Strong passwords do not use common words that are easy to guess. A strong password includes sufficient length and mix of special characters, numbers and uppercase and lowercase letters. Therefore, defending brute-force attacks by increasing password length or increase password complexity, increases time to brute-force crack.

- *Cross-site scripting attack*: Represent attacks, which use third-party web resources to run scripts in the attacked web browser or scriptable application. In the cross-site scripting attack type attacker injects a payload with malicious JavaScript, or VB Script, or Active X, or Flash, into a website's database. When a website user request a page from the website, the website transmits the page with the attacker's payload as part of the HTML body to the user's browser, which executes the malicious script. Cross-site scripting attacks can also exploit vulnerabilities that can enable an attacker to steal cookies, log key strokes, capture screenshots, discover and collect network information and remotely access and control the attacked computer system or network of the user.

- *Data destruction*: Destroying data stored on electronic/digital media so that it is completely unreadable and cannot be accessed or used anymore. However, that data destruction is not the same as destroying the media on which data is stored which is a physical destruction.

- *Data manipulation*: Threat event attack form of an indirect type of sabotage by altering data, compromising indirectly compromise a project that decisions are based on bad data, which have the potential to cause a great damage later.

- *Dictionary search attack*: A password cracking technique. An attacker attempts to search in a targeted system each word in the dictionary for correct passwords. Password dictionaries exist for a variety of topics and combinations of topics. Rather than trying to input every possible permutation, an attacker using a dictionary approach would attempt all the permutations in tits predetermined library. Sequential passcodes, like 12345, and static passcodes, like 00000, tested in the dictionary attacks. If the five-digit permutation particularly is unique, the dictionary attack likely would not guess.

- *Denial of Service (DoS) attack*: Threat event attack type that shut down a machine or network, making it inaccessible to its intended user(s). The two general DoS attack methods are accomplished by flooding the target with traffic for the server to buffer, causing them to slow down and eventually stop, or sending it information that triggers a crash. The most popular flood attacks are as follows:

  - *Buffer overflow attack*: Send more traffic to a network address than the system is able to handle.

- *Internet control message protocol (ICMP) flood attack*: Threat event attack is also known as smurf attack or ping of death attack. Normally ICMP echo-request and echo-reply messages are used to ping a network device to diagnose:

  (a) Health and connectivity of it
  (b) Connection between the sender and the network device to diagnose the health and connectivity of it
  (c) Connection between the sender and the network device

  This means the ICMP flood attack flooding the targeted network devices by sending spoofed packets that ping every computer at the targeted network with request packets, the network is forced to respond with an equal number of respond packets which result in amplified network traffic.

- *SYN flood attack*: Form of DoS attack in which the threat event sends a succession of SYN requests to a targeted system in an attempt to consume enough server resources to make the target system unresponsive to legitimate traffic that the targeted system cannot complete the handshake. Continues until all open ports saturated with requests and none available for legitimate users to connect

- *Distributed Denial of Service (DDoS)*: Threat event attack to disrupt normal traffic of a targeted server, service or network by overwhelming the target or its surrounding infrastructure with heavy Internet traffic

- *Drive-by download attack*: In drive-by download attacks the malicious code is downloaded from a website via a browser, application or integrated operating system without user's permission or knowledge. In this attack form, hackers look for insecure websites and embed a malicious script into Hypertext Transfer Protocol (HTTP) or Hypertext Preprocessor code on one of the pages. The script install malware directly on the user's computer system who visits the website or it might re-direct the user to a website controlled by the hacker(s). Drive-by-downloads can occur when visiting a website or viewing an email message or a pop-up window. Unlike other types of threat event attacks, a drive-by-download attack does not rely on a user to click on anything, e.g., open a malicious email attachment to activate the download to become infected. Just accessing or browsing a website can start a download to enable the attack. Thus, a drive-by-download attack can take advantage of an app, operating system or web browser that contains security flaws due to unsuccessful updates or lack of updates. It is an unintended download of a malicious code on a computer system or another device. Unlike other types of threat event attacks, a drive-by-download attack does not rely on the user to do anything to actively enable the attack.

- *Eavesdropping attack*: This attack type occur through the interception of network traffic. Attacker can obtain passwords, credit card numbers and other confidential information the user(s) send over the network. Eavesdropping can be passive or active:

  - *Passive eavesdropping*: A threat event actor detects the information by listening to the message transmission in the network.

- *Active eavesdropping*: A threat event actor actively grabs the information by disguising himself as friendly unit and by sending queries to transmitters, called probing, scanning or tampering.

Detecting passive eavesdropping attacks is often more important than spotting active ones, since active attacks require the attacker to gain knowledge of the friendly units by conducting passive eavesdropping before.

- *Emotet malware*: Sophisticated Trojan that steal data and load other malware. Emotet benefits from insecure passwords: a reminder of the importance of using strong passwords to protect against cyber-threats.
- *Guessing*: A password cracking technique. An attacker is able to guess a password without the use of tools. If the threat event actor has enough information about the targeted victim or the victim is using a common enough password, the attacker be able to come up with the correct characters.
- *Intellectual property theft*: Threat event attack through digital technologies and Internet file sharing networks, robbing ideas, inventions, and creative expressions that can include everything from trade secrets and proprietary products and parts to movies, music, and software.
- *Man–in-the-middle attack*: Threat event attack when a perpetrator positions him in traffic between a targeted user and an application, either to eavesdrop or to impersonate one of the parties, making it appear as normal exchange of information is underway to finally steal personal information. Critical is that the targeted user is not aware of the man in the middle.
- *Malvertising*: Threat event attack in which perpetrators inject malicious code into legitimate online advertising networks. The code typically redirects users to malicious websites.
- *Malware attack*: Refers to malicious software installed without consent. It can attach itself to legitimate code and propagate and can replicate itself across the Internet. The most common types of malware are as follows:

  - *Macro viruses*: Infect applications such as Microsoft Word or Excel, attached to an application's initialization sequence. When the application is open, the virus executes instructions before transferring control to the application. The virus replicates itself and attaches to other codes in the computer system.
  - *File infectors*: Viruses usually attach themselves to executable code, such as . exe files. Located code install virus. Another version of a file infector attack associates itself with a file by creating a virus file with the same name, but an . exe extension. When the file is open, the virus code will execute.
  - *System or boot-record infectors*: Virus attaches to the master boot record on hard disks. When the system start, it looks at the boot sector and loads the virus into memory, where it can propagate to other disks and computers.
  - *Polymorphic viruses*: Conceal themselves through varying cycles of encryption and decryption. Encrypted virus contains an associated mutation engine, initially decrypted by a decryption program. The virus proceeds to infect an area of code. The mutation engine then develops a new decryption routine and the virus encrypts the mutation engine and a copy of the virus with an

algorithm corresponding to the new decryption routine. The encrypted package of the mutation engine attached to new code, and the process repeats. Such viruses are difficult to detect but have a high level of entropy, because of the many modifications of their source code. Antivirus software or free tools like Process Hacker can use this feature to detect them.

– *Stealth viruses*: Take over system functions to conceal them. They do this by compromising malware detection software so that the software report an infected area as being uninfected. These viruses conceal any increase in the size of an infected file or changes to the file's date and time of last modification.

– *Trojans of Trojan horses*: Program that hides in a useful program and usually has a malicious function. A major difference between viruses and Trojans is that Trojans do not self-replicate. In addition to launching attacks on a system, a Trojan can establish a back door exploited by attackers. Programming a Trojan is required to open a high-number-port so the hacker can use it to listen and then perform an attack.

– *Logic mombs*: Type of malicious software appended to an application and triggered by a specific occurrence, such as a logical condition or a specific date and time.

– *Worms*: Differ from viruses in that they do not attach to a host file, but are self-contained programs that propagate across computer systems or networks. Worms commonly spread through email attachments; opening the attachment activates the worm program. A typical worm exploit involves the worm sending a copy of itself to every contact in an infected computer's email address list. In addition to conducting malicious activities, a worm spreading across the Internet and overloading email servers can result in Denial of Service (DoS) attacks against nodes on the network.

– *Droppers*: Program used to install viruses on computer systems. A not with malicious code infected dropper might not be detected by virus-scanning software. A dropper can also connect to the Internet and download updates to virus software that is resident on a compromised computer system.

– *Ransomware*: Type of malware that blocks access to the victim's data and threatens to publish or delete it unless a ransom is paid. While some simple computer ransomware can lock the system in a way that is not difficult for a knowledgeable person to reverse, more advanced malware uses a technique called crypto viral extortion, which encrypts the victim's files in a way that makes them nearly impossible to recover without the decryption key. A dangerous new trend of ransomware actors have begun ex-filtrating sensitive data from the targeted organization and threatening to publish them if the ransom is not paid.

• *Negligent and malicious insiders*: Employees, associates, and/or affiliates who have legitimate access to an IT system in organizations. Maliciously focused on what assets are at risk of leaving the organization through the IT environment as well as threat events entering the organization through the same means.

- *Password attack*: Attack mechanism to authenticate users to an information system, obtaining passwords is a common and effective attack approach. Access to a person's password obtained by looking around the person's desk, sniffing the connection to the network to acquire unencrypted passwords, using social engineering, gaining access to a password database or outright guessing. The last approach can be done in either a random or systematic manner:

  - *Brute-force password guessing*: Using a random approach by trying different passwords and hoping that one work. Applying some logic by trying passwords related to the person's name, job title, hobbies or similar items.
  - *Dictionary attack*: A dictionary of common passwords used to attempt to gain access to a user's computer system or network. One approach is to copy an encrypted file that contains the passwords, apply the same encryption to a dictionary of commonly used passwords, and compare the results.
  - *Password cracking*: Use an application program to identify an unknown or forgotten password to computer systems or networks resources. With this information a threat event actor obtain unauthorized access, to undertake a range of cybercriminal activities, such as identify theft and fraud.

- *Phishing*: Form of a fraud in which a threat event actor masquerades as a reputable entity or person in email or other communication channels. The threat event actor uses phishing emails to distribute malicious links or attachment that can perform a variety of functions, including the extraction of login credentials or account information from attacked computer systems or networks to gain sensitive, confidential information such as usernames, passwords, network credentials, and more, by posing as a legitimate individual or institution (see Sect. 2.1.1). Some of the more common types of phishing attacks include the following:

  - *Clone phishing attacks*: Use previously delivered but legitimate emails that contain either a link or an attachment. Attackers make a copy (clone) of the legitimate email, replacing any number of links or attached files with malicious links or malware attachments. Since the message appears to be a duplicate of the original, legitimate email, targeted individuals or public and private organizations can often be tricked into clicking the malicious link or opening the malicious attachment.
  - *Evil-twin Wi-Fi phishing attack*: An evil twin is a rogue wireless access point that masquerades as a legitimate Wi-Fi access point, that a threat event actor can gather personal or corporate information without the end-user's knowledge. When targeted individuals or organizations connect to the evil twin Wi-Fi network, the threat event actors gain access to all transmissions to or from the targeted individuals or organizations devices, including user IDs and passwords. Threat event actors can also use this vector to target individuals or organizations devices with their own fraudulent prompts for system credentials, which appear to originate from legitimate systems.
  - *Pharming phishing attacks*: Type of phishing that depends on domain name system cache poisoning to redirect users from a legitimate site to a fraudulent

one, and tricking users into attempting to log in to the fraudulent site with personal credentials.

- *Spear phishing attack*: Directed at specific individuals or organization, using information specific to the targeted individuals or organization gathered more successfully represent the message as being authentic. Spear phishing emails might include references to co-workers or executives at the targeted organization, and the use of the targeted individual's name, location, or other personal information.
- *Voice phishing attack*: Known as vishing, it is a form of phishing that occurs over voice communications media, including Voice over IP (VoIP) or Plain Old Telephone Service (POTS).
- *Whaling phishing attacks*: Type of a spear phishing attack that specifically targets senior executives within an organization with the ability to authorize payments, often with the objective of stealing large sums. Attackers preparing a spear phishing campaign search their targeted individuals or organizations in detail to create a more genuine message, as using information relevant or specific to a target increase the chances of the attack being successful.

- *Pre-phishing attack*: Threat event attacks that try to uncover names, job titles and email addresses of potential victims, and information about their colleagues and the names of key employees in their organizations. This information then used creating a believable email. Targeted attacks, including those carried out by APT groups, typically begin with a phishing email containing a malicious link or attachment.
- *Rainbow attack*: A password cracking technique. This attack approach involves using different words from the original password in order to generate other possible passwords. Malicious threat event actors keep a list called rainbow table. This list contains leaked and previous cracked passwords, which make the overall password cracking method more effective.
- *Ransomware*: Type of malware that typically lock the data on a targeted computer system or user's files by encryption. The attack demand payment before the ransomed data is decrypted and access returned to the targeted user. In recent years, over 50 different ransomware variants exist with names such as

  - *Cryptlocker*: Blocks access to computer systems while attackers require payment before unlock access
  - *Crypto*: Encrypt all or some files on a computer system while attackers require payment before handling over a decryption key.
  - *Double extortion*: Cybercriminals targeting and crafting attacks to increase the hight of theirs ransom.
  - *Data exfiltration*: Occurs when data is unauthorized copied, transferred, or accessed from a server or an individual's computer system in order to steal data or to extort a ransom.
  - *KeRanger*: Malware that arrives via a Trojan app, which is uncommon for crypto ransomware that usually infects targeted systems through malicious links.

- *Locky*: Malware that encrypts essential files on computers in order to force ransom payment.
- *Log4Shell*: Software vulnerability in Apache Log4j 2, a popular Java library for logging error messages in applications, which enables a remote attacker to take control of a device on the Internet, if the device is running certain versions of Log4j 2.
- *TeslaCrypt*: Ransomware Trojan gained immediate notoriety as a menace to computer games. Among other types of target files, it tries to infect typical gaming files: game savers, user profiles, record replays, and others.
- *Third party risk*: Likelihood that an organization will experience an adverse event, like data breaches, operational disruptions, reputational damages, when choose to outsource certain services or use software built by third-parties to accomplish certain tasks.
- *WannaCry*: Also known as WannaCrypt, WannaCryptor, and WannDecryptor, it is one of the first example of a worldwide ransomware attack. It began with a threat event attack on May 12, 2017, which affected hundreds of thousands of industrial organizations' computer systems in as many as 150 countries. WannaCry spread using EternalBlue, an exploit leaked from the National Security Agency (NSA)- EternalBlue enables attackers to use a zero-day vulnerability to gain access to computer systems. It targets Windows computers that use a legacy version of the Server Message Block (SMB) protocol.
- *Zero-days risk*: Recently found unknown or undocumented program problems, referring to software and computer systems vulnerabilities, viruses, worms, malware and attacks that exploits these vulnerabilities to take control of the targeted computer systems and networks.

In this regard, ransomware attacks crossed from the digital world into the physical world. In this regard, the ransomware attack against Colonial Pipeline resulted in panic and gas shortages. Another ransomware attack affected JBS Foods, the world's largest meatpacker, which affected an already strained supply chain. Third-party risk also garnered increased attention with attacks against Kaseya software and Kronos, highlighting the systemic risk of vulnerability or incidents.

As soon as unsuspecting user activates ransomware, it contacts a control server that sends it a randomly generated Advanced Encryption Standard Key (AESK). AESK is a symmetric key encryption cipher, which means that the same key used to encrypt the data used to decrypt. Hence, the AESK used to encrypt essential files on the local hard drive and drives in the network and in the connected cloud. From this point of view, the data is completely under the control of the cyber-criminal attacker, the hacker, who now requests an immediate ransom to restore the files and not to publish them. Thus, industrial public and private organization make use of virtualization, cooperative mobility management, file synchronization and to protect their computer systems, tablets, smartphones, and other devices, from infection with ransomware. Moreover, data of organizations then quickly restored in case of a successful ransomware attack. Furthermore, avoiding ransomware payment, if a computer system hit by a ransomware attack, can be done by air gapping as part of a 3-2-1 backup strategy, which means having an offline copy of the latest backup for recovery and not to pay ransom.

- *Rogue software*: Type of Internet fraud using computer malware to trick users into revealing financial and social account details or paying for bogus products. Rogue software misleads users into believing there is no virus on the computer system and aims to convince them to pay for a fake malware removal tool that actually installs malware on the computer.
- *Romance scamming*: Cybercriminals who pretend to be in love in dating portals, chat rooms, and apps, pretend to be looking for a partner in order to persuade their victims to disclose personal information.
- *Spyware*: Software installed on computing devices without end user's knowledge, which can violate end user's privacy and has abused potential. The software can be difficult to detect; often the first indication that a computing device infected with spyware is a noticeable reduction in processor or network connection speed and in mobile devices data usage and battery life time.
- *Structured query language (SQL) injection attack*: Injection with database-driven websites occurs, when a threat event attacker executes a SQL query to a database via the input data from the client to server. These SQL commands are inserted into the data-plane input to run predefined SQL commands, Therefore, a successful SQL injection exploit read sensitive data from the database, modify (insert, update or delete) database data, execute administration operations (such as shutdown) on the database, recover the content of a given file, and, in some cases, issue commands to the operating system.
- *Wiper attacks*: Malware with the sole intention of destroying computer systems and/or data assets, usually causing great financial and/or reputation damage.
- *Zero-day exploit*: Threat event attack occurs on the same day a weakness is discovered in software. Weakness exploited before a security patch to fix the flaw becomes available from the software creator. If this happens, there is little protection against a threat event attack because the software flaw is so new.

2.1.3 Cybersecurity Residual Risk Rating, Likelihood, and Consequence Levels

The manifold of threat event attacks affects a large group of mission-critical computer systems, networks and infrastructure resources, and others. That could result in denying essential services, stealing sensitive data and business assets, causing various types of damage, disruption, and others. One interesting approach defending threat event attacks is the honeypot approach, decoy systems method to lure attackers, by mimic likely targets of cybercriminal attacks, used to detect cybercriminal attacks or deflect them from a legitimate target. Further use is to gain information about how cybercriminals operate. Thus, the honeypot approach projects itself onto the following features:

- Keep cybercriminal attacker's away from accessing mission critical and crucial computer systems, networks, infrastructure resources, and others
- Collect information about cybercriminal attacker's activities
- Encourage cybercriminal attacker's to stay on computer systems, networks, and others to steal fake information filled in so that the intrusion detection system can gain information about cyber-criminal attacker's activities to respond on

Therefore, cyber-secure computer systems, networks, infrastructure resources, and other digital systems against potential threat event attacks is essential and requires management of cybersecurity risk assessment. Once a major threat event attack is identified, an analysis carried out to determine the likelihood and impact of successfully exploiting the vulnerabilities through this threat event attack, if it occurs. This process called risk quantification. However, understanding the potential consequences (impacts) of cybersecurity risks on computer systems, networks, infrastructure resources, and other digital systems requires a solid understanding of the risk, which means

$$risk = (threat\ event * vulnerability) * impact$$

or more in general

$$risk = (probablility\ of\ attack) * (expected\ losses\ through\ that\ attack)$$

to decide about the potential likelihood and consequences. Thus, risk management becomes an important issue that evaluates risk occurrences [10]:

- *Risk acceptance*: Formal acknowledgement of the presence of a risk with a commitment to monitor it
- *Risk avoidance*: Elimination of the cause of the risk
- *Risk mitigation*: Reduction of the probability of risk occurrence or of its impact

However, cybersecurity risk-models often suffer from vague, non-qualified output, based on partial information or on unfounded assumptions. Therefore, threat intelligence provides an approach that directly support risk models through defined risk measurement to be more transparent about their assumption, variables and outcomes, supporting answering questions such as:

- How often was this threat event attack measured?
- Is the trend of this threat event attack up or down?
- What kind of damage, technical- and financial-wise, has this threat event attack?
- Which threat event actors use and how do they target the business?
- Which vulnerabilities does the threat event attack exploit, and are these vulnerabilities present in the industrial, public and private organizations?

Table 2.3 Definition of risk levels and potential consequences

Risk level	Consequences (Impact)
High risk	*Not acceptable risk level:* Identified threat event attacks classified defend them successfully and in real-time. Potential standard tactics like phishing to get access to targeted computer systems, networks, and others, intruding malicious modifications in targeted objects
Medium risk	*Moderate risk level:* Monitoring identified threat event attacks with consideration whether necessary measures done to defend a potential threat event attack. Potential threat event attacks at this risk level use malware like ransomware or infect EthernalBlue vulnerabilities
Low risk	*Acceptable risk level:* Observing identified threat event attacks to discover changes that could increase the risk level. Potential threat event attacks at this risk level are password violations or insider threat events

With this in mind, a simple classification scheme for distinct cybersecurity risk levels is possible developed that assume the levels high, medium and low, as illustrated in Table 2.3 [12].

A detailed analysis of well-known and documented threat events like Stuxnet [13] on critical computer systems, networks, infrastructure resources, and others might reveal vulnerabilities, interactions, or dependency signatures. These threat events belong to the cybersecurity risk level high, which made use or zero-day vulnerabilities that require detailed knowledge of the process, and thoroughly testing of the attack on a copy of the actual targeted system.

In case of such recognizable signatures, further investigation take action to identify specifications, severity, impact, counter-measures, and if possible, developing a methodological approach to reveal them before a threat event attack happens. With this clear and structured understanding of interactions and dependencies of vulnerabilities by threat event attacks, developing a cybersecurity framework in terms of essential business assets becomes possible. If different types of risks exist, an associated shared risk chain by data breaches through threat event attacks happen. To avoid threat event attacks, cybersecurity approaches evolve to reveal before threat event attacks happen. Thus, a methodological approach should use qualitative measures for likelihood levels, e.g., described as frequency values, which means how easy it is to exploit threat event attacks, shown in Table 2.4.

Table 2.5 classifies consequence levels against the appropriate impact of threat event attacks.

2.1.4 Cybersecurity Risk Management and Quantifying Cybersecurity Risk

Industrial organizations rely on emerging technologies to run their business successfully. Risks through threat event attacks, inherent with using emerging technologies

Table 2.4 Likelihood levels compared to frequency and ease of misuse

Likelihood	Frequency of threat event attacks	Ease of misuse
High likely	Threat event attacks occur very often means more frequently than $\geq$ 10% of the time per threat event attack	Made possible by: Lack of system knowledge Performing careless/ wrong system usage
Likely	Threat event attacks occur often means frequently in between >1% and <10% of the time per threat event attack	Made possible by: Minor lack of system knowledge Performing careless/ wrong system usage
Possibly likely	Threat event attacks happen means less frequently in between >0.1% and <1% of the time per threat event attack	Made possible by: Regular system knowledge Security aware performing system usage
Unlikely	Threat event attacks occur rarely means <0.1% of the time per threat event attack	Made possible by: Good knowledge about the system Security aware performing system usage

Table 2.5 Definition of likelihood levels and possible threat event impact

Consequence level	Threat event attack impact
Catastrophic	Threat event attacks classified as severe level risk with successful devastating impact and loss of trust
Severe	Threat event attacks classified as serious level risk with serious level of successful impact and loss of trust
Moderate	Threat event attack classified as low level risk with low level of successful impact that may influence trust
Small	Threat event attack classified as no dangerous or marginal risk

in the digital transformation era, comes into action in many forms and cause IT outages, malicious code intrusion, stealing secret data, and others. Therefore, cybersecurity risk management becomes an important narrative. Cybersecurity risk management is the process identifying assessing and controlling threat event attacks to organizations critical and crucial computer systems, networks, infrastructure resources, and others. Cybersecurity risks stem from innumerous sources, including technological issues, strategic management errors, legal liabilities, financial uncertainties, and others. However, between financial and reputational damages, the fallout of threat event attacks is extremely difficult to recover. Cybersecurity risks that impede business objectives require a purposeful strategy defending them. There are two approaches, the top-down and the bottom-up approach. In a top-down approach business organizations identifies mission-critical processes and activities with their internal and external partners, to determine the conditions that potentially could impede them. The bottom-up approach starts with the threat event attack and considers their potential impact on critical assets in business organizations. Therefore, planning and implementing a successful cybersecurity risk management support business organizations, considering the full range of cybersecurity risks they face. This process includes, as described in [14], the steps in Table 2.6.

Cybersecurity risk management contains risk assessment as an integral part of an organizations risk management process. By considering a risk assessment, organizations would be able to identify risks to business organizations, to analyze the identified risk(s), and to evaluate the risk(s) impact on the organization. Risk identification is about identification and creation of an inventory of all assets to the organizations business. This means analyzing the elements that make up each risk to determine the likelihood of it, to examine the relationship between risk(s) and the

Table 2.6 Risk management process components, characteristics, and capabilities

Risk management process components	Characteristics and capabilities of the risk management process
Framing risk	Establish a risk context, describing the risk-based decision environment, to create a risk management strategy that addresses how organizations assess risk(s), respond to risk(s) and monitor risk(s). Risk management strategy establish a foundation managing risk and delineates the boundaries for risk-based decisions in the organization
Assessing risk	Address how organizations assess risk(s) in the context of the organizational risk frame. Purpose of risk assessment is identifying: (i) threat event attacks to organizations or threat event attacks directed through organizations by other organizations or nation; (ii) organizations internal and external vulnerabilities; (iii) likelihood that occur harm. Determination of risk, typically a function of degree of harm and likelihood of harm is the result
Response to risk	Address how organizations respond to risk once that risk is determined based on results of risk assessment. Purpose of risk response is to provide consistent, organization-wide response to risk in accordance with organizations risk frame by: (i) develop alternative courses of action responding to risk; (ii) evaluate alternative courses of action; (iii) determine appropriate courses of action consistent with organizational risk tolerance; (iv) implement risk response based on selected courses of action
Monitoring risk	Address how organization monitor risk over time. Purpose of risk monitoring is: (i) determine ongoing effectiveness of risk responses; (ii) identify risk-impacting changes to organizational information systems and environments in which systems operate; (iii) verify planned risk responses implemented and information security requirements derived from and traceable to organizational missions and/or business functions, federal legislation, directives, regulations, policies, standards, and guideless are satisfied

cascading input they could have on an organizations business [15], and determining likelihood, based on historical or expected occurrences. In this regard, business organizations use measures of time or frequency of historical or expected occurrences of threat event attacks to determine their likelihood in cybersecurity risk(s), as shown in Table 2.4. The consequences refer to the magnitude of harm to a business organization, resulting from the threat event attack. Therefore, risk evaluation is about determining and understanding the significance of risk levels to determine and prioritize compromising risks, as shown in Table 2.7.

The shading in the matrix visualizes the existing different risk levels. Based on the acceptance criteria in Table 2.3, the risk level *High* indicates an unacceptable severe risk level. Any threat event attack obtaining this level treated in order to reduce this risk to an acceptable level. Only risk level *Catastrophic* describes an unacceptable catastrophic risk that must immediately treated if identified, to minimize a malicious impact. Thus, identifying risks, it is impotent to understand that cybersecurity risks are harmful events. Therefore, the conditions for cybersecurity, or minimizing the risk impact, require a risk level analysis. The first step in this analysis is to find out that the respective risk scenario has a positive or negative impact on the organizations capability to conduct their business. The risk level itself refers to the likelihood and impact of the identified risk in order to support grading risks. The risk matrix in Table 2.7 is useful, providing a visual representation of the nature and impact of the identified risks, showing that a severe risk has a high likelihood for harmful impacts. To achieve a High Cybersecurity Risk Degree (HCRD), an existing risk level R_{LE} have to be less than a reasonable risk level R_{LR}. Therefore, the expected value of *HCRD* depends on the ratio of risk level R_{LE} and risk level R_{LR}

$$E(HCRD) \leftrightarrow R_{LR} < R_{LE}.$$

However, a threat event actor may perform a threat event attack (TE_A), which refers to the actual risk level (R_{LA}) representing the attack risk condition

$$\{a_R : R_{LA} \in TE_A\}$$

where a_R represents the attack risk.

Table 2.7 Risk matrix as a function of defined risk levels and input

Likeli-hood	Consequence				
	Catastrophic	Severe	Moderate	Small	Insignificant
Highly Likely	Catastrophic	High	Medium	Medium	Low
Likely	High	High	Medium	Medium	Low
Possible	Medium	Medium	Medium	Low	Low
Unlikely	Medium	Medium	Low	Low	Low
Rare	Low	Low	Low	Low	Low

Let's assume that a threat event actor only has a small amount or resources r to support his threat event attack effort. Then threat event attack (TE_A) that a threat event actor (T_{EA}) is able to perform to the subset

$$T_E T_{EA} \subseteq a_R$$

If a threat event attack occurs, its success μ refer to the membership relation $\mu \in [0, 1]$. In case a successful threat event attack happen, $\mu = 1$. In case an unsuccessful threat event attack happen, $\mu = 0$. There are several methods evaluating the threshold μ of threat risk levels, which may result in an alarm. However, before an alarm is active, an alert happens, to warn of a danger or problem, typically with the intention observing it to discover intrinsic changes that could result in increasing or decreasing the risk level. Alerting means that the risk level is at likelihood level unlikely, as shown in Table 2.3, and typically, alerting has the possibility to neglect it or deal with it, which is considered as accepted marginal risk level (R_{LAM}). Let's assume the targeted system consists of several sub-systems for which the different sub-systems be analyzed. This analysis result in the probability of the sub-system threat event risks, as shown in Table 2.8, marked by light white boxes.

The vulnerability of threat event attacks refers to cyber attacker's goals and belief. However, it is unsatisfactory only knowing what attacker's beliefs and desires is, because it is also important to know how strong they are. Beliefs vary in strength, proved not only in decision theory, but also in AI, statistics, and others. Thus, the theory of belief functions provides a way using mathematical probability theory in subjective judgment. In this regard, it is a generalization of the Bayesian theory of subjective probability. Using the Bayesian theory to quantify judgments about a

Table 2.8 Risk matrixes as a function of system or sub-system risk levels

Likeli-hood	Consequence				
	Catastrophic	Severe	Moderate	Small	Insignificant
Highly Likely	Catastrophic	High	Medium	Medium	Low
Likely	High	High	Medium	Medium	Low
Possible	Medium	Medium	Medium	Low	Low R_{LAM} system 1 has a warning
Unlikely	Medium	Medium	Low	Low R_{LAM} system 1 component 4 has a warning	Low R_{LAM} system 1 component 3 has an alert
Rare	Low	Low	Low	Low R_{LAM}' system 2 has alert	Low R_{LAM} system 5 has alert

question, one must assign probabilities to the possible answers to that question. In this context, the theory of belief functions is more flexible; it allows deriving degrees of belief for a question based on probabilities. These degrees of belief may or may not have the mathematical properties of probabilities, but how much they differ depend on how closely the questions are related [16]. With this in mind, Bayesian models derive from the assumption that rational degrees of belief satisfy the mathematical conditions of a probability function. In the era of artificial intelligence, big data, and digital transformation, the Bayesian method gains the ability to solve real business problems like the Analytics Project Life Cycle Management (APLCM), studying forecasting solutions. Among others, this means that the credence assigned to a status in a decision-making problem must add up to 1. That can be introduced as Degree of Belief (*DoB*) or credence (*Cr*), which obey the probability calculus. The idea beyond is that belief comes in degrees based on the observation being more certain to some objectives than others. Therefore, a Degree of Belief is a rational and objective to give available evidence that may also rise to contradictory results.

Assuming a simple example with two questions, q_1 and q_2, with probabilities for q_1 and derived *DoB* for q_2. Let's also assume that q_1 and q_2 had only two possible answers: yes and no, then no formal notation is required. Moreover, we assume that more answers are possible. In this context a notation is required for each set of possible answers, as well as a notation for the possibilities for q_1 and the *DoB* for q_2, representing the constraints that an answer to q_1 may put on the answer to q_2. Let's also assume that one of the answers is correct, but not known which one. Assuming that the set of answers S_A is a frame for q_1, the question for which probabilities exist, and S_B be the frame for q_2, the question of interest, while $p(s)$ is the probability of element s of S_A. For $p(s)$ and a subset A of S_B, the Degree of Belief $DoB(A)$ is given, which means that A contains the correct answer to q_2. Against this background, a decision-making problem-solving approach evaluates each form by the weighted average of the utility of all possible outcomes, weighted by the likelihood of the relevant status, as given by the cyberattackers *DoB* or *Cr*. For a simple example averages can be calculated as follows for the numbers $x_1, x_2, \ldots, x_n$, and then the average of the numbers is

$$\frac{x_1 + x_2 + \ldots + x_N}{N} = \frac{1}{N}x_1 + \frac{1}{N}x_2 + \ldots + \frac{1}{N}x_N.$$

Each number in this simple example has the same weighted average; but the weight can also be different for different numbers. Assuming a threat event attack leads to outcomes

$$O_1, \ldots, O_N$$

for the system status

$$S_{S1}, \ldots, S_{SN}$$

Let's also assume $DoB(s)$ denote the threat event attack Degree of Belief in Ss_1 and $DoB(Ss_2)$ is the Degree of Belief in Ss_2, and finally $DoB(Ss_N)$ as the Degree of Belief in Ss_N. Let $E(O_1)$ denote the expected outcome of O_1 for the first identified threat event attack, $E(O_2)$ the expected outcome of O_2 for the second identified threat event attack, and finally $E(O_N)$ the expected outcome for O_N for the n-th identified threat event attack. Then the expected threat event attack vector (EC_{TA}) of identified threat event attacks is:

$$EC_{TA} = D_0B(S_{S1})E(O_1) + D_0B(S_{S2})E(O_2), \ldots, D_0B(S_{SN})E(O_{2N}),$$

which can also be written in summarizing form as:

$$EC_{TA} = \sum_{i=1}^{N} D_0B(S_{Si})E(O_i)$$

Furthermore, it should be noted that from practical experience subjectivism give the same probabilities to frequently repeated threat event attacks, where the probability is characterized as relative frequency of threat incidents in the long term as shown for instance in Table 2.4.

For practical reasons, affiliations in cybersecurity risk management published several documents what organizations must do to manage cybersecurity risks. One of the common sources is the ISO/IEC 31000:2009 standard Risk Management – Principles and Guideline (RMPG) developed by the International Organization for Standardization (ISO) and International Electrotechnical Commission (IEC) [17]. Any business organization can use ISO's five-step risk management process:

1. Identify the risk.
2. Analyze the likelihood and impact of each one.
3. Prioritize risk based on business objectives.
4. Treat (or respond to) the risk conditions.
5. Monitor results and adjust as necessary.

Based on the ISO 31000 best practice principles risk management should meet the following objectives:

- Adaptable to change
- Activities embedded in organization's overall decision-making process
- Explicitly addressing uncertainties
- Integral part of the overall organizational process
- Processes continuously monitored and improved upon
- Process creating value for the organization
- Systematic and structured processes
- Taking into account human's factors, including potential errors
- Transparent and all-inclusive processes
- Using best available information

Table 2.9 Threat event actor chain of action and countermeasures

Action chain of threat event actors	Countermeasures
Enlightenment Identify target Search for vulnerabilities	Monitoring and logging Situational Awareness Cooperation and Exchange of Knowledge
Attack target Exploit vulnerabilities Defeat remaining controls	System design by security Standard controls (e.g., ISO 27001) Penetration testing
Achieve goals Disruption of systems Extraction of data Manipulation of information	Proactive response strategy to cybersecurity incidents Proactive business continuity strategy Proactive disaster recovery strategy Cybersecurity Insurance

Following the ISO best practice principles, a simple three-step threat event actor chain of action and countermeasures is required that can be set up as shown in Table 2.9.

Another Risk Management Framework (RMF) is the National Institute of Standards and Technology (NIST) NIST RMF, which provides a comprehensive, flexible, repeatable and measurable 6-step process that any organization can use to manage information security and privacy risk for organizations and systems. The steps are as follows:

- Impact analysis to classify organizations IT system and the data it process, stores, and transmits.
- Choose a starting point for the IT system's baseline security measures based on the security categorization. Adjust and supplement it as necessary based on an organizations risk assessment and include local circumstances.
- Implement the security measures and describe how to use in the IT system's operational context.
- Determine the degree to which the security controls implemented, operating as intended, and providing the expected result with respect to meeting the security requirements for the IT system by an adequate evaluation of the security controls.
- Determine the risk that the operation of the IT system poses to organizations operations, assets, people, other organizations, and decide whether that risk is acceptable before approving its use.
- Regularly assess the efficiency of the security measures in the IT system, note changes to the IT system or its operational environment, execute security impact evaluations on the resulting changes, and report the security status of the IT system to designated organizations officials.

Beside this two risk management frameworks, other common cybersecurity frameworks introduced in Chap. 5. The following three approaches also important to enable cybersecurity risk management at a certain level:

- *Multifactor Authentication (MFA)*: Authentication method that requires the user to provide two or more verification factors to gain access to a resource such as an application, online access, or a Virtual Private Network (VPN). MFA is a core component of an Identity and Access Management (IAM) policy. Rather than only asking for a username and password, MFA requires one or more additional verification factors, which decreases the likelihood of a successful cyberattack. Hence, MFA improve organizations security posture on all public-facing insider (employee) services and portals as well as restricting internet-facing protocols such as Remote Desktop Protocol (RDP) and Server Message Block (SMB) protocol to inhibit unauthorized access to an organizations environment. The SMB protocol is a client-server communication protocol that enable applications and their users to access files on remote servers and to other resources. The protocol can also communicate with server programs configured to receive SMB client requests.

- *Next-Generation Endpoint Security*: Incorporate real-time analysis of user and system behavior to analyze executables using artificial intelligence (AI) and machine learning techniques, an a tighter integration of network and device security to provide more comprehensive and adaptive protection that traditional endpoint security solutions. This improve organizations security posture by utilizing advanced endpoint protection across their environment, which leverage AI and machine learning to identify anomalies and perform heuristic analysis, in addition to conducting antivirus and anti-malware activities in real time.
- *Privileged Account Management (PAM)*: Improve organizations security posture by managing and auditing account and data access by privileged users. A privileged user has administrative access to critical systems who securely con-tainerizes and rotates credentials of privileged accounts such as local or domain administrators, service- and database accounts as well as set up and delete user accounts and roles. This helps organizations make sure that employees have only the necessary levels of access to do their jobs. PAM enables security teams to identify malicious activities linked to privilege abuses and take action to remedi-ate before the threat event actor is able to deploy the malicious activity or exfiltrate data from the organizations environment.

2.2 Threat Intelligence

Threat intelligence, also known as cyber threat intelligence (CTI), is information gathered from a variety of sources about current or potential threat event attacks against organizations. The information obtained is analyzed, redefined, and orga-nized to minimize and mitigate cybersecurity risks. Building on that, threat intelli-gence is evidence-based knowledge, including context, mechanisms, indicators, implications and actionable advice about an existing or emerging menace or hazard to assets. Therefore, the main purpose of threat intelligence is to make the various risks, organizations may face, visible and defendable. These risks are, besides internal risks, external threat event attacks, such as zero-day attack, APT, and others. For this purpose, threat intelligence includes in-depth information and specific content about threat events, such as who is attacking, capabilities and motivation of threat event actors, the IoC, and others, which provides a better insight into the threat event landscape and their latest Tactics, Techniques and Procedures (TTP). With this information, any organization can become proactive in configrating its cybersecurity measures to identify, detect, protect and defend against the most damaging threat event attacks. Therefore, threat intelligence is an essential part of any organizations cybersecurity strategy, gathering data prior to malicious threat event attacks, making decisions about how best to secure their IT resources, business assets, and others. Using threat intelligence in cybersecurity protection, based on the latest observed threat event attacks, is required to collected data by a third party, shared between individual organizations, or shared between groups of organizations. In this case, the risk depend on threat event actors, specialized in finding exploits and continuously developing malware platforms to improve their craft especially to find out how stealthy their malware infect and/or operate. Malware is a specifically designed software disrupting, damage or gain unauthorized access to a computer system, network, infrastructure resource, and others. Vulnerabilities in these devices allow the malware performing not permitted actions on them, like running arbitrary code. Such malicious actions affect the confidentiality, integrity and availability of critical and crucial digital systems.

As reported in [18], zero-day-malware exist in the wild for over 300 days before identification. Hence, moving toward more proactive threat intelligence is required, based on forward learning to capture the essence of the evolving nature of threat event attacks. Machine learning (see Chap. 8) and big data analytics (see Sect. 1.2.5) gathering and analyze massive amounts of threat event attack data from technical sources, open sources, closed dark sources, and threat event attack research. By correlating these insights with internal available sources, industrial, public and private organizations can realize faster and better-informed security decisions. Therefore, threat intelligence is widely being seen as the domain of data scientists, a profession investigating complex data from the business perspective. Data scientists make predictions supporting organizations to take accurate decisions with regard to their cybersecurity issues. For this reason, they must have a solid foundation in computer science, network communication, data analytics, applied mathematics like statistics and stochastics, modeling skills, and business informatics, to deal with the different business and IT requirements. Data scientists identify threat event objectives to reveal value in cybersecurity after resolving it, treated as a separate option in the cybersecurity framework, which depends on collection, processing and analytics of data. Therefore, threat intelligence can be broken down in four subcategories, each of which has its own usage, techniques, and challenges:

- *Strategic Threat Intelligence*: A non-technical risk-based approach, used by managers and operators making decisions. Allow insight in mission critical and crucial risk areas, enabling managers and operators to act in conjunction with action lines, signatures in threat event actor's tactics and targets as well as in geopolitical trends. In [19] the usefulness of the Violent Extremist Risk Assessment (VERA) method, applied to five groups of cybercriminal attackers, determines1 whether it is more applicable to threat event actors, who work on their own or as part of a group. The method enables a systematic and structured assessment of actors with regard to the impact of their risk. Therefore, threat intelligence tools on a structured professional judgment method are developed. They take individual characteristics of cases into account and finally based of the structured professional judgment method. This allows various possible courses of case to be included in the planning of measures. Hence, the strategic category of threat event attacks has tremendous value for business decision-making; however, it is only one aspect out of the broader threat intelligence options. Benefits of the Strategic Threat Intelligence are as follows:

– Understanding the organizations risk(s)
– Safeguard of organizations security posture
– Actionable insight in organizations security policies

• *Tactical Threat Intelligence*: Gain information to determine how a threat event actor typically cyberattacks a software infrastructure and network, and using this intelligence to detect similar potential cyberattacks and reduce the probability or effects of such events. This approach to cybersecurity is proactive, ensuring all relevant parties are fully updated on recent developments or trends. Specifically, Tactical Threat Intelligence refers to Tactics, Techniques, and Procedures (TTPs) focusing on the strengths and weaknesses of organizations networks and its ability to prevent cyberattacks to support threat event response teams to understand how their organizations might be attacked and what are the best ways to defend against or mitigate those threat event attacks. Benefits of Tactical Threat Intelligence are as follows:

– Structured and proactive cybersecurity system
– Make complex data more digestible
– Improve responsiveness to cyberattacks
– Adaptable framework to withstand a range of cyberattacks

Thus, Tactical Threat Intelligence help answer questions such as what tactics, techniques and procedures the cyberattacker may have access to and how they can be countered.

• *Technical Threat Intelligence*: Refers to technical threat indicators, e.g., malware hashes, a commonly shared form of threat intelligence practiced by sharing host-based indicators for malicious code that often create file names and hashes, C to Integer Program (C2IP) addresses, and many others. A hash of a file means compute the cryptographic checksum of the file. Hashes are the output of a hashing algorithm, essentially to generate a unique, fixed-length string, the hash value, for any given piece of data or message. The hash value is of great help to cybersecurity research teams defending malware, sharing IoC, and others. Using hash values, cybersecurity researchers can reference malware samples and share them with others through malware repositories like:

– *VirusTotal*: Perform file searches through datasets in order to identify files that match certain criteria like hash, antivirus detections, metadata, submission file names, file format structural properties, file size, and others.

– *VirusBay*: Web-based collaboration platform that connects professionals of security operation centers with malware researchers. That enables targeted organizations to collaborate with malware researchers on Indicators of Compromise (IoC) and the creation of a threat event attack report. In return, the researchers gain access to malware samples for analysis to improve detection knowledge that are published.

– *Malpedia*: Free service offered by Fraunhofer FKIE [20]. The primary goal of Malpedia is to provide a resource for rapid identification and actionable context, when investigating malware. Openness to curated contributions ensures accountable levels of quality in order to foster meaningful and reproducible research. Malpedia has Application Programming Interfaces (APIs) for different malware such as: Emotet, Yara, MISP Risk, Qakbot, Drivex, Conti, Sandworm, and others.

– *MalShare*: Collaborative project to create a community driven public malware repository that works to build additional tools for the benefit of the security community at large [21].

– *MISP Risk*: Free and open source threat sharing platform, supports information sharing of threat intelligence, including cybersecurity indicators, and financial fraud or counter-terrorism information. It includes multiple subprojects to support the operational requirements of analytics to improve the overall quality of information shared [22].

• *Operational Threat Intelligence*: Collect information from a variety of sources and past threat event attacks, used to anticipate the nature and timing of future threat event attacks. Security and incident response teams use operational intelligence to change the configuration of certain controls, such as firewall rules, detection rules of threat event attacks and access control. It supports proactively pursue threat event attack hunting before compromise and after recovering, knowing better where to investigate. It also can improve response times as the information provides an idea how and where what to look for. Hence, Operational Threat Intelligence provides a coherent framework to analyze and prioritize potential threat event attacks and vulnerabilities in organizations threat environment, thereby linking the possibility and impact of cyberattacks to their strategic implication.

The four categories of threat intelligence strategic, tactical, technical and operational can be classified into low and high level objects and short term and long-term usage as shown in Fig. 2.1 after [23].

An essential issue of these subcategories is how to gain the respective knowledge to threat event response and threat event defending teams, to enable them to answer the question "Who will potentially benefit from threat event attacks?" Answering such kind of questions are essential information sources of information. Common sources of information used for strategic threat intelligence are as follows:

• *Policy documents*: Can be documents from nation-states, nongovernmental organizations, and others

	Strategic Threat Intelligence	Tactical Threat Intelligence
↑ **Long term use**	High level information on changing risk(s) *High level executives and Management*	Information on cyber-attackers TTPs *IT Service Administrators and SOC Managers*
Short term use ↓	**Operational Threat Intelligence** Information on specific incoming cyberattacks *Security Manager and Network Defender*	**Technical Threat Intelligence** Specific Indicator of Compromise (IoC) *Security Operations Center (SOC) Staff*
	← **High Level**	**Low Level** →

Fig. 2.1 Cyber Threat Intelligence categories

- *News*: From local and national media, industry- and subject-specific publications, and other subject-matter experts
- *White papers, research reports, and other content*: Generated by security organizations and companies offering security products
- *Cyber threat and cyber attack databases*: Developing databases becomes a major task of threat intelligence.

In this context pre-reconnaissance refers to information gathered before malicious threat event attacks interacts with defended computer systems, networks, infrastructure resources, and others in organizations, detecting a vulnerability or a threat event attack and possibly repelling it.

Detecting hostile threat event attacks depends on the number and type of appropriate actions, obtained from publicly available data, found for example in the

- *National Vulnerability Database (NVD)*: US government repository of standards-based vulnerability management. A vulnerability is a mistake in software code that provides a cyberattacker with direct access to a system or network. Data represented in the NVD using the Security Content Automation Protocol (SCAP). This data enables automation of vulnerability management, security measurement and compliance. NVD includes databases of security checklist references, security-related software flaws, misconfigurations, product names, and impact metrics [24]. This service let users stay up to date on the latest vulnerabilities identified by popular automation such as "Get an email when a new vulnerability is published," "Track new vulnerabilities' in a spreadsheet," "Get a weekly digest of new vulnerabilities," and others. The NVD provides analysis on Common Vulnerabilities Exposures (CVE), the catalog of known cybersecurity threats. This provides various other pieces of information relevant to the vulnerabilities, used by organizations to prioritize the vulnerabilities and the patches they should be deploying to keep their IT infrastructure safe.

- *Common Vulnerabilities and Exposures (CVE) Database*: The program launched by MITRE, a nonprofit organization that operates research and development centers sponsored by the federal government, enables to identify and catalog vulnerabilities in software or firmware into a free dictionary for organizations to improve their security. The dictionary list entries, descriptions, an identification number, and at least one public reference for publicity known cybersecurity vulnerabilities. A CVE number uniquely identifies one vulnerability from the list that administrators can access technical information about specific threat events across multiple CVE-compatible information sources. Entries are including the NVD [24]. CVE is sponsored by US-CERT, within the Department of Homeland Security office of Cybersecurity and Information Assurance. MITRE maintains the CVE dictionary and public website. CVE is free to use and publicly accessible. An example of a CVE ID is CVE-2020-16891, which includes the CVE prefix, the year that the CVE ID is assigned or the year the vulnerability is made public and sequence number digits.
- *Relationships*: CVE List feeds NVD that builds upon the information included in CVE entries to provide enhanced information for each entry such as fix information, severity scores, and impact ratings. As part of enhanced information, NVD also provides advanced searching features such as by OS; by vendor name, product name, and/or version number; and by vulnerability type, severity, related exploit range, and influence [25, 26].

These databases financed by the U.S. Department of Homeland Security, the US Office of Cybersecurity and Communications, and the Computer Emergency Readiness Team (CERT). They support understanding the severity of the current cybersecurity threat landscape. Recent years have seen the adoption of open standard languages and protocols. The MITRE Cooperation has published a document with the title "Standardized Cyber Threat Intelligence Information with the Structured Threat Information eXpression (STIX^{TM})" [27], which reflects ongoing efforts to create, evolve, and refine a community-based development sharing and structuring threat event attack information.

MITRE Cooperation, a nonprofit organization to operate research institutes on behalf of the US Government, created by separation from the Massachusetts Institute of Technology (MIT). The Structured Threat Information eXpression (STIX^{TM}) is an evolving, collaborative community-driven activity, used to exchange cyber threat intelligence (CTI). It enables public and private organizations to share CTI with one another in a consistent and machine-readable form. This allows security communities understand better what computer-based attacks they are most likely to see, and to anticipate and/or respond to those threat event attacks faster and more effectively. In this context, STIX^{TM} designed to improve many different capabilities, such as collaborative analysis of threat event attacks, automated threat event exchange, automated detection and response, and others [29]. Therefore, STIX^{TM} defines and develops a language and serialization format to represent structured threat event attack information to convey the full range of threat event attack information and strives to be fully expressive, flexible, and extensible, automatable, and as human-readable as possible. STIX^{TM} built upon feedback and active participation from industrial, public and private organizations and experts across a broad spectrum of industry, academia, and government. This includes consumers and producers of information of threat event attacks in security operation centers, Computer Emergency Response Teams (CERT), threat intelligence activists, security executives and decision makers, as well as numerous currently active information-sharing groups, with a diverse set of sharing models. CERT is a group of experts who respond to cybersecurity incidents. These teams deal with the evolution of malware, viruses, and other threat events. Against this background, the Transport Protocol Automated eXchange of Indicator Information (TAXII^{TM}) enable sharing of security information. These open standards allow secure sharing of information of threat event attacks. However, the advancement in threat event attacks urgent require new and outward-looking collaborative approaches to defend cybersecurity risks. Threat intelligence and threat information sharing are on the cutting-edge of novel approaches with high potential shifting the balance of power between threat event actor's and defenders of threat event attacks. However, a core requirement for maturing effective threat intelligence and threat event attack information sharing depends on the availability of an open-standardized structured representation for information of threat event attacks. In this regard, STIX^{TM} is a community-driven effort to provide such a representation adhering to guiding principles to maximize expressivity, flexibility, and extensibility. All parties interested in becoming part of the collaborative community discussing, developing, refining, using and supporting STIX^{TM} are welcome to join the community (see STIX^{TM} website: http://stix.mitre.org/).

Furthermore, developing a Threat Intelligence Model (TIM) will be another helpful approach in analyzing threat event actor's potential intention. This allows identifying what kind of information to collect, investigate single and multiple cybersecurity incidents, to achieve threat intelligence. In this context, TIM based on capabilities referred to as preventive and detective capabilities, forming some kind of profiling approach. As described in [28], this model contains the following elements at the detective level:

- *Identifying the identity of a threat event actor*: This refers to the name of a person, a public and private organization, a nation state, and others. Sometimes, identity linked to other threat event actor's without actual attribution or even location of their operations. However, it is important to connect multiple threat event attacks to the same threat event actor in order to determine any strategy, Tactics, Techniques, and Procedures (TTPs) expected to be used.
- *Motivation*: Driving force that enables actions in the pursuit of specific goals. Motivation derived from the benefits achieving a goal. Threat event actor's goals may change, but motivation most of the times stays the same. Knowing threat event actors motivation narrows down to which targets that actor may focus, helping defenders focus their mostly limited defense resources on the most likely

scenarios of the threat event attacks, as well as shapes intensity and persistence of a threat event actor [29]. However, motivation can be different which means ideological, military, financial, and others.

- *Goals*: According to [30] a goal is "a cognitive representation of a desired endpoint that impacts evaluations, emotions, and behaviors." Thus, a goal consists of an overall end state and behavior objectives and plans needed for attaining it. Establishing a goal guides behavior (strategy). A threat intelligence goal defined as a tuple based on actions and objects, but work is required to create a consistent taxonomy at an adequate level of detail [31]. It is well known that typical forms of goals are stealing Intellectual Property (IP), damage infrastructure; embarrass competitors, and others.
- *Strategy*: A nontechnical high-level description of a planned threat event attack. There are typically multiple different ways attackers can achieve its goals, and the strategy defines which approach the attacker should follow. Assuming that introducing a formal taxonomy, describing relationships between motives, goals, and strategies, would be advantageous for the advancement of threat intelligence, as well as risk assessment processes.
- *Tactics, Techniques, and Procedure (TTP)*: Characterize adversary behavior in terms of what they are doing and how they will do it.

  - *Attack patterns*: Type of TTP describing ways adversaries utilize to compromise targets.
  - *Malware*: Type of TTP referring to software inserted into a computer system or network with the intent of compromising the target in terms of the CIA Triad (see Sect. 1.6.2).
  - *Infrastructure*: Type of TTP referring to threat event actor's resources available to perform threat event attacks. Examples of adversarial infrastructures include Command and Control (C2) servers, malware delivery sites, phishing sites, and many others.

- *Tools*: Threat event actor's install and use tools within the attacked network. The tools often modified so that a tool detect and analyze that a previous security incident might be similar, but not the same in new threat event attacks. In addition, tools might be non-malicious software like vulnerability scanners, network scanning tools, and many others used for malicious reasons.
- *Indicators of Compromise (IoC)*: This element represents detective mechanisms describing how to recognize malicious or suspicious behavior. Examples of IoCs include the following:

  - Anomalous spikes of requests and read volume in industrial, public and private organization files
  - Dubious log-ins, access, and other network activities that indicate probing or brute-force attack
  - Irregular or suspicious activities, such as increases in data reads
  - Large amounts of compressed files and data unexplainably found in locations where they shouldn't be

- Network traffic that traverses in unusually used ports
- Surf traces that are not indicative of human behavior
- Suspicious activity in administrator or privileged accounts
- Tampered files, Domain Name Servers (DNS) and registry configurations as well as changes in system settings, including those in mobile devices
- Unusual password request and traffic going in networks
- Unknown files, applications, and processes in the system

These and other unusual activities enable cybersecurity team staff monitoring a computer system, network, infrastructure resource, and others to detect malicious intruders early as part of an intrusion detection process. To create an IoC it is desirable to classify indicators in a number of ways like Atomic Indicators, Behavioral Indicators, and Computed Indicators, which is often referred to as ABC Indicators [32].

- *Atomic indicators*: Value of atomic indicators is limited due to the short shelf life of information that can include file hashes, domain names, IPs, and others. Atomic indicators are the type of data and information that has the longest history in threat intelligence and many threat intelligence efforts are based upon them.
- *Behavioral indicators*: Combining other indicators or other behaviors to form a profile. They can be a combination of computed indicators such as geolocation of IP addresses, MS Word attachments by magic number, base64 encoded in email attachments, behaviors such as targeted sales forces, atomic indicators such as adversary activity on their own, covert Command-and-Control (C2) channel indicator, and others. However, C2 indicators are critical to detect and analyze. Behavioral indicators also be referred to as TTPs, which means that identifying and defending threat events is based on behavioral indicators and thus a profiling-driven investigation, to match threat event behavior with probably known information.
- *Computed indicators*: Are indicators such as hashes of malicious files, specific data decoded custom C2 protocols, back doored office documents, website reputation, and others [32].

• *Target*: Organizations, companies, sectors, nations, individuals, and others.

At the preventive level of threat event attacks the following element is of importance:

• *Courses of action*: Refer to techniques and procedures of the target to mitigate the threat event actor's to achieve their goals. This call for measures taken to prevent or respond to threat event attacks.

Mastering the tactics and tools of the Advanced Persistent Threat (APT) hacker referred to in [32], reveals the mindset, skills, and effective attack vectors, needed to compromise any target of choice. APT hacking discusses the strategic issues that make all organizations vulnerable and provides noteworthy empirical evidence. However, APT hackers methodology for systematically targeting and infiltrating organizations and its IT systems, can be understand, as mentioned in [33]. A unique, five-phased tactical approach to APT hacking is presented in [34] for real-world examples and hands-on techniques, used immediately to execute very effective attacks. Review empirical data from actual attacks conducted by unsophisticated and elite APT hackers one can learn the APT hacker methodology – a systematic approach designed to ensure success, avoid failures, and minimize the risk of being caught, with regard to

- *Reconnaissance*: Mission undertaken to obtain information about the activities and resources of a threat event actor. Perform in-depth reconnaissance is used to build a comprehensive understanding of the targeted computer system, network, infrastructure resources, and others. In the military domain reconnaissance produces combat information.
- *Obtain nontechnical data*: About targeted computer systems, networks, infrastructure resources, and others, as particular attack area, and others.
- *Use of social engineering*: Term used for a broad range of malicious activities accomplished though attacker's interactions, to compromise a specific computer system, network, infrastructure resource, and others, to identify and attack wireless targeted devices. It uses psychological manipulation to trick users into making security mistakes or giving away sensitive information.
- *Spear phishing*: Email spoofing attack that infiltrates targeted organizations facilities, seeking unauthorized access to sensitive information. It attempts are not typically initiated by random hackers, but are more likely conducted by perpetrators for financial gain, trade secrets, or military information to obtain access to assets and compromise digital devices.

2.2.1 Problem of Known-Knowns, Known-Unknowns, and Unknown-Unknowns

The analysis of ABC Indicators tries to seek most static of ABC Indicators but also adversarial behavior, because they often reveal themselves. In this regard, profiling ABC Indicators typically begin with the most potential assumed cyber-criminal attacker (also known as actor or adversary) as first step and then dovetails into potential defending strategic options. These options refer primarily to *Known-Knowns* (see Table 2.10) with regard to previous threat event actor's behavioral indicators as threat event attack specific signature out of a crowd of indicators to track. However, profiling in this regard is some kind of stochastic process and one

Table 2.10 Cyber threat risk level and respective security model

Cyber-threat event risk level	Security model
Known-Knowns (KK)	Information security (IS)
Known -Unknowns (KU)	Cybersecurity (CS)
Unknown-Unknowns (UU)	Cyber resilience (CR)

never knows for sure whether the tracked *Known-Knowns* are identical signatures again behind the actual threat event actor with the initiated threat event attack. Finally, it is only an expectation whether or not the same attacker (actor, adversary) or even attacker group is truly at the other end of threat indicator behavior every time. The attacker group, also called hacker group, can exist out of individual groups, but all should share the same goals [32]. Based on the received profile it is possible to facilitates predicting future activity and detecting it in the context of *Known-Unknowns* (see Table 2.10).

Known-Knowns in Table 2.10 are threat event attacks in the relation with the CIA Triad (see Sect. 1.6.2), based on threat event information that refer to methods used in information security (see Sect. 1.6). Risks considered, as *Known-Knowns* are existing risks, documented as facts. Detecting *Known-Knowns* describe an obvious context being aware about all identified risks organizations defenders are certain about, and have solutions that best allay those risks. Risks that don't fit this attribution represent *Known-Unknowns*, which refer to objects *Known*, and the one not known yet belong to *Unknowns*. Thus, *Known-Unknowns* describes a complicated context due to yet not knowns, which referred to as *Unknowns, Known* to as existing, but not enough is known about what the unknown is. This raise the question "How gain knowledge to quantify accurately their potential impact"? Finally, *Unknown-Unknowns* characterize a risk situation not clear whether a risk appears or not. Thus, *Unknown-Unknowns* refers to threat event attacks not identified yet by anyone, stated as unforeseeable and unpredictable, which requires ongoing exploration. For this type of risk, no one know how to find out is a risk intruded or not, because there is not enough knowledge to ask the right questions to identify and close the knowledge gap to risks. Therefore, *Unknown-Unknowns* are risks that may be real, but no one knows how it will show up. Thus, this threat event level is believed as impossible to understand or imagine it in advance. Threat events at this level assumed as highly dangerous because no defense strategy exists. A probable attempt to address this problem is attempting to identify as many risks as possible to turn as many *Unknown-Unknowns* into *Known-Knowns*. While this may be a worthwhile endeavor, it does not address the fundamental unknowability of some challenges.

The goal of observation or measures of *Known-Knowns* threat event attacks is to defend and protect data, assets and IT resources against known threat event attacks, which can be simple so far they belong to the CIA triad and cyber threat intelligence (CTI) approaches. Therefore, well-written documentation how to mitigate threat event attacks has to be available within the CTI community. In some cases, specific configured mitigation tools allow detecting and even mitigating the threat event attacks automatically. Google and Amazon have implemented changes in their Domain Name System (DNS) services that inspect Server Name Indication (SNI) fields to detect domain fronting. To detect domain fronting, industrial, public and private organizations need the capability to inspect Transport Layer Security (TLS) traffic between internal networks and external hosts.

Another way to mitigate threat event attacks is to follow up different security alerts such as Common Vulnerabilities and Exposures (CVE) and National

Vulnerability Database (NVD) that often also contains information how to mitigate the vulnerability. However, achieving a reasonable quality in CTI is a tricky task, which can become a major problem, when industrial, public and private organizations security analysts try to collaborate enhancing quality by create, share, improve and use CTI to improve their cyber-defense strategy [18, 35, 36]. Nevertheless private and public organizations need to have their own strategic cybersecurity CTI defense approach for building up cybersecurity resilience [18, 37]. Such an approach calls for the cybersecurity characteristics as shown in Table 2.10, which deals with the issue how to handle known and unknown information.

For *Known-Unknowns* threat event attack identification and observation becomes more difficult and decisions required for CTI due to threat event attacks, which is much harder to achieve. The reason lies in the variety of APTs and non-CIA Triad threat event attacks. The later requires an enhanced and systematic application of the CIA Triad to all assets of the industrial, public and private organizations and their environments as an informal description of a cybersecurity solution approach. This requires an End2End (E2E) real-time monitoring to detect malware by analyzing network traffic to extract network behavioral indicators across different protocols and network layers. This approach refers to different observation methods such as transaction, session, flow and conversation windows, and others. A feature selection method is used to identify the most meaningful features to reduce the data dimensionality to a tractable size. Finally, evaluating supervised methods to indicate whether traffic in the network is malicious, to assign the unknown to the known malware features to discover new threat events [38].

The detection problem of unknown threat event attacks be a known topic in many different areas in computer systems, networks and infrastructure resources security. Network and malware attacks are among the most common threat event vectors [39], and are focal point in the dissertation of Duessel [40]. Machine learning technique (see Chap. 8) is concerned solving the problem of unknowns by minimizing the expected risk.

Against this background, the category of *Unknown-Unknowns* requires a solution for unpredictable implications making devices cyber-resilient. Cyber-resilience is the ability of a computing system to recover fast in case it experience attack conditions. It require continues effort and deal with many aspects of IT security, Disaster Recovery (DR), Business Continuity (BC) and computer forensics. Unfortunately, there exist only a few methodological approaches to deal with *Unknown-Unknowns*. One of which is computer or digital forensic that also take into account digital profiling; the other approach is machine learning-based techniques (see Chap. 8). Digital Forensic is the application of scientific investigation techniques to digital crime (cybercrime) and threat event attacks) but also a crucial aspect of law and business in today's digital transformative era. Therefore, a digital process of preservation, identification, extraction, and documentation evidence, based on finding evidence from digital systems and devices by forensic teams by the court of law is required.

2.2.2   *Digital Forensic and Threat Intelligence Platforms*

The scope of digital forensic is to conduct a structured examination through profiling, and at the same time documenting a chain of evidence, so that it is possible to determine exactly which expectations took part through cybercriminal events. This can help answering the question "What attacker group (also termed actor or adversary group) potentially may be responsible for the respective cybercriminal threat event attack. Against this background, digital profiling be understood as process gathering and analyzing information about an individual that exists online. Hence, digital profiles include information about personal characteristics, behavior, affiliation, connections, interactions, and others. Digital profiling is used in a manifold of areas. In industrial, public and private organizations, digital profiling is used to identify suspect insiders (employees) or outsiders (see Table 1.15) to protect organizations from an inside as well as outside cybercriminal attacks. To determine whether the identified insider (employee) really poses a risk to his organization requires scrutinizing his online behavior as digital profile. In a low-profile incident case, typically information gathered through corporate email, logs and social media content, connections, and posts. In a more high-profile incident case, investigators might employ specific surveillance technologies for a more complex profile of the individual attacker.

Forensic approaches use various techniques and proprietary forensic applications to examine the infiltrated computer system, network or infrastructure resources, searching for hidden folders and unallocated space on volumes of deleted, encrypted or corrupted files. Finding evidence to the unknown will carefully documented in a Forensic Investigation Profile Document (FIPD). Determining a forensic profile for unknown threat event attacks contains several steps such as:

- *Identifying and documenting*: Highest probability of a potential nature and purpose for an unknown cybercriminal threat event attack(s)
- *Identifying and documenting*: Highest probability of a potential unknown cybercriminal threat event infection mechanism(s)
- *Identifying and Documenting*: Highest probability of how an unknown cybercriminal threat event attack(s) interact with potentially targeted host computer systems, networks or infrastructure resources
- *Identifying and documenting*: Highest probability of the profile and sophistication level of unknown cybercriminal threat event attack(s)
- *Identifying and documenting*: Highest probability of the extent of infection and compromise of the potentially targeted computer system(s), network(s) or infrastructure resource(s) by unknown cybercriminal threat event actor(s)

Based on procedure specific digital forensic incident response programs one can be build up combining the respective cybersecurity tools and approaches with regard to the different cybercriminal types of threat event attacks to respond effectively. The portfolio of tools and approaches in digital forensic requires specific skills such as:

- Reverse-engineering of malware
- Detecting malicious files and software code
- Discovering, searching computer systems memory for infections and malicious code
- Digital documents of infections and threat event attacks

These tools and approaches used before and after a security breach, including endpoint detection and response monitoring, cybersecurity information and incident management, log analyzers, threat intelligence databases, penetration testing, firewalls, intrusion detection and prevention systems (see Chap. 3), machine learning (see Chap. 8), and others. Therefore, application-specific forensics deals with forensic issues, unique to a specific application. Although the overarching forensic process is similar, but questions like "What extraction method would be appropriate in a particular application"? result in different answers [41]. The application-specific forensic deal with different evidence collection approaches, examination, analysis and reporting, as shown in Table 2.11 [32].

Finally, the sum off all evidences form the puzzle analyzed to identify how to transform the unknown threat event attack to an understandable new threat event attack of a cybercriminal attacker. For this reason, there are several efforts available to find a uniform method in the documentation and incident report:

- *Open Indicators of Compromise (OpenIoC)*: Open framework that currently exist for organizations that want to share threat event attack information internally and externally in a machine-digestible format, developed by the US cybersecurity company MANDIANT in 2011 [43]. MANDIAT, written in eXtensible Markup Language (XML) can easily customized for additional intelligence so that incident responders can translate their knowledge into a standard format. This allows organizations to describe the technical characteristics that identify a known threat event attack, an attacker's methodology, or other evidence of compromise. Thus, organizations can leverage this format to share threat event attacks related to latest IoCs with other organizations, enabling real-time protection against the latest threat events attacks [44].
- *Structured Threat Information eXpression (STIX^{TM})*: Standardized structured language describing CTI developed by MITRE Corporation and the Open Standards Open Source (OASIS) Cyber Threat Intelligence Technical Committee, which supports automated information sharing for cybersecurity situational awareness, real-time network defense, and sophisticated threat event attack analysis.

In STIX^{TM} terminology, an individual or group involved in malicious cybercriminal activity is a $Threat\ Event\ Actor$. A set of activity ($Threat\ Events$) carried out by $Threat\ Event\ Actors$ using specific techniques (TTP) for some particular purpose is termed a $Campaign$. Such activity might fit along the lines of stealing information from customers or targeting a particular business sector. When data collected on various related intrusion attempts ($Threat\ Events$), it may initially not include enough information for characterizing attributions of the

Table 2.11 Application-specific digital forensic approach, modifies after [42]

Digital forensic phases	Application-specific forensic context
Collection	Forensic processes require data, obtained by a collection process. Collection means identifying, labeling, recording, and acquiring data from identified sources, containing digital evidence, following procedures that preserve data integrity and maintain a credible chain of custody. Data collection process is essential to trustworthiness of the outcome of the entire process, and usually follows standard mechanisms to check for exactness and to ensure that the primary data used for the forensic investigation not subjected to unethical modifications that could question its integrity
Examination	Data collected requires examination by forensically processing collected data, using a combination of automated and manual methods. During examination, data of interest assessed and extracted methodically preserving its integrity through the entire process. Examination of data of interest includes looking for deletion history, abnormal login sequences, irregular file operations, unconventional file naming and extensions, exceptional data traffic statistics, unidentifiable user accounts, and others. Focus of the examination stage is simply to look for events of interest, showing anomalies, variances, deviations, and other signs of abnormal occurrences that indicate a pattern or profile of a threat event issue. Assume proper examination of the signs of abnormal and irregular occurrences essential appear as pattern and the examination phase produces a documentation of the patterns in addition to operations and actions carried out
Analysis	After examining data of interest, a detailed security analysis follows. The analysis phase focusses on using legally justifiable methods and techniques to scrutinize or analyze the examination results, in order to derive useful trends and information that addresses objectives of performing the collection and examination in the first place. The analysis phase interprets the process examination to deduce an explanation, and infers credible intelligence based on methodological observation of the trends. Essentially the analysis phase derive a meaning through examination of collected data
Reporting	The reporting phase generates the results and outcome of the analysis, including a clear description of the methods adopted and actions carried out. Reporting also provides justification explaining choice of tools and procedures adopted; and determines any other follow-up action(s) required including lessons learnt, e.g., forensic examination of additional data sources, securing identified vulnerabilities, improving existing security controls, and others to forestall future occurrence. Reporting includes recommendations for improvement covering policies, procedures, tools, and identified areas that could strengthen future forensic tasks

actor causing them. In this case, for cross-incident analysis of the "who" and "why", the preferred method is to begin by defining a *Campaign* for that activity, with a placeholder *Threat Event Actor* identity until additional information becomes known. As more information evolves for characterizing the responsible threat event actor's, the *Threat Event Actor* placeholder incrementally fleshed out [45]. In Fig. 2.2 the STIX[TM] relationship is illustrated after [26]. IoC contain a pattern used to detect suspicious or malicious threat event activity. Threat event activity belongs

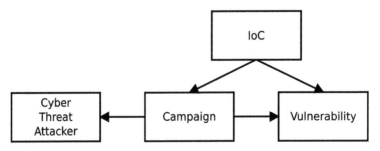

Fig. 2.2 STIX^{TM} Relationship Example

to individuals, groups, or organizational attackers, believed operating with malicious intent. *Campaign* of attackers behavior describes a set of malicious activities or threat event attacks that occur over a time-period against a specific set of targets, and/or vulnerabilities. This is assumed as mistake in software that can be directly used by a hacker to gain access to a computer system, network, or infrastructure resources.

STIX^{TM} adopted by numerous communities and information sharing organizations is an international standard, developed for distribution by TAXII^{TM}, but redistributed elsewhere. STIX^{TM} designed to allow users to describe threat event attacks [26, 46, 47] with regard to:

– Motivation
– Fitness
– Skills
– Reaction

• *Trusted Automated eXchange of Indicator Information (TAXII^{TM})*: Application layer protocol for the communication of threat event information in a simple and scalable manner, used to exchange CTI over HTTPS. TAXII^{TM} enable industrial, public and private organizations to share CTI using an Application Programming Interface (API), a set of services and messages as requirements for clients and servers, supporting common sharing model collections and channels, as illustrated in Fig. 2.3 after [26]. A collection is an interface to a logical repository of CTI objects provided by a TAXII^{TM} Server. It allows operation teams to host a set of CTI data that requested by users or groups: TAXII^{TM} clients and servers exchange information in a request-response model, and a channel, maintained by a TAXII^{TM} server. This allows operation teams to push data to many users groups and/or users to receive data from many teams of threat event operation: TAXII^{TM} clients exchange information with others TAXII^{TM} clients as publish-subscribe model. Note: The TAXII 2.0 specification reserves the keywords required for channels but does not specify channel services. Defining channels and their services be aviated in a later version of TAXII^{TM}.

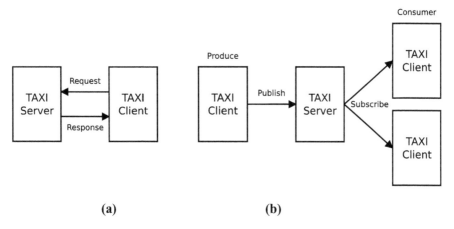

Fig. 2.3 TAXII^{TM} collections (**a**) and channels (**b**)

TAXII^{TM} is a free and open framework that standardizes the automated exchange of information of threat event attacks through services and messaging, designed to integrate with existing sharing agreements, including access control limitations. TAXII^{TM} transitioned to OASIS, which supports automated information sharing for cybersecurity situational awareness, real-time network defense, and sophisticated threat event analysis. Developed to support STIX^{TM} information, which is done, based on a common exchange model [46]. The three main models of TAXII^{TM} are as follows:

- Hub and spoke: Single store of information
- Source/subscriber: Single source of information
- Peer-to-peer: Several groups exchange information with each other

TAXII^{TM} defines four services. However, users or groups can select and implement as many as they need and combine with each other for different exchange models [47]:

- *Discovery*: Way to discover which services a unit supports and how to interact with them
- *Collection management*: Way to learn about data collections and request subscriptions to them
- *Inbox*: Way of receiving content (push messaging)
- *Query*: Way of requesting content (pull message delivery)

Several companies that have business in cybersecurity like Anomaly, EclecticQ, Fujitsu, Hitachi, IBM Security, New Context, NC4, ThreatQuotient, and TruSTAR demonstrating how STIX and TAXII^{TM} are being used to prevent and defend against intruded threat event attacks by enabling threat intelligence to be analyzed and shared among trusted partners and communities [48].

2.2.3 Threat Event Profiling, Threat Intelligence, Threat Lifecycle

Threat event attack profiling described by a linear sequence of phases, as shown in Fig. 2.4.

However, important is not that this is a linear sequence but rather how far along a threat event actor has progressed in his threat event attack process due to the corresponding damage, and identifying what must be performed to prevent the threat event attack impact. Furthermore, Fig. 2.4 shows that if all phases completed, compromising code is present at a targeted computer system, network, infrastructure resource, and others that should not be there, which need defending intrusion of threat event attacks to avoid malicious impacts, such as exfiltration of sensitive data or business assets, and others. In this context, the Command-and-Control phase (C2) of the threat event attack represents the period after which treat event actors exploit to leverage a computer system, network or infrastructure resource. This clue to the importance to analyze threat event attacks taking into account the different indicators to the phases, as shown in Table 2.12. Assuming that a threat event actor attempt to use the zero-day attack to compromise computer systems, networks or infrastructure resources, the Command-and-Control phase (C2) act as for past threat event attacks by the same threat event actor. Different proxy IP addresses used to relay a threat event attack, but the intrusion detection tools used should not change between these threat event attacks. The immutable or infrequently changing properties of threat event attacks by a threat event actor form threat event actor's behavioral profile. This profile refer to the knowledge that capturing, knowing, and detecting the specific intrusion form that facilitates discovering of other threat event attacks by the same threat event actor, even if many other aspects of the threat

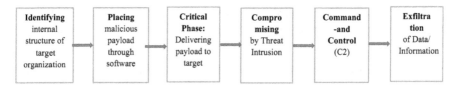

Fig. 2.4 Threat event attack progression sequence

Table 2.12 Cybersecurity operations versus capabilities

Cybersecurity operations	Capabilities
Triage	Determine importance and urgency of reaction to incoming alerts
	Decision making if alert is legitimate and should be tracked
First response	Determine scope of incident identified
	Identify infected and vulnerable devices
	Recommend actions to contain the effects
Investigation	Determine strategic weaknesses in defending threat event incidents
	Recommend actions to prevent recurrences

event attack change. However, analysis of threat event attacks quickly becomes complicated, and makes profiling the cyber threat actor more complex, because the analysis of attack progression requires knowing the threat event attack Indicator of Compromise Lifecycle (IoCL). However, the IoCL has an intrinsic cycle which means that with the discovery of known threat event attacks IoC start the revelation of new ones. Therefore, knowledge is of importance to decide if a detected threat event attack IoCL will further be active or not and has therefore to be taken into consideration. In this regard IoCL refers to answering the intrinsic question "How to collect, analyze and disseminate data sets or data packets of threat event attacks and vulnerabilities incidents"?

Against this background, CTI emerged in order to support cybersecurity practitioners in recognizing the indictors of threat events, extract information about the threat attack method, and responding to threat event attacks accurately in a timely manner. However, a salient problem in CTI is shortening time improving decision-making due to the increasing volume of possible threat event alerts, which has to be fast. Alerts provide timely information about current security issues, vulnerabilities, and exploits. Many alerts classified as unknown because not investigated and mark a type of uncertainty, captured by their ontologies (see Chap. 4) or incidents not been thought of. To reduce response time of threat event attacks, response must become less reactive and more proactive. This requires additional methods, e.g., filter out false alarms, speed up simplifying analysis of threat event attacks.

If suspicious incidents detected through decision-making, Cyber Threat Intelligence Algorithms (CTIA) executed immediately to reduce the impact and severity of threat event attacks. These kind of algorithms based on machine learning (see Chap. 8) through static malware analysis, classification algorithms, data sets and data packets analysis, as well as feature selection algorithms, and others [49]. CTIA deal with cybersecurity operations, classified as illustrated in Table 2.12.

Reducing response time to threat event attacks require that response must become pro-active to detect probable threat event attacks and periodization, strengthening response to threat event attack(s) through threat intelligence-based detection and defense algorithms of threat event attacks. The objective of these algorithms is to accurately discover suspicious incidents based on the analysis of actual measured data to defend threat event attack vulnerability in real-time. However, the objective is hard to achieve, especially in terms of accuracy required.

If threat intelligence-based detection and defense algorithms generate inaccurate results, the outcome negatively affect the performance of the entire Threat Intelligence Defense System (TIDS). In this regard, the inaccurate detection algorithm may generate large numbers of false alarms, which can lead to unnoticed detection of live attacks or intrusions, and unmanageable numbers of alarm notifications that may overwhelm security nodes [50]. Consequently, research has explored new algorithms and methods aiming to increase the performance and accuracy of intrusion detection systems, see [51–55]. Beside this, industrial, public and private organizations have established Threat Response Operation Teams (TROT) gathering threat event data to protect themselves from incoming threat event attacks to maintain a strong cybersecurity posture. In this regard, TROT also sharing information along with the data collected internally. Industrial, public and private organizations need external information to have a comprehensive view of the landscape of threat event attacks. The information about threat event attacks comes from a variety of sources; including sharing communities, open-source and commercial sources, which spans many different levels and timescales. Immediately actionable information are often interpreted as low level IoC, such as known malware hash values or Command-and-Control (C2) IP addresses, where an actionable response can be executed automatically by a cybersecurity system [50].

Moreover, threat intelligence refers to more complex threat information acquired or inferred through the analysis of existing information. Information such as the different malware types used over time within threat event attacks or the network of threat event actors involved in threat event attacks is a valuable information and can be vital to understand and predict threat event attacks, threat development, as well as informing law enforcement investigations. Moreover, there is need for effective Threat Intelligence Management Platforms (TIMP) to facilitate the generation, refinement, vetting of data, post sharing, and others. Designing such a platform, some of the key challenges are working with multiple intelligence sources, combining and enriching data for deeper knowledge, determining intelligence relevance based on technical constructs, and organizational input, delivery into organizational workflows and finally into technological products [56].

2.2.4   Threat Intelligence Sharing and Management Platforms

A Threat Intelligence Management Platform support teams responsible for threat event attack defense, security operation centers, as well as threat intelligence analysts, responsible for threat event response, risk and vulnerability management. They not only have to react to events and warnings, but also predict threat event attacks and act more proactively. To achieve this goal, TIMP serves as a central source of information for all data of threat event attacks from internal and external sources. This simplifies collaboration, structured decisions, proactive measures and faster detection and response. In [57] an open source overview about Threat Intelligence Management Platforms is given. This overview includes the Collective Intelligence Framework), the Collaborative Research in Threats, the Model-Based Analysis of Threat Intelligence Source (MANTIS) framework, the Malware Information Sharing Platform (MISP), and the Soltra Edge Platform for Automated Standard Threat Intelligence Sharing (Soltra Edge) and concludes that the market for Threat Intelligence Sharing Platforms is still developing. Soltra Edge takes threat event information from any source in any format, and aggregates and distributes threat intelligence within and outside industrial, public and private organizations in STIX^{TM} standard format.

Brown et al. [56] discuss the community requirements and expectations of an all-encompassing Threat Intelligence Management Platform (TIMP), based on studies on some threat intelligence sharing platforms. As the landscape of threat event attacks and the organizations environments will change, TIMP helps to stay up to date with potential threat event attacks, predict them, and adopt to ongoing assessments on threat event attacks, as well as strengthening countermeasures. In addition, a subsequent phase concept is required for classifying the lifecycle of threat intelligence as introduced in [57], focusing on:

- *Direction*: Set platform algorithm goals for threat intelligence. If a threat event attack is identified, the needs for information to distinct requirements can be targeted.
- *Collection*: Gathering information to identify the most important threat intelligence platform requirements.
- *Processing*: Executing and transformation of information into an easy to use format.
- *Analysis*: Turns data into intelligence that can perform decisions.
- *Dissemination*: Get intelligence output to places it need to go.
- *Feedback*: Regular feedback is required to make sure that industrial, public, and private organizations threat intelligence platform addressed quite well.

The Malware Information Sharing Platform (MISP) allows effective sharing of malware information. It automatically imports data in intrusion detection systems for faster and better detection, collects threat intelligence from public sources, enriches data and represents the Threat Information in the Comma Separated Value format [58]. The purpose of Comma Separated Values is to present data uniformly to exchange it between TIMP systems. Table 2.13 show an example how Commy Separated Values can occur in a TIMP. TIMP widely adopted to assist organizations in creating and participating in inter-organizational sharing efforts to protect against threat event attacks.

A possible Comma Separated Value export of Table 2.13 in a hypothetical code may look like as shown in Table 2.14.

The first line in the hypothetical code contains the designation of the data fields and lines 1–3 of the corresponding data records. Under normal consideration of a full word document notation of Table 2.13 a semicolon is required, as a separator for the individual data fields. The individual data field is enclosed in quotation marks. However, the question is still paramount: "Should a TIMP build or bought, to

Table 2.13 Comma Separated Value example to TIMP

Threat type	Threat type known as	Date of detection	Number of incidents
Benign	Phishing	10.01.2021	22
Vicious	Zero-day	12.01.2021	2
Vicious	Malware	15.01.2021	15
. . .	. . .	. . .	. . .

Table 2.14 Comma Separated Value export of Table 2.13

Line	Data fields			
	Threat type	Threat type known as	Date of detection	Number of incidents
1	Benign	Phishing	10.01.2021	22
2	Vicious	Zero-day	12.01.2021	2
3	Vicious	Malware	15.01.2021	15
...	...	...	...	...

support, identifying, prioritizing, and act on the most relevant threat event attacks to industrial, public and private organizations business, to cyber secure it against threat event attacks"? In this context, some important considerations required to decide: build or buy. In the one hand, the answer depends on available skilled human resources, because it can be difficult for organizations to find the required skilled staff with the required experience and the essential knowledge necessary, to build up such a central software platform and keep this resource in the long-term continuously up-to-date. On the other hand, this kind of software development is more then only software. The software developer(s) must be familiar with software engineering, agile Development Processes (DevOps), Security Processes (SecOps) and cybersecurity technologies and tools, a combination that is extremely rare to find ad hoc. This means that the required skilled employees from the different domains have to be at hand: software developer(s), big data expert(s), machine learning specialist(s), cybersecurity expert(s), quality assurance expert(s), all of which being able to aggregate and carry out internal essential analyzes of malware data, vulnerability data and risk management strategies, to name a few. Against this background, it could be much easier for organizations to buy a TIMP framework. This is a possible solution in case that buyable TIMP meets more than 80% of the specific requirements of the organizations, and costs for support and upgrades are appropriate because it's all about minimizing threat event attacks and thus cybersecurity risks. However, it is difficult to compare directly the risk minimization performance between a self-developed TIMP framework solution and a commercial TIMP framework. As a result, there is no standard solution or best practice for build or buy. A general workflow of a TIMP can receive and process threat event information from differ torrent sources in order to a better understanding of threat event attacks and recommend an appropriate set of actions [59–61]. These sources include, but not limited to, commercial threat event intelligence providers information sharing communities, such as CiSP1, and internal sources. A simplified conceptual ecosystem model of a TIMP is introduced in [60]. This ecosystem takes action by developing one or more threat event attack response systems including network or endpoint controllers, and security elements or data center controllers [60, 61]. Ultimately, it's all about minimizing threat event attack risk(s). However, it is difficult to directly compare the performance and risk minimization performance between a self-developed TIMP solution and a commercial TIMP solution. As a result, there is no standard solution or best practice for build or buy.

2.3 Exercises

2.3.1 Threats

What is meant by the term *Threat*?

Describe the main characteristics of Threats.

What is meant by the term *Threat Event*?

Describe the main characteristics and capabilities of Threat Events.

What is meant by the term *Threat Event Attack* ?

Describe the main characteristics and capabilities of Threat Event Attacks.

What is meant by the term *Threat Intension*?

Describe the main characteristics and capabilities of Threat Intensions.

What is meant by the term *Threat Actor*?

Describe the main characteristics and capabilities of Threat Actors.

What is meant by the term *Cybercrime*?

Describe the main characteristics, capabilities, and efforts of Cybercrime.

What is meant by the term *Cybercriminal*?

Describe the main characteristics, capabilities and efforts of a Cybercriminal.

What is meant by the term *Advanced Persistent Threat?*

Describe the main characteristics, capabilities, usage of Advanced Persistent Threats.

What is meant by the term *Denial of Service Attack*?

Describe the main characteristics and capabilities of Denial of Service Attacks.

What is meant by the term *Malware Attack*?

Describe the main characteristics of the five most common types of Malware Attacks.

What is meant by the term *Phishing Attack*?

Describe the main characteristics and the five most common types of Phishing Attacks.

What is meant by the term *Ransomware Attack*?

Describe the main characteristics and capabilities of the five most common Ransomware Attacks.

What is meant by the terms *Likelihood*?

Describe the main characteristics and capabilities of the Likelihood.

What is meant by the term *Impact*?

Describe the main characteristics and capabilities of the Term impact by the respective usage.

What is meant by the term *Consequence Level*?

Describe the main characteristics and capabilities of Consequence Levels in the context of a threat event attack.

What is meant by the terms *Internal Threat Event Intension*?

Describe the main characteristics and capabilities of the Internal Threat Event Intension.

What is meant by the terms *External Data Breaches*?

Describe the main characteristics and capabilities of External Data Breaches.
What is meant by the terms *Password Guessing*?
Describe the main characteristics and capabilities of Password Guessing
What is meant by the terms *Drive-by Download Attack*?
Describe the main characteristics and capabilities of the Drive-by Download Attack.

2.3.2   Threat Intelligence

What is meant by the term *Threat Intelligence?*
Describe the main characteristics and capabilities of Threat Intelligence.
What is meant by the term *National Vulnerability Database?*
Describe the main characteristics and capabilities of the National Vulnerability Database.
What is meant by the term *Tactics, Techniques and Procedures?*
Describe the main characteristics and capabilities of Tactics, Techniques and Procedures.
What is meant by the term *Indicators of Compromise?*
Describe the main characteristics and capabilities of Indicators of Compromise.
What is meant by the term *ABC Indicators?*
Describe the main characteristics and capabilities of ABC Indicators.
What is meant by the term *Digital Forensic?*
Describe the main characteristics and capabilities of Digital Forensic.
What is meant by the term *Threat Intelligence Management Platform?*
Describe the main characteristics and capabilities of the different Threat Intelligence Management Platforms.
What is meant by the term *Operational Threat Intelligence?*
Describe the main characteristics and capabilities of Operational Threat Intelligence.
What is meant by the term *Common Vulnerabilities and Exposures*?
Describe the main characteristics and capabilities of Common Vulnerabilities and Exposure.
What is meant by the term *Attack Profiling*?
Describe the main characteristics, capabilities, advantages and disadvantages of Attack Profiling.
What is meant by the term *Threat Intelligence Sharing*?
Describe the characteristics and capabilities of Threat Intelligence Sharing.
What is meant by the term *Residual Risk Rating* ?
Describe the main characteristics and capabilities of Residual Risk Rating.
What is meant by the term *Risk Management* ?
Describe the main characteristics and capabilities of the Risk Management Process.
What is meant by the term *Action Chain of Threat Event Actors*?
Describe the main characteristics and countermeasures of Action Chain of Treat Event Actors.
What is meant by the term *STIX*?

Describe the main characteristics and capabilities of STIX.
What is meant by the term *TAXII*?
Describe the main characteristics and capabilities of TAXII.

References

1. McMillan, R.: Definition: Threat Intelligence. Gartner Research 2013
2. https://www.executech.com/insights/top-15-types-of-cybersecurity-attacks-how-to-prevent-them/ (Accessed 12.2022)
3. Lehto, M., Phenomena in the Cyber World. In: Cyber Security: Analytics, Technologies and Automation, Lehto, M., Neitaanmaki, P. (Eds.), Spinger Publ. 2015
4. M. Goodman "Future Crimes", Penguin Random House, 2016
5. T. J. Holt, B. H. Schell, "Hackers and Hacking", ABC-CLIO Press, 2013
6. M. Sikorski, A. Honig, "Practical Malware Analysis", No Starch Press, 2012
7. https://www.imperva.com/learn/application-security/insider-threats/ (Accessed 12.2022)
8. https://techtarget.com/searchcontentmanagement/definition/brandjacking (Accessed 12.2022)
9. Shaw, E., Ruby, K.G., Post, J.M.: The Insider Threat to Information Systems: The Psychology of the Dangerous Insider. In: Security Awareness Bulletin, Vol. 2, pp. 1–10, 1998
10. https://www.pratum.com/services/it-risk-management/risk-assessment (Accessed 12.2022)
11. Miller, B., Rowe, D.: A Survey SCADA of and Critical Infrastructure Incidents. In: Proceedings 1st ACM Annual Conference on Research in Information Technology, pp. 51–56, 2012
12. Möller, D.P.F.: Cybersecurity in Digital Transformation: Scope and Applications. Springer Nature 2020
13. Langner, R.: Stuxnet: Dissecting a Cyberwarfare Weapon. In: IEEE Security and Privacy, Vol. 9 No.3, pp. 49–51, 2011
14. Information Security: Guide to Conducting Risk Assessment. NIST Special Publication 800-30, CODEN: NSPUE2, 2012
15. Tucci, L.: What is Risk Management and Why it is so Important. Tech Target Report, 2021
16. Shafer, G.: Perspectives on the Theory and Practice of Belief Functions. In: International Journal of Approximate Reasoning, Vol. 4, pp. 323–362, 1990
17. https://www.iso.org/obp/ui/std:iso:3100ed-2v1:en (Accessed 12.2022)
18. Robertson, J., Diab, A., Martin, E., Nunes, E., Paliath, V., Shakarian, J., Skakarian, P.: Darkweb Cyber Threat Intelligence Mining. In: Cambridge University Press, 2017
19. Borum, R., Felker, J., Kern, S., Demnesen, K., Feyes, T.: Strategic Cyber Intelligence. In: Information and Computer Security, Vol. 23, No. 3, pp. 317–332, 2015
20. https://malpedia.caad.fkie.fraunhofer.de/usage/tos (Accessed 12.2022)
21. https://malshare.com/about.php (Accessed 12.2022)
22. https://www.misp-projectorg/galaxy.html (Accessed 12.2022)
23. https://socradar.io/what-is-operational-cyber-threat-intelligence/ (Accessed 12.2022)
24. https://nvd.nist.gov/ (Accessed 12.2022)
25. https://cve.mitre.org/ (Accessed 12.2022)
26. https://www.mitre.org/sites/default/files/publications/stix.pdf (Accessed 12.2022)
27. https://oasis-open.github.io/cti-documentation Accessed 12.2022)
28. Mavroeidis, V., Bromander, S.: Cyber Threat Intelligence Model: An Evaluation of Taxonomies, Sharing Standards, and Ontologies within Cyber Threat Intelligence. In: Proceedings European Intelligence and Security Informatics Conference, pp. 91–98, 2017
29. Casey, T.: Understanding Cyber Threat Motivations to Improve Defense. Intel White Paper, 2015

30. Fishbach, A., Ferguson, M.J.: The Goal Construct in Social Psychology. In: A. W Kruglanski, E. T. Higgins (Eds.) Social Psychology: Handbook of Basic Principles, pp. 490–515, The Guilford Press, 2007
31. Bromander, S., Josang, A., Eian, M.: Semantic Cyber threat Modeling. In: STIGDS, pp. 74–78, 2016
32. SANS, Security Intelligence: Attacking the Cyber Kill Chain. https://digital-forensics.sans.org/blog/2009/10/14/security.intelligence-attacking-the-kill-chain/ (Accessed 12.2022)
33. Wrightson, T.: Advanced Persistent Threat Hacking: The Art and Science of Hacking any Organization. McGraw-Hill, 2010
34. van Haaster, j., Gevers, R., Spengers, M.: Cyber Guerillas. Elsevier Publ., 2016
35. O. Al-Ibrahim, A. Mohaisen, C. Kamhoua, K. Kwiat, L. Njilla, "Beyond Free Riding: Quality of Indicators for Accessing Participation in Information Sharing for Threat Intelligence". Technical Report University at Buffalo and Air Force Research Lab, 2017 https://arxiv.org/abs/1702.00552 (Accessed 12.2022)
36. C. Sillaber, C. Sauerwein, A. Mussmann, R. Breu, "Data Quality Challenges and Future Research Directions in Threat Intelligence Sharing Practice". In: Proceedings ACM Workshop on Information Sharing and Collaborative Security, pp. 65–70, 2016
37. G. Sharkov, "From Cybersecurity to Collaborative Resiliency" In: Proceedings ACM Workshop on Automated Decision Making for Active Cyber Defense, pp. 3–9, 2016
38. D. Bekerman, B. Shapira, L. Rkach, A. Bar, "Unknown Malware Detection Using Network Traffic Classification". In: Proceeding IEEE Conference on Communications and Network Security (CNS), pp. 134–142, 2015
39. Fogla, P., Sharif, M., Perdisci, R., Kolesssnikov, O., Lee, W.: Polymorphic Blending Attacks. In: Proceedings 15th UNSENIX Security Symposium, pp. 241–256, 2006
40. Duessel, P.: Detection of Unknown Cyber Attacks Using Convolution Kernels over Attributed Language Models. PhD Thesis University of Bonn, Germany, 2018
41. Zia, T., Liu, P., Han, W.: Application Specific Digital Forensic Investigative Model in Internet of Things (IoT). In: Proceedings ACM-ARES Conference, 2017 https://doi.org/10.1145/3098954.310404 (Accessed 12.2022)
42. Okreafor, K., Djhaiche, R.: A Review of Application Challenges of Digital Forensic. In: International Journal of Simulation Systems, Science & technology, pp.36.1–36.6, 2020. https://doi.org/10.5013/IJSSST.a.21.02.35
43. https://cyware.com/educational-guides/cyber-threat-intelligence/what-is-open-indicators-of-compromise-openioc-framework-ed9d (Accessed 12.2022)
44. Lock, H.Y.: Using IOC (Indicators of Compromise) in Malware Forensi. SANS Institute, 2019
45. https://stixproject.github.io/documentation/idioms/campaign-v-actors/ (Accessed 12.2022)
46. https://www.anomali.com/de/what-are-stix-taxii (Accessed 12.2022)
47. Struse, R., Wunder, J., Davidson, M., Jordan, B.: TAXI^TM Version 2.0 Working Draft 02. OASIS Open, 2017
48. https://www.oasis-open.org/news/pr/cybersecurity-companies-demo-support-for-stix-and-taxii-standards-for-automated-threat-intel?platform=hootsuite (Accessed 12.2022)
49. Dehghantanha, A., Dargahi, M. (Eds.): Cyber Threat Intelligence. Springer Publ., 2015
50. Thames, L., Schaefer, D.: Cybersecurity for Industry 4.0 and Advanced Manufacturing Environments with Ensemble Intelligence. In: L. Thames and D. Schaefer, (Eds.): Cybersecurity for Industry 4.0 – Analysis for Design and Manufacturing, pp. 243–65, Springer Nature 2017
51. Anderson, J.P.: Computer Security Threat Monitoring and Surveillance. Contract 79F296400, 1980
52. Axelsson, S.: Intrusion Detection Systems: A Survey and Taxonomy. Technical Report, Department of Computer Engineering, Chalmers University of Technology, 2000
53. Ghorbani, A.A., Lu, W., Tavallee, M.: Network Intrusion Detection and Prevention Concepts and Prevention. Springer Publ. 2010

54. Khor, K.C., Ting, C.Y., Amnuaisuk, S.P.: From Feature Selection to Building of Bayesian Classifiers: A Network Intrusion Detection Perspective. In: Am. J. Appl. Sci. Vol. 6, pp. 1949–1960, 2009
55. Zhang, J., Porras, P., Ullrich, J.: Gaussian Process Learning for Cyber-Attack Early Warning. In: Proceedings SIAM International Conference in Data Mining, pp. 255–264, 2008
56. Brown, S., Gommers, J., Serrano, O.: From Cybersecurity Information Sharing to Threat Management. In:Proceedings of the 2nd ACM Workshop on Information Sharing and Collaborative Security, pp. 343–49, 2015
57. Poputa-Clean, P.: Automated Defense _ Using Threat Intelligence to Augment Security. Technical Report, SANS Institute InfoSec, 2015
58. Pace, C.: The Threat Intelligence Handbook – A Practical Guide for Security Teams to Unlocking the Power of Intelligence. Cyberedge Press 2018
59. Gupta, B.B, Agrawal, D.P., Wang, H. (Eds.): Computer and Cybersecurity: Principles, Applications, Algorithms, and Perspectives, CRC Press, 2019
60. Zibak, A., Simpson, A.: Cyber Threat Information Sharing: Perceived Benefits and Barriers. In. Proceeding ARES Conference, 2019. https://doi.org/10.1145/3339252.3340528 (Accessed 12.2022)
61. Appala, S., Cam-Winger, N., McGraw, D., Verman, J.: An Actionable Threat Intelligence System: Using a Publish-Subscribe Communications Model. In: proceedings 2nd ACM Workshop on Information sharing and Collaborative Security, pp.61–70, 2015. https://doi.org/10.1145/2808128.2808131 (Accessed 12.2022)

Chapter 3
Intrusion Detection and Prevention

Abstract Intrusion detection and prevention are security measures used to detect and prevent cybersecurity risks to computer systems, networks, infrastructure resources, and others. Intrusion detection and prevention systems automatically detect and respond to cybersecurity risks in order to reduce potential risks through threat event attacks. They use different methods for a successful execution. In this context, the signature-based approach that corresponds to known threat event attacks is used, or the anomaly-based detection that compares definitions of what activity is considered normal against observed threat event attacks, to identify significant deviations. Other methods are the stateful protocol analysis, which compares predetermined profiles of general accepted definitions of benign protocol activities for each protocol state against observed events, to identify deviations, or the hybrid system approach that combines some or all of the other methodologies to detect and respond to cybersecurity risks, and others. However, the need of intrusion detection and prevention systems architectures require distinguished decisions to the essential methodology used and the deployed system architecture. Against this background, this chapter seeks to offer a clear explanation of respective methodologies and comparing theses methodologies with regard to effectivity and efficiency. This requires (i) a discussion regarding the importance of intrusion detection and prevention to combat against threat event attack risks, malicious threat event attacks, by logging information about them and attempt to stop this, and (ii) reporting the identified malicious threat event attacks to the cybersecurity response team. Furthermore, investigation of threat event attacks is done, because threat event actor's seeking out computer systems, networks, and infrastructure resources to exploit vulnerabilities and to attack, causing serious problems for threat event attacks for the targeted industrial, public, and private organizations. Therefore, Intrusion Detection and Prevention Systems (IDPSs) are a valuable approach in keeping information systems secure against malicious threat event attack risks by monitoring, analyzing, and responding to possible cybersecurity violations against computer systems, networks, or infrastructure resources. The violations may result from attempts by unauthorized intruders that try to compromise the computer systems, networks, infrastructure resources, and others. These intruders can be privileged internal users that misuse their authority, or external single cyberattackers or attacker-groups. In this context, Chap. 3 introduces in Sect. 3.1 in the specific background of intrusion detection methods and in Sect. 3.1.1 in the specific characteristics and capabilities of the different intrusion detection forms and their advantages and disadvantages. Thus, anomaly detection is part of Sect. 3.1.2, while Sect. 3.1.3 refers to misuse intrusion detection, and Sect. 3.1.4 focuses on advantages and disadvantages of anomaly and misuse intrusion detection forms. Section 3.1.5 refers to the Specification-based Intrusion Detection, which combines the strength of anomaly and misuse detection, and Sect. 3.1.6 refers to the characteristics of intrusion detection types. The focus of Sect. 3.1.7 is on intrusion detection systems and its architecture. In this sense, Sect. 3.2 focusses on intrusion prevention, whereby Sect. 3.2.1 describes the intrusion prevention system, while Sect. 3.2.2 focuses on the architecture of the intrusion prevention system. Section 3.3 refers to the intrusion detection and prevention system architecture and the respective performance measures as constraints for the proof of concept approach. Section 3.4 introduces the intrusion detection capability metric, which includes the necessity developing the respective detection approach to detect known and unknown threat event attacks. Finally, Sect. 3.5 summarizes the intrusion detection and intrusion prevention approaches, concerning a stable and resilient system operation. Section 3.6 contains comprehensive questions from the topics intrusion detection and intrusion prevention methodologies and architectures, while reference section refers to references for further reading.

3.1   Intrusion Detection

An intrusion is a threat event attack intruded by a threat event actor or a group of actor's in the attempt to gain access to valuable data or information assets to disrupt or manipulate normal business operation. As introduced in [1], intrusions are actions that attempt to bypass cybersecurity mechanisms of information systems to threaten their confidentiality, integrity, and availability, the CIA Triad (see Sect. 1.6.2). In this regard, confidentiality means that information is not made available or disclosed to unauthorized individuals, entities, or processes, whereas integrity means data or information not been altered or destroyed in an unauthorized manner. Finally, availability refers to the fact that a system assures that the required data or information is available and usable upon demand by authorized users. An intrusion may be caused by a threat event actor or a group of actors to assess the targeted information system through the Internet, or exploits any security flaw of a third party (Middleware) application. In case threat event attacks come from outside origins, they are called external threat event attacks or cyberattacks. Threat event attacks that involve unauthorized internal users, e.g., employees, attempting to gain and misuse non-authorized access privileges, are called insider cyberattacks [2]. These possibilities result from the exponential growth of Internet-based facilities that corresponds with a number of digital systems, connected to computer systems, networks, infrastructure resources, and others, providing and serving online services to users. This allows an infinite amount of threat event actor's to access unauthorized to intrude threat event attacks, e.g., traffic flooding, worms, Denial of Service (DoS) attacks, and others [3], which simultaneously enhance the need for proactive intrusion detection methods. However, there are a number of ways to differentiate threat event attacks with respect to one another, a priori. Two or more threat event attacks may be related in the following way [4]:

- Threat event attack TEA_1 may be a subclass of threat event attack TEA_2.
- Threat event attack TEA_1 may be an another view of threat event attack TEA_2, e.g., TEA_1 may be the same as TEA_2, but viewed from a network perspective, rather than a process perspective.
- Threat event attacks TEA_1, TEA_2 .... TEA_m may enable threat event attack TEA_n.
- Threat event attacks TEA_1, TEA_2 .... TEA_m may combine to form a composite treat event attack $CTEA$.

Similarly, the intrusion detection methods may yield a way to differentiate between treat event attacks. The intrusion detection methods are related in the following way:

- One Intrusion Detection Method may be relatively more sensitive than another to attack TEA_1, as compared with threat event attack TEA_2.
- One Intrusion Detection Method may operate at a different level of abstraction than another.
- One Intrusion Detection Method may use a composite analysis, already taking input from many different sources. If this is the case with a commercial, proprietary intrusion detection system, its combination may not be expressible using the format already defined.

Thus, intrusion detection is the process of observing and tracking threat event attacks on computer systems, networks, infrastructure resources, and others, to identify signs of threat event attacks, monitor threat event activities, and others. Against this background, the main goal of intrusion detection is avoiding unauthorized access and misuse of computer systems, networks, infrastructure resources, and others by insiders, e.g., organizations employees, as well as external (outsiders) penetrators, with their cybercriminal threat event activities. In this regard, intrusion detection becomes a challenging task due to the proliferation of digital resources with their increased connectivity in the era of digital transformation, which gives outsiders a greater opportunity to access and make it easier for them to intrude their threat event attacks.

In this context, the main task of an Intrusion Detection System (IDS) is to identify and detect patterns and/or signatures of cybersecurity breaches monitored. Monitoring means to monitor data flow or data packets in computer systems, networks, infrastructure resources to analyze them for patterns and/or signatures of suspicious activity, and finally report detected cybersecurity incidents to the responsible cybersecurity team for further action. Therefore, IDS is designed and installed to deter or mitigate the damage caused by hackers (see Table 1.15). Thus, IDS is based on hardware and software components to detect attempts compromising the CIA Triad (see Sect. 1.6.2), addressing the cybersecurity needs of computer systems, networks and infrastructure resources, and others. In advanced IDS that detect documented threat event attacks, the IDS is also able to learn to detect new threat event attack forms. In this sense, intrusion detection, documentation, and reporting is a primary IDS function.

In contrast, the prevention of detected intrusions is part of the Intrusion Prevention System (IPS), which takes actions to respond to malicious activities or anomalous behavior. The IDS is typically placed inline, e.g., at a spanning port of a switch, or on a hub in place of a switch, and other, passively monitoring the presence of threat event attacks and provide notification for active intrusion defense services, in case of identified and detected suspicious incidents.

3.1.1 Significant Intrusion Detection Methods

If a cybersecurity breach is identified as violation that intends to corrupt the actual executed code by intruding malicious code, then IDS sends out an alert to the IPS about the detected treat event attack. In this regard, the IPS responds to conquer the detected malicious incident, preventing it from continuing to the intended destination. Depending of placement and deployed cybersecurity methodology, several IDS types are present. In this regard, IDS and IPS are valuable assets in a cybersecurity strategy for computer systems, networks, infrastructure resources, and others. IDS monitor and IPS possibly prevent attempts to intrude or compromise computer systems, networks, infrastructure resources, and others. There are three significant types of IDS [5–7]:

- *Host-based IDS (HIDS)*: Run on all computer systems in the network and other resources of the organizations network to monitor the network for malicious traffic. Their characteristics and capabilities are summarized in Table 3.1.
- *Network-based IDS (NIDS)*: Deployed and intended to manage networks only when chances of vulnerabilities are high and refer to the desired specification of the NIDS. Their characteristics and capabilities are summarized in Table 3.2.
- *Specification-based IDS (SIDS)*: Describe the desired behavior through functionalities and security policies. Their characteristics and capabilities are summarized in Table 3.3.

After detecting an intrusion incident, the generated alert transferred to a response team that responds to the alert and takes appropriate action defending the intrusion to conquer the intruder. Nevertheless, the problem with intrusion detection is that an incident caused by a threat event attack is one out of a number of different types of threat event attacks. In one case, an unauthorized user might steal passwords to masquerade identity to the attacked computer system, network, or infrastructure resource. In another case of threat event attacks, threat event actors are people who are legitimate users, who abuse their privileges, or people who use prepacked exploit scripts that are often found on the Internet to attack computer systems through a network. Therefore, a combination of Intrusion Detection and Intrusion Prevention

Table 3.1 Characteristics and capabilities of HIDS

IDS type	Characteristics and capabilities
HIDS	HIDS are host-based ID that represent an application, which monitors computer systems, networks, or infrastructure resources for suspicious activities. The activities monitored include intrusions created by external actors and internally by misuse of resources or data. In this sense, a HIDS use rules and policies in order to search for log files, flagging those with events or activities the rules have determined as potential malicious behavior. HIDS is divided into four types [8]: Connection analyzing IDS: Monitors computer systems, networks, or infrastructure resource connection attempts to and from a host. File systems IDS: Monitors computer systems, networks, or infrastructure resource checking the integrity of files and directories. Kernel-based IDS: Detect malicious activity on a kernel level. Log-file analyzing IDS: Analyze log-files for patterns, indicating suspicious activities. Configured and implemented correctly, they are useful tools in information security. HIDS's main function is monitoring internal risks, which means identifying and detecting anomalies and compromising activities. Thus, a HIDS monitors the behavior of applications on the host observing their interaction with the underlying operation system. In practice, security relevant interactions typically are system calls and their work schemes, examining the trace of system calls by each application [9]. Most HIDS use a combination of two methods [10]: Anomaly detection-based HIDS: Approach that analyze a trustworthy behavior, using machine learning to flag malicious behavior to determine when something is probing a computer system, network, or infrastructure resource prior to a real attack. The success of this HIDS type depends on the degree of distribution across the computer systems, networks, or infrastructure resources and the level of training provided by the IT administrators. Signature detection-based HIDS: Searching for a previously known signature, or a specific intrusion event. However, threat actors can make small changes to the threat event so that data cannot keep up in real time. Most IDS needs regular update to keep up with regular and known threat event attacks. As long as the IDS database is up to date, the IDS work in an appropriate manner. HIDS residing on trusted computer systems, networks, and infrastructure resources are in close nature to authenticated users of these components. If one of the users attempts an unauthorized activity, the HIDS usually detect and collect the most pertinent information as fast as possible. In addition to detect unauthorized insider activities, HIDS is also effective detecting unauthorized file modifications. Another example for HIDS is rouge program detection that accesses computer systems, networks, or infrastructure resources in a suspicious way, discovering that the rouge program has modified the registry in a harmful way or the program served and receives applications or operation systems audit logs. Advantages of HIDS are to monitor user and file access activity, including file accesses, changes to file permissions, attempts installing new executables, using system logs containing events that have actually occurred to determine whether a threat event occurred or not, and others like near real-time detection and response. Usage of switched network protocols does not affect a NIDS. Disadvantages of HIDS referred harder to manage, because configuring information and management has to be arranged for every host. It also is not well suited to detect network scans or other forms of surveillances. It also is not optimized to detect multihost scanning, and is vulnerable to both, direct threat events and to apply threat events against the host operating system.

Table 3.2 Characteristics and capabilities of NIDS

IDS type	Characteristics and capabilities
NIDS	IDS become a challenging task due to the proliferation of computer networks and the increased connectivity of computer systems, which give outsiders a greater opportunity in accessing and intruding threat event attacks. It also makes threat event attacks easier without identification or threat event attacks. However, NIDS differs from an HIDS in that way that it usually is placed along a local Area network (LAN) wire. It attempts to discover unauthorized and malicious access to a LAN by analyzing traffic that traverses the LAN to multiple hosts. In this regard, the NIDS observes the traffic at specified points in the network, checking traffic packet by packet that pass through the network in real time to detect intrusion patterns. NIDS can examine the activity at any layer of the network such as network layer, transport layer, and application layer protocol. Network-based systems are best detecting unauthorized outsider access, bandwidth theft, denial of service (DoS) attacks, and others. The packets that initiate bandwidth theft attacks are detected best by NIDS. There are many algorithms to detect malicious traffic, but they generally read inbound and outgoing packets and searches for any suspicious patterns. Any alert generated by an NIDS allows notification of the organization cybersecurity team or take active actions such as blocking the source internet protocol (IP) address. When unauthorized users log in successfully, or attempts to log in, they are tracked by the HIDS. Therefore, detecting unauthorized users is accomplished by the NIDS before their logon attempt. Some NIDS are *Dragon* [11]: Include Dragon sensor that monitors network packet for traffic that indicate network misuse or abuse. Dragon server manages data from Dragon blocks like sensor. Network flight recorder (NFR) [12]: Utilize auditing program to extract an extensive set of features that describe each network connection or host session, applying data mining to learn rules that accurately capture the behavior of intrusions and normal activities. The rules used for misuse and anomaly detection. LITNET-2020 [13]: New annotated network benchmark dataset obtained from real-world academic networks. Dataset presents real-world examples of normal and under attack network traffic. Advantage of NIDS is that some well-placed IDS monitor a large network. Typically, NIDS are passive entities that listen on a network LAN without interfering with other operations. NIDS advantages are difficulties processing all packets in a large or busy network. Therefore, it may fail to detect a threat-event launched during periods of high packet traffic.

in one System (IDPS) is required, which consists of devices that collect data about the computer systems, networks, infrastructure resources, and others, being monitored, which will be processed and evaluated to enhance detection quality in threat event attack detection. Thus, Table 3.4 shows a comparison between advantages and disadvantages of HIDS, NIDS, and SIDS.

However, any definition of an intrusion detection type is imprecise as security policy requirements do not always integrate well-defined set of actions. Against this background, intrusion detection is stated as a general methodology to detect intrusions by threat event attacks. Therefore, IDS can be located inline the legitimate data or packet traffic to any computer system, network, infrastructure resource, and others to monitor all internal legitimate data or packet traffic, as shown in the generic IDS

Table 3.3 Characteristics and capabilities of SIDS

IDS type	Characteristics and capabilities
SIDS	In SIDS, the correct behaviors of critical and crucial objects are manually abstracted as security specifications, which are compared with the actual behavior of the objects. Intrusions that usually cause objects to behave in an incorrect manner detected without exact knowledge about them [14]. Thus, any sequence of operations executed outside the SIDS specifications considered as a security violation. Specification-based detection applied to privileged programs, applications, and several network protocols. Various formalisms for SIDS have been suggested to capture operation sequences of SIDS, either manually with the parallel environment grammar [15], regular expressions to events [16], abstract state machine languages [17] or automatically with techniques such as inductive logic-programming [18], software fault-tree as well as colored Petri-Nets [19]. SIDS approaches have also been integrated to automated techniques [20] or combined with anomaly-based methods [21, 22] to keep high detection accuracy while decreasing the cost of manually defining the specifications [23] and pattern matching models [24].

Table 3.4 Comparison between intrusion detection systems

IDS type	Advantages	Disadvantages
HIDS	Accurate in intrusion detection	Higher cost
	Able to detect encrypted threat event attacks	Cause performance issues or resource logging
	Does not require additional hardware	
NIDS	Low cost	High fluctuations in network traffic cause packets to be lost
	Detect network attacks such as denial of service (DoS) attacks	Requires more CPU power & resources in large-scale LAN
		Unable analyzing encrypted packets
SIDS	Medium cost	Manual capturing of operation sequences of SIDS
	High detection accuracy	Manually defining detection specification

model in Fig. 3.1. Positioning the IDS module in the legitimate network traffic make use of both a Firewall and IDS.

Since, an IDS generates a large amount of data, traffic, and events in its logs, the main feature for the IDS is generating alerts on threat event attacks of interest and danger. Thus, effective IDS should have a low rate of false positives and false negatives, as described in [25], and shown in Table 3.5.

Intrusion detection systems use many different methodologies to detect incidents. Their characteristics and capabilities are summarized in Tables 3.1, 3.2 and 3.3. The primary classes of detection methodologies are [26] as follows:

- *Anomaly-based Intrusion Detection (AID)*: Depend typically on normal patterns, classifying any deviation from normal as malicious. It compares what activity is

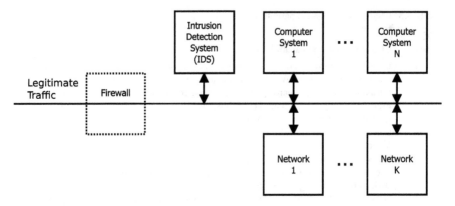

Fig. 3.1 Integration of an intrusion detection system in a computer system and network environment

Table 3.5 True and false positives and negative

Rate	Positive	Negative
True	Alerts when there is malicious data and/or traffic flow	Silent when data and/or traffic flow is rare or unlikely
False	Alerts when data and/or traffic flow is rare or unlikely	Silent when malicious data and/or traffic flow occurs

considered normal against observed threat event attacks to identify significant deviations, using profiles developed by monitoring the characteristics of typical activities over a period-of-time.

- *Specification-based Intrusion Detection (SpID)*: Operates in a similar way as AID. It employs a model of valid behavior in a form of specifications to the development, which requires user guidance.
- *Signature-based Intrusion Detection (SiID)*: Relies on pre-specific signatures of threat event attacks. Therefore, it compares known threat event signatures with observed threat event attacks to identify incidents that are very effective, detecting known threat event attacks, but largely ineffective in detecting unknown threat event attacks and many variants of known threat event attacks (see Sect. 2.2). SiID cannot track and understand the state of complex communications, so it cannot detect most of threat event attacks that comprise multiple threat event attacks.

However, there are two major categories of intrusion detection methods used by most IDSs [25–29] that are broadly classified as:

- *Anomaly Intrusion Detection (AID)*: Protects computer systems, networks, or infrastructure resources against malicious incidents.
- *Misuse Intrusion Detection (MID) or Signature Intrusion Detection (SiID)*: Protects computer systems, networks, or infrastructure resources against patterns that

describe suspect, based on patterns that describe suspect collections of sequences of activities or operations that can be possibly be harmful.

3.1.2   Anomaly-Based Intrusion Detection

A behavior that is neither nominal nor normal is an anomalous behavior. Therefore, Anomaly-based Intrusion Detection (AID) is a key element to identify items, incidents, or observations that do not confirm to an expected pattern or other signatures in a dataset or data packet. Examples of anomalies are data points not following a particular distribution, the occurrence of incorrect or infrequent values, or the repeated presence or absence of particular events, and others. Thus, AID classifies deviations from normal behavior that indicate the presence of intentionally or unintentionally accomplished threat event attacks or cybercriminal fraud. Therefore, AID focusses on detecting unusual activity patterns in the observed data, whereby the AID system must be able to distinguish between normal and anomalous behavior. AID detection approaches are based on models of normal data sets or data packets to detect deviations from the monitored normal data sets or data packets. Beside this, anomaly-based detection algorithms must also be capable to detect new types of threat event attacks based on intrusion incidents as deviations from normal data transmission usage. In this sense, it is suitable to divide anomaly AID systems into static and dynamic AID-systems:

- *Static Anomaly Detection System (SADS)*: Based on the assumption that there is a portion of the object monitored not constant (static) as it should be. This type of AID system checks the static portion of the computer system, network, infrastructure resource, and others and reports whenever it deviates from its normal (regular) behavior. This occurs when an error has occurred or a threat event attack has altered the static portion of it. Therefore, static AID systems define one or several static bit strings to monitor the desired state of the inspected system to archive a representation of that state. Periodically, SADS compares the archived state representation to a similar computed representation, based on the current state of the same static bit string(s). Any detected difference represents an error or intrusion. Hence, SADS checks for data sets or data packets integrity. In this regard, virus-specific checkers belong to SADS;
- *Dynamic Anomaly Detection System (DADS)*: Includes a behavior, defined as a sequence or a partially ordered sequence of distinct incidents. Therefore, DADS examine and extract basic traffic features. After pre-processing, the goal of the DADS algorithm is to train the AID-system with normal (regular) data to model normal network traffic from the given set of normal (regular) data. Thereafter, the task of the DADS is to determine whether the test data belong to normal (regular) or to an abnormal (irregular) behavior to given new test data [30–32]. In this regard, the generic AID-system architecture is composed of the submodules:

- Pre-processing.
- Anomaly detection.
- Alerting/reporting.

Furthermore, AID can be either time dependent or time independent. Time-dependent AID focusses on anomaly detection in temporal data sets, using time data, e.g., time between threat event attacks, time of occurrence, or threat event attack ordering. Time-independent AID ignores temporal information and focuses on the detection of anomalies on individual data points, e.g., inside the multivariate data set of an event or data aggregations [28].

AID systems often use Self-Organizing Maps (SOMs) algorithm to model signatures of normal (regular) data or packet traffic to determine if a computer system, network, infrastructure resource behavior is normal (regular) or abnormal (irregular). Alerting and reporting handles communication based on a decision support system. The SOM algorithm belongs to the category competitive learning model, commonly successfully used for various clustering problems [33]. The SOM-based method belongs to unsupervised learning (see Sect. 8.1.1.2) to map nonlinear statistical relationships between high-dimensional input data into two-dimensional space, called output space. SOMs efficiently place similar signatures to adjacent locations in the output space and provide projection and visualization options for high dimensional data. Training the SOM algorithm, the first step is to enumerate and normalize input vectors that be accomplished in a pre-processing submodule. The input data sets for training contain normal data sets and data set obtained from threat event attacks or abnormal signature data packets, e.g., for network traffic. The anomaly-based detection submodule extracts pre-processed data sets out of the normal data sets used, to train the SOM. After a successful training phase, the SOM is ready to use for classification, normal or abnormal behaviors.

In this regard, activities in AID-systems are monitored that periodically generated signatures capturing their behavior. When monitored input data sets or packets are processed, the AID-system periodically generates a value to indicate a normal or abnormal behavior. In case of an anomaly, where too much deviation from the normal signatures occurs, the decision system of the intrusion detection reports an alert due to an identified threat event intrusion, which can lead to false positive alarms (see Table 3.5). This means that a monitored regular functional operation mistakenly identified as threat event attack, due to an error, which result in a fault stop of the regular functional system operation. This is likely a negative impact on computer system, network, and infrastructure resource function, depending on conditioning or sensitivity of the AID-system. False positives are events reported as malicious, in reality they are not. Therefore, logging is an important issue of AID to record intrusion-related activities, by means not to determine what kind of intrusion is in the first place.

Furthermore, a capability lack in intrusion detection enables a threat event actor to attempt threat event attacks until a successful one is identified. However, AID allow to identify a threat event attack before the threat event attack is likely successful. A simple rule-of-thumb says if system traffic not reasonably generated by a legitimate user of the application, it is almost certainly a threat event attack. Once a threat event intrusion identified by the AID-system, the threat event attack alert to the Intrusion Prevention System (IPS), which is next in line, to respond appropriately to the alert. Typically, this contains logging off the user from the system intruding the anoma-lous traffic, invalidating the ac-count, recording information for the cybersecurity team, or patching the root cause vulnerability. Against this background, there are three important findings the actual alert might cause:

- Almost certainly a threat event attack.
- Not sure whether it is a threat event attack or not.
- Almost a certainly legitimate input.

which requires some extra information, because this is a more policy decision, organizations should be aware about in their requirements phase over the cyberse-curity lifecycle. To find an appropriate solution, the safest rule is to assume that everything except legitimate traffic is a threat event attack. However, this causes probably false alerts that can be false positive as well as false negative, which are hard to deal with. In case of false positive, the AID system identifies an incident as threat event attack. Nevertheless, a false positive is a false alert. A false negative incident is the most serious and dangerous alert when the AID system identifies an incident as acceptable, whereas this incident actually is a threat event attack. Therefore, a false-negative alert is an incident, where the AID system fails to detect a real threat event attack. This is the most dangerous threat event alert, because it pretends a regular computer system, network, or infrastructure resource behavior, albeit a threat event attack took place. In this regard, in Table 3.5, the true and false positives and negatives are summarized. Against this background, anomaly-based detection include the following advantages and disadvantages:

- *Advantage of Anomaly Intrusion Detection*: No predefined rules for detection of intrusions are required; detects new threat events;
- *Disadvantages of Anomaly Intrusion Detection*: False-positive alert can arise, leading to user's inconvenience. Establishment of regular profile usage is required but often is hard to achieve.

Furthermore, anomaly detection techniques also include beside the foregoing described methods the following types of models [33]:

- *Statistical Models (SM)*: These techniques make use of different kinds of models, such as:

  - *Operational Model or Threshold Metric*: Monitors data or data packet traffic incidents that occur over a period of time and raise the alarm if incidents are higher than a threshold x.
  - *Markov Model or Marker Model*: Monitors computer systems, networks, infrastructure resources, and others at fixed intervals and keeps track of its state probability at a given time interval Δt_{SP}. A change of the state occurs when an incident happens, and thus, the behavior is detected as anomalous, if

the probability of occurrence of that state is high. Markov-based techniques are extensively used in HIDS.

- *Standard Deviation Model*: If an incident falls outside the set of interval, e.g., higher or lower the statistical moment, or mean, or standard deviation, or any other correlation, this incident is said being anomalous.
- *Multivariate Model*: Based on major difference between mean and standard deviation model with regard to correlations among two or more metrics.
- *Time Series Model (TSM)*: Interval timers together with an incident counter or resource measurement are key components in this model.

Beside these models, other tools and techniques are used to detect anomalies based on simple statistical techniques up to complex machine learning algorithms, depending on the complexity of data and sophistication. Some are simple yet powerful statistical techniques, used for initial screening. While complex algorithms can be inevitable to use, sometimes simple techniques serve better:

- *Z-score*: Probably the simplest one yet an useful statistical measure for anomaly detection.
- *Modified Z-score*: Relies on the median for calculating the Z-score.
- *Interquartile Range (IQR)*: Split data in quartiles.
- *Boxplot*: Provides better graphical representation of IQR, but also provides additional information.
- *Histogram*: Shows the shape of the values, or distribution, for a continuous variable.

• *Cognitive Models (CM)*: These techniques make use of different kinds of methods such as finite state machine method, description scripts that can describe signatures of attacks on computer systems, networks or infrastructure resources, as well as adept process management model, which is trained by a huge amount of knowledge of patterns with known threat event patterns, obtained from a knowledge community.

• *Cognitive-based Detection Techniques (CDT)*: This technique makes use of an audit data classification technique, based on sets of predefined rules, classes, and attributes identified from training data sets of classification rules, parameters, and procedures inferred. Different kinds of methods exist under this headline such as boosted decision tree, support vector machine, as well as artificial neural networks, and others, as illustrated in Fig. 3.2 after [30]. However, beside the foregoing mentioned methods of intrusion detection models, a new methodological approach is coming up based on the computer immunology approach. However, applying computer immunology to basic computational models is subject of research, based on the widely applied basic notions of antigen and antibody models.

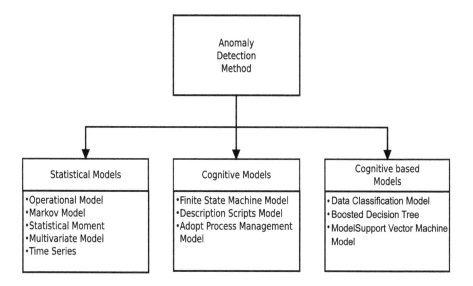

Fig. 3.2 Classification of Anomaly-based Intrusion Detection techniques

3.1.3 Misuse-Based Intrusion Detection

A multitude of technical safeguards is available to detect and prevent threat event attacks. Concepts behind are broadly categorized in the Anomaly-based Intrusion Detection (AID), described in Sect. 3.1.2, Misuse-based Intrusion Detection (MID), described in Sect. 3.1.3, and Specification-based Intrusion Detection (SpID), described in Sect. 3.1.5.

Misuse-based detection methods intended to recognize known threat event signatures described by rules. Rule-based solutions are divided into

- *Blacklist-based Method (BM)*: Blacklisting has been deployed as a key element in anti-virus and security software suites, typically in the form of a virus database of known digital signatures, heuristics, or behavior characteristics, associated with viruses and malware that have been identified in the wild. It can be differentiated into:
  - *Signature-based Approach*: Enables to detect threat event attacks based on specific threat event patterns, e.g., malicious byte sequences.
  - *Heuristic-based Approach*: Enables detection of unknown threat event attacks, based on expert-based probabilistic rule sets that describe malicious indicators.
- *Whitelist-based Method (WM)*: Whitelisting draws up a list of acceptable entities that enable access to a computer system, network, or infrastructure resource and blocks everything else. Based on a zero-trust principle it essentially denies all, and allows only what is necessary. It usually includes policies that enable threat

event detection on the deviation from a pre-defined negative baseline configuration, e.g., IP whitelists.

The majority of intrusion detection systems are classified as MID. In this regard, MID tries to match actual data and traffic flow activity with stored signatures of known exploits or threat event attacks. This means MID use a priori knowledge of threat event attacks to investigate for threat event attack traces, based on well-defined signatures of input incidents. Assuming that the state transition of the respective components leads to attacked state when exercised with the intrusion pattern, weaknesses in computer systems, networks, infrastructure resources, and application software exploited. The objective of MID is to frame the intrusion detection problem as a signature or pattern-matching problem to develop efficient algorithms for such a matching problem. Indeed, simply specifying intrusion patterns without the initial state specification is often insufficient to capture intrusion scenario in total.

Other primary approaches to MID are [24]:

- *Expert Systems*: Code knowledge about threat event attacks is based on if-then implication rules.
- *Model-based Reasoning System*: Combines models of misuse with evidential reasoning to support conclusions about the occurrence of misuse.
- *State Transition Analysis*: Represents threat event attacks as a sequence of state transitions of the monitored computer system, network, or infrastructure resource.
- *Key Stroke Monitoring*: Uses user key strokes to determine the occurrence of a threat event attack.

Another MID method makes use of signatures of threat event attacks, described by rules. These rule-based methods divided into blacklist- and whitelist-based approaches.

To overcome challenges with existing approaches, e.g., limited ability to detect unknown threat event attacks by signature-based methods and lack of detection accuracy by behavior-based methods, interest has grown in the cybersecurity community to utilize machine learning as alternative and more accurate approach (see Chap. 8).

3.1.4 Disadvantages of Anomaly and Misuse Intrusion Detection

A primary disadvantage of AID is that statistical measures of user behavior be trained gradually. Threat event actors, who knows be monitored, train computer systems, networks, or infrastructures resources over a length of time to the point where intrusive behavior is considered being normal (regular). MID that is simpler than AID, is immune to such training. If signatures of a threat event attacks are developed, even major variations to the same threat event attack signatures be detected, but this is simpler in MID compared to AID.

Another disadvantage is that these methods only look for known vulnerabilities and are currently of little help in detecting unknown intrusions. However, modern approaches enhanced the capability in AID by using a Self-Organizing-Map (SOM) structure to model normal behavior, whereby the deviation from the normal behavior is classified as a security violation, and hence a threat event attack [28].

SOM from a general perspective is an Artificial Neural Network (ANN) to analyze and visualize high dimensional data set, which belongs to the category of unsupervised learning (see Sect. 8.1.1.2) to map nonlinear statistical relationships between high-dimensional input data sets into two-dimensional space, the output space. A SOM efficiently place similar patterns to adjacent locations in the output space and provides projection and visualization options for high dimensional data sets [31]. In a two-dimensional use case the ANN can be arranged either on a rectangular or a hexagonal lattice. The artificial neurons that are adjacent belong to the 1-neighborhood N_{il} of the artificial neuron i. A neighborhood function determines how strongly the artificial neurons connected to each other. In this regard, the neighborhood functions and the number of artificial neurons determine the accuracy and the generalization capability of the SOM mapping.

An overview of state-of-the art AID and MID methods are provided is given in the references [32, 33].

3.1.5   Specification-Based Intrusion Detection

The concept of Specification-based intrusion detection (SpID) was introduced by Ko in 1997 [14]. It describes the desirable behavior through its embedded functionalities and through the used security policy. Any sequence of operation executed outside the SpID specifications is considered as a security violation [24]. In this regard, SpID proposes a promising alternative that combines the strengths of MID due to accurate detection of known threat event attacks and AID with regard to the ability to detect novel treat event events. For this purpose, SpID use specific manually developed behavioral specifications characterizing legitimate behavior as a basis to detect threat event attacks. As this method is based on legitimate behaviors, it does not generate false alarms when unusual, but legitimate behaviors are encountered. Since it detects threat event attacks as deviation from legitimate behavior, it has the potential to detect previously unknown threat event attacks [17]. The promise of SpID argued for some time was the question of whether these benefits be realized in practice, which remained open. Some of the questions that arise in this context are [17]:

- How much effort is required to develop behavioral specifications?
- How do these efforts compare with that required for training AbID-systems?
- How effective is the approach in detecting novel threat events?
- Are there classes of threat events detected by SpID-systems that can't be detected by AbID detection or vice-versa?
- Can it achieve false alarm rates that are comparable to MbID?

3.1.6 Intrusion Type Characteristics and Detection

In order to classify breaches by threat event attacks, a scheme based on intrusion types is presented in [34, 35] and summarized in Table 3.6, which introduces some important intrusion types, their characteristics, and detection possibilities.

Let A_1, A_2, ..., A_n be n measures used to determine if an intrusion by a threat event attack is occurring to a computer system, network or infrastructure resource at any given time, whereby each A_i measures a different aspect of the computer system, network or infrastructure resource with

$$A_i = \begin{cases} 1 & implying \ that \ the \ measure \ is \ anomalus \\ 0 & otherwise \end{cases}$$

where A_I is an $i \ x \ 1$-dimensional vector. Let H be the hypothesis that the computer system, network or infrastructure resource is currently undergoing an intrusion by a

Table 3.6 Intrusion types, characteristics and detection

Intrusion type	Characteristics	Detection
Attempted break-in	Breaking into a system generates a high abnormally rate of password failures with regard to a single account or the system as a whole.	Atypical behavior profiles or violations of security constraints
Denial of service (DoS)	An intruder be able monopolize a resource might have abnormally high activity with regard to the resource, while activity for all other users is abnormally low.	Atypical use of system resources (e.g., network)
Inference by legitimate user	User attempts to obtain unauthorized data from a database through aggregation and inference might retrieve more records than usual.	Atypical behavior profiles using I/O resources
Leakage by legitimate user	User trying to leak sensitive documents, log into the system at unusual times, or route data to remote printers not normally used.	Atypical usage of I/O resources
Masquerading or successful break-in	A log into a system or network through an unauthorized account or password might have different login time, location, or connection type from that of account's legitimate user. Intruder's behavior may differ considerably from that of the legitimate, e.g., a user using most of his time browsing through directories and executing system status commands, whereas the legitimate user might edit, compile, or link programs.	Atypical behavior profiles or violations of security constraints
Trojan horse	A program is substituted for a legitimate program.	Atypical CPU time or I/O activity
Virus	Causes an increase in the frequency of executable files rewritten or storage used by executable files.	Atypical CPU time or I/O activity

threat event attack. The reliability and sensitivity of each anomaly measure A_i is determined by

$$p(A_i = 1|H)$$
$$and$$
$$p(A_i = 1|/H).$$

The combined belief in H is

$$p(H|A_1, A_2, \ldots, A_n) = p(A_1, A_2, \ldots, A_n|H) \times \frac{p(H)}{p(A_1, A_2, \ldots, A_n)}$$

which requires the joint probability distribution of the set of measures conditioned on H and $/H$. In [36], covariance matrices used to account for the interrelationships between measures. If the measures A_1, A_2, ..., A_n are represented by vector A, then the compound anomaly measure is determined by

$$A^T C^{-1} A$$

where C is the covariance matrix representing the dependence between each pair of anomaly measures A_i and A_j.

This intrusion detection method broadened by intrusion prevention, the process of performing intrusion detection and defending possible detected incidents. The objective introducing intrusion detection and prevention methods is their primary focus on identifying possible threat event attacks; logging information about them; attempting to defend them; reporting them to responsible cybersecurity administrators in the industrial, public, and private organizations; and documenting existing threat event attacks. Therefore, intrusion detection and prevention methods have become an essential narrative to the cybersecurity infrastructure of nearly every mission-critical and crucial computer systems, networks and infrastructure resources, and others. The type of an Intrusion Detection and Prevention System (IDPS) extracted by the types of potential threat event attacks that they monitor and the ways in which they deployed, as shown in Table 3.7.

3.1.7 Intrusion Detection System Architecture

Securing critical and crucial computer systems, networks, or infrastructure resources is a very important objective in cybersecurity, because threat event actors try to gain access to sensitive business assets. Therefore, specific protective activities are of essential importance such as encryption and other security approaches that enable a cyber-secure data transmission. This includes verifying that the components are working as desired and not anomalous. That requires monitoring for security issues

Table 3.7 Intrusion detection and prevention system types

IDPS type	Characteristics
Host based	Monitoring characteristics of a single host and events occurring within that host for suspicious activity
Network based	Monitoring network traffic for particular network segments or devices and analyzing network and application protocol activity to identify suspicious activity
Network behavior analysis	Examines network traffic identifying threat event attacks that generate unusual traffic flows, such as distributed denial of service (DDoS) attacks, certain forms of malware, and policy violations (e.g., client system providing network services to other systems)
Wireless	Monitoring wireless network traffic and analyzing it to identify suspicious activity involving the wireless networking protocols themselves

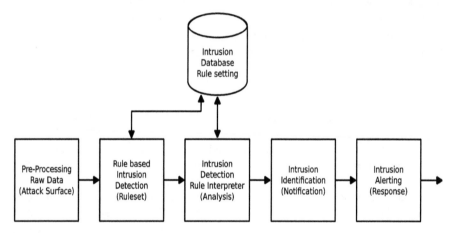

Fig. 3.3 Block diagram of a generic Rule based Intrusion Detection System

to perform regular vulnerability assessments, respond appropriately to vulnerabilities and testing and deploying of an Intrusion Detection System (IDS). In such IDS data sets stored and processed directly, and the output of which is fed into a rule-based system that take further action to defend detected intrusions. Therefore, monitoring is required to identify causes of threat event attacks and alert. Thus, an alert list of threat event attacks is created that refer to the targeted computer system, network or infrastructure resource, the threat event actor is attempting to intrude, to alert about the specific threat event attack and block the attacked entity, alert and proceed the alert for alerting response, as shown in Fig. 3.3 [37, 38].

Figure 3.3 illustrate that the generic Rule-based Intrusion Detection System (RIDS) contains a pre-processing component for raw data, transforming the data into a suitable form for further data processing in the RIDS. Pre-processing offers a manifold of methods to use as described more in detail in Sect. 3.2. It aims to reduce the size of data is to find relations between data, normalize data, remove outliners and extract data features. It includes several techniques such as data cleaning, integration transformation and reduction [38]. Pre-processing technique feature extraction pulls out data significant in some particular context to detect normal and abnormal activities. In this context, RIDS enables estimating relations between data that describe how the output depends on the input.

The block shown in Fig. 3.3 contains beside the pre-processing block three important components:

- *Rule Base*: A rule is an ordered pair of strings. Set of rules that govern decision such as known threat event attack, normal event activity, known malicious activity, and others. A rule base typically has a format of source, destination, service, action;
- *Database*: Collect and organizes all data, and be updated regularly. Moreover, the database contains stored known threat event patterns, signatures or uncertain data that compared with actual threat event attacks for intrusion detection. In case of a cybersecurity incident case an alert will be generated.
- *Rule Interpreter*: Learning kernel, based on an inference engine for decision-making with regard to normal and anomalous activities in data that successfully match against threat event attack related patterns or signatures in the database or a combination of several uncertainty sources. If anomalous activity is found, the rule interpreter checks the rule base by comparing the ordered pair of signatures of each rule until one is identified, which can successfully be matched against the known threat event attack related patterns or signatures in the database in order to detect the incident point and type of intrusion.

To achieve this goal, the block diagram model, shown in Fig. 3.3, also contains an adaptive Expert System (not shown in the block diagram). The Expert System solves intrusion detection problems by applying knowledge, generated based on expertise in the field of an application, such as decision support to identify the source of intrusion incidents to suggest best possible prevention techniques and suitable controls for the different types of threat event attack intrusions.

Furthermore, from Fig. 3.3 it can be seen that the RIDS uses a cybersecurity audit such as a vulnerability scan and alerting and reporting mechanisms. For this purpose, the intrusion database has stored the known malicious incidents for future intrusion detection of threat event attacks. Moreover, rules are defined and stored in the Rule Set-based Intrusion Detection Engine (RSIDE) of the RIDS, while intrusion points and types is passed to the Expert System to evaluate that data with known malicious incidents, stored in the intrusion database, to detect the threat event attack source, using a backward chaining approach. In advanced intelligent intrusion detection systems the intelligent RIDS suggests the appropriate prevention technique after detecting the threat event attacks. In this regard, the integrated Expert System approach in the RIDS permits the incorporation of human experience in the RIDS, and then utilizes that knowledge to identify incidents that match the defined characteristics of misuse and cyberattack. The expert system used in the RIDS recognizes intrusions of threat event attacks by comparing them with the sequence of regular actions, stored in the set of rules. These rules replicate the normal (regular) activities in order to conclude whether or not the identified data refer to a malicious intrusion scenario.

Rule based analysis, as shown in the block diagram in Fig. 3.3, relies on sets of predefined rules that repeatedly applied to a collection of implemented facts. Facts represent conditions that describe a certain situation in the audit records or directly from system activity monitored. Rules represent heuristics that define a set of actions executed in a given situation to describe known intrusion scenario(s) or generic techniques. The rule that fires has identified a malicious incident, which causes an alert that takes further action. An alert requires as an operational routine immediate response to defend the identified threat event intrusion. Therefore, the RIDS represent a core element to control the data flow between cyberattack surfaces and mission-critical and crucial devices or components. However, the described generic intrusion detection model in Fig. 3.3 has to expand its rule set-based approach by integrating intelligent signature detection methods due to the i) growing complexity of computer systems, networks and infrastructure resources and ii) the ever growing complexity of threat event attacks, to effectively and efficiently defending threat event attacks. Otherwise, this result in providing invalid, unexpected, or random data that requires expanding the *Rule based Intrusion Detection (Ruleset)* block. This is an important fact if intrusion detection is embedded at interfaces that cross a trust boundary, because the violation refers to vulnerabilities that have not been validated before crossing the boundary. As reported in [39], numerous static, dynamic, and hybrid solutions are available for analyzing patterns and signatures in program codes and the behavior of program execution, to identify the presence of malicious threat event attacks, helping to disable them. In real-time systems, used for mission-critical tasks, intrusion incidents be detected through static timing analysis.

In [40], mechanisms for time-based intrusion detection described to detect the execution of unauthorized threat event attacks in real-time computer systems, networks or infrastructure resources. This intrusion detection utilized information obtained by static timing analysis. For real-time systems, timing bounds on code sections are available as they are already determined prior to the schedule analysis. In [40] it is demonstrated how to provide micro timings for multiple granularity levels of application code. Through bound checking of these micro timings, techniques be developed, to detect intrusion incidents (i) in a self-checking manner by the application and (ii) through the Operating System Scheduler (OSS), which is a novel contribution in the real-time system domain. Another important IDS issue is the application of ANN techniques for the anomaly and misuse detection in computer systems, networks and infrastructure resources. ANN is a non-linear statistical data modeling method that tries to simulate the functions of biological neural networks. ANN consists of an interconnected collection of simple processing elements, artificial neurons and processes information in a connectionist approach to computation, which means its treatment is to transform a set of inputs to a set of searched outputs, through a set of simple processing units, or nodes, and connections between them. Generally, an ANN is an adaptive system that changes its structure in response to external or internal information that flows through the network during the learning phase. Subsets of the units are input nodes, output nodes, and nodes

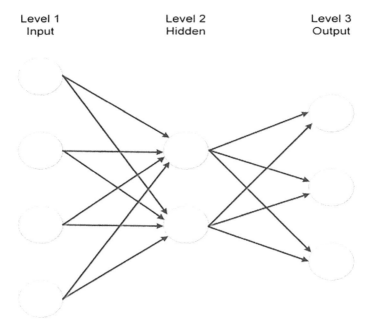

Fig. 3.4 Architecture of an artificial neural network (ANN) with its layers: input, hidden, output

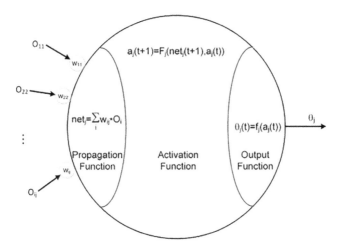

Fig. 3.5 Node structure of an artificial neural network

between input and output, the hidden layers. The connection between two units has some weight, used to determine how much one unit will affect the other, as illustrated in Figs. 3.4 and 3.5 [26].

ANNs with one hidden layer belong to the category of shallow learning. ANN typically based on:

- *Architecture Body*: Specifies variables involved in the ANN and their topological relationships;
- *Activity Rules*: Represent local rules that define how the activities of the artificial neural neurons change in their response to each other;
- *Learning Rules*: Specify the way in which the ANN's weights,

$$w_{i,j}; i,j = 1, \ldots, m, n,$$

change with time. Usually learning rules depend on the activities of the ANN neurons. They may also depend on the target values supplied by a training phase and on the current value of the weights, $w_{i,j}$, as shown in Fig. 3.5.

Figure 3.5 shows the following conditions of the ANN:

- Input Information oij: Pass along the weighted connections wij to neuron j;
- Propagation Function $netj$: Also called net input, summarizes the weighted input information according to a predefined rule, often a simple summation of all weighted ones

$$net_j = \sum_i w_i * o_i$$

- Activation Function Fj: Determines based on the grid input and a predefined threshold value Θj the state of the neuron aj

$$a_j = F_j(net_j, \Theta_j)$$

- In simple ANN it is usually a sigmoid threshold function. When the net input $netj$ exceeds the threshold, the neuron is activated;
- Output Function θj: Uses the state of the neuron aj to calculate the output value θj, which in turn is passed on to a neuron and is part of the input information there:

$$\theta j = f_j(a_j)$$

- The identity used most often, so that the state of the neuron is used as the output value.
- Activation aj: Represents the current state of neuron j. Binary numbers most frequently used, e.g., with $0 \equiv inactive$ and $1 \equiv active$ or $\{-1; 1\}$. When using the logistic function or the hyperbolic tangent, aj is assigned real values;
- Threshold Θj: Determines from which network input value the neuron is activated.

In this context, AAN are massively distributed parallel entities made up of processing units (nodes), as shown in Fig. 3.5. They have the capability to store experimental knowledge and making it available for use in monitoring abnormal behavior of applications in the cyber-space of digital transformation. The nodes are effective against hidden attacks of threat event actor's. The ability dealing with uncertain and partially true data makes ANN attractive in intrusion detection. In this regard, some IDS exploited as Pattern Recognition Technique, implemented by using a Feedforward ANN that be trained accordingly. During training, the ANN parameters optimized to associate outputs with corresponding input patterns. When an ANN is used, it identifies the input pattern and tries to output the corresponding class. In case a connected record has no output associated with, it is referred to as input. The neural network gives the output that corresponds to a taught input pattern that is least different from the given pattern. Applying these techniques for IDS two from the manifold of ANN types are of interest:

- *Multilayer Feedforward ANN*: Are non-parametric regression methods, which approximate the underlying functionality in data by minimizing the loss function;
- *Self-Organizing Map (SOM)*: Promising technique in cluster analysis. SOMs are popular unsupervised training algorithms. A SOM tries to find a topological mapping from the input space to clusters. SOM employed for classification problems.

3.2   Pre-processing in Intrusion Detection Systems

Raw data obtained from different sources are often not direct usable for data analysis in their gathered form, because data may contain certain errors, noise or have missing data values, and others. Therefore, raw data require pre-processing to transform them into a suitable for the respective application domain for further analysis. In this context, data pre-processing methods consist of data cleaning, data transformation and data reduction, as summarized in Table 3.8.

Beside the data pre-processing methods in Table 3.8, two other methods are often used [43–45]:

- *Histogram-based Pre-processing*: Unsupervised approach that does not use class label(s). The approach distribute attribute values into ranges, whereby the values divided into equal ranges in the equal width histogram because of the equal frequency histogram each part has the similar part of data;
- *Entropy-based Pre-processing*: Used for discretization of data. It's a top down and supervised approach [46] that use class information to reduce the size of data. To discretize a numerical attribute X it must choose the value of X with minimum entropy as a splitting point. This step is repeated recursively to get hierarchically discretization [47]. Let's assume a set of tuple(s) exist with a number of attributes and classes $C1$ and $C2$ for attribute X. To discretize tuples T to an attribute X assumes that all tuples with class $C1$ are in one part and tuples with class $C2$ fall into another class. The expected information to classify is a tuple in T based on X.

Table 3.8 Raw data pre-processing methods

Raw data pre-processing	
Pre-processing type	Characteristics and capabilities
Data cleaning	Raw data contain incomplete data records, noise values, outliners and inconsistent data. Data cleaning is use to 　Find missing value(s) in unrecorded value(s) through: 　　Fill the missing value(s) manually 　　Use a global constant to fill the missing value 　　Use the attribute means to fill the missing value(s) 　　Use the most probable value(s) to fill the missing value(s) Smooth noise data because noise is a random error or variable tin a measured variable. Noise is a raw data that also means there are outliners, derivate from the normal but noise can be corrected using the methods described in [41, 42]
Data transformation	Means transforming raw data into forms suitable for the data mining process, which involves: 　Aggregation: Process applying statistical metrics like means, median, variance, necessary to summarize the data 　Generalization: Includes replacing the lower level data (primitives) by higher level using hierarchical concepts 　Normalization: Adjusting data values into a specific range such as between 0 and 1 or 1 and −1; 　Smoothing: Removes noise from data. Methods used are clustering, regression and binning
Data reduction	Used to reduce dataset(s) into smaller volumes to maintain the integrity of the original dataset(s). Data reduction methods are: 　Attribute subset selection: Reduce dataset(s) size by removing redundant features or dimension and irrelevant attributes 　Data cube aggregation; construct data cube applying operations of aggregation(s) of data without losing the necessary information for data analysis 　Dimensionality reduction: Also known as data compression that issues mechanisms of encoding to reduce size of data set(s) 　Numerosity reduction: Replace or map data to an alternative or smaller representation of data

In this context, data preprocessing is required in knowledge discovery tasks such as network-based intrusion detection (NIDS), which attempts to classify network traffic as normal or anomalous. For this purpose, various formal process models developed for knowledge discovery and data mining. They include dataset creation, data cleaning, integration, and feature construction to derive new higher-level features, feature selection to choose the optimal subset of relevant features, reduction, and discretization techniques. The most relevant ones in NIDS are [48]:

- *Dataset Creation*: Method involves identifying representative network traffic for training and testing. These datasets be labeled indicating whether the connection is normal or anomalous. Labelling network traffic can be a very time consuming and difficult task;
- *Feature Construction*: Method aims to create additional features with a better discriminative ability than the initial feature set, which bring a significant

improvement to machine learning algorithms. Features can be constructed manually, or by using data mining methods such as sequence analysis, association mining, and frequent-episode mining;

- *Reduction*: Method commonly used to decrease the dimensionality of the dataset by discarding any redundant or irrelevant features. This optimization process called feature selection, and commonly used to alleviate the curse of dimensionality. Achieve data reduction with feature extraction, which transforms the initial feature set into a reduced number of new features. Principal Component Analysis (PCA) is a common linear method used for data reduction.

Preprocessing converts network traffic into a series of observations, where each observation is represented as a feature vector. Observations optionally labelled with its class, such as regular (normal) or irregular (anomalous). These feature vectors are then suitable as input to data mining or machine learning algorithms. Machine learning (see Chap. 8) is the use of algorithms, which evolve according to the labeled data instances (observations) provided to it. The algorithms are able to generalize from these observations, hence allowing future observations automatically classified. Machine learning is widely used in anomaly-based NIDS [48].

3.3 Intrusion Detection Capability Metric

Evaluation of threat event attack intrusion detection is fundamental to measure the objective effectiveness of an Intrusion Detection System (IDS) in terms of ability to correctly classify detected incidents as normal (regular) or malicious incidents. Measuring IDS detection capability is essential to decide about the detection accuracy of threat event intrusion(s). In this regard, measuring the detection capability of threat event intrusion(s) is required in an Intrusion Detection System Architecture (IDSA). However, several metrics available that enables measuring different aspects of an IDSA, but none of them objectively measures the IDSA detection capability metric of threat event intrusion(s) detection [49]. Therefore, it is difficult to determine, which IDSA is more effective to detect objectively malicious intrusions in terms of only False Positive Rate (*FPR*) than any other, i.e., probability the IDSA outputs an alert when no threat event attack occurs, and True Positive Rate (*TPR*), i.e., probability that the IDSA outputs an alert when threat event attacks happen.

Let's assume IDSA1 detect 10% more threat event attacks then IDSA2, but IDSA2 generate 10% less false alerts, which raise the questions: Which IDSA fit better, IDSA1 or IDSA2?, and How to determine a unified objective metric that enables calculating the Intrusion Detection Capability (*CID*) of an IDSA? In [50] a metric is suggested that enables identifying the IDSA configuration for threat event intrusion detection, which fit best identifying input data or packet streams being normal (regular) or malicious, generating a trustable alert if truly malicious. For this purpose, an abstract model for intrusion detection of threat event attacks is developed, which is referred to as Butterfly scheme, shown in Fig. 3.6.

Fig. 3.6 Butterfly scheme of abstract intrusion detection model

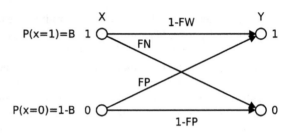

As can be seen from Fig. 3.6, the processing of the data follows a butterfly pattern, connecting the inputs x (left) to the outputs y that depends on them (right) through a butterfly step. The general associated system of equation is

$$x'(k) = x(k) + ax(k + Q)$$
$$x'(k + Q)) = x(k() + bx(k + Q)$$

with a and b are the respective signal flow metric coefficients.

The assumptions for the specific intrusion detection model of threat event attacks based on random variables as follows

X be equal 1

if a threat event attack intrusion happen, and

X be equal 0

in normal (regular) case. Thus, the IDSA output alerts if random variable

Y equals 1

while a threat event attack intrusion happen, and random variable

Y equals 0

which means no alert because no threat event attack happen. Therefore, the IDSA output generated corresponds to each input.

Let's assume data or packet streams encoded for formal reasons as follows:

$$X = 1$$

for malicious incidents, and

$$X = 0$$

for normal (regular) data or packet streams, examined for every data or packet stream in the IDSA to detect outputs that indicates a malicious incident

$$Y = 1$$

or not

$$Y = 0.$$

In the abstract intrusion detection model, $p(X = 1)$ represents the base rate B, the prior probability of a threat event attack intrusion detected by the IDSA, representing the *Known-Knowns* (see Table 2.10).

Let's assume a threat event attack intrusion has a probability $p(Y = 0|X = 1)$ and considered as normal (regular) by the IDSA. This represents a False Negative Rate *(FNR)* denoted by the symbol β, identifying there is no alert /A, when there is an intrusion of a threat event attack *I*.

Let's assume a normal (regular) event has a probability $p(Y = 1|X = 0)$ and misclassified as threat event attack intrusion incident by the IDSA. This represents a False Positive Rate *(FPR)* denoted by the symbol α, identifying there is an alert *A*, when there is no threat event intrusion /I.

However, evaluating an IDSA data is required to calculate symbols a, β and B. The model in Fig. 3.6, introduced in [48], finally enables to calculate the Intrusion Detection Capability *(IDC)* metric, to identify the best IDSA configuration for detection of intrusions of threat event attacks, to evaluate different types of IDSA with regard to fit best to *IDC*. The *IDC* is defined in [50] as follows: "Let X be the random variable representing the IDSA input and Y the random variable representing the IDSA output. *IDC* is defined as:"

$$IDC = \frac{I(X;Y)}{H(X)} = \frac{H(X) - H(X|Y)}{H(X)}$$

where $I(X;Y)$ represent the mutual information of X and Y, $H(X)$ is the entropy of X, and $H(X|Y)$ is the conditional entropy of X, after Y is known. The mutual information measures the reduction of uncertainty of the IDSA input, knowing the IDSA output and normalized, using the entropy, i.e., the original uncertainty of the input. Hence, *IDC* represents the uncertainty reduction ratio of the IDSA input, to estimate the IDSA output. Its value range is [0,1]. Therefore, a larger *IDC* value means that the respective IDSA have a better capability classifying input threat event attack intrusion(s) accurately."

The model in Fig. 3.6, can also be interpreted as follows: Let's assume $\tilde{X}$ is a stochastic binary vector that represents the assumed correct assessment of the input data and/or packet stream $\tilde{S}$, i.e., the correct indication whether each stream is a threat event attack intrusion or not. Thus the threat event attack intrusion detection algorithm developed has to be a deterministic function acting on $\tilde{S}$, yielding a bit-string $\tilde{Y}$ that should ideally be identical for $\tilde{X}$. In this regard the IDSA makes correct guesses about the unknown $\tilde{X}$, based on the input stream $\tilde{S}$. The actual number of required binary guesses is $H(\tilde{X})$, the so called real information content of $\tilde{X}$. Against this background, the number correctly guessed by the IDSA is

$$I\left(\tilde{X}; \tilde{Y}\right).$$

Therefore,

$$I\left(\tilde{X}; \tilde{Y}\right) = H\left(\tilde{X}\right)$$

is the fraction of correct guesses for correct cyber threat attack intrusion detection events".

Let's now abstract the model in Fig. 3.6 for IDSA with input (X) and output (Y) for which *IDC* be assumed as function of the foregoing mentioned basic variables B, α, and β. Furthermore, let's assume two cases for B as follows:

$$B = 0$$

and

$$B = 1,$$

which means that for the first case the IDSA input is 100% true to normal (regular) and for the second case the IDSA input is 100% true to a threat event attack intrusion incident, which yield

$$H(X) = 0.$$

These two cases be defined for *IDC* = 1. Therefore, the *IDC* metric be expressed using the Bayes theorem so that base rate B is assumed as part of prior data or information stream about the IDSA operational threat event intrusion detection. In this regard, the Bayesian detection rate corresponds to different prediction rates as shown in [51]:

- *True-Positive Prediction Value (TPPV)*: Probability of a treat event attack intrusion when the IDSA output alerts, expressed by the term $TPPV(A_O|I_E)$ with $A_O=$ alert, $I_E=$ intrusion incident.
- *False-Positive Prediction Value (FPPV)*: Probability of an output alert when no threat event attack intrusion occur, expressed by the term $FPPV(A_O|/I_E)$ with $A_O=$ alert, $/I_E=$ no intrusion incident.

- *True-Negative Prediction Value (TNPV)*: Probability of no treat event attack intrusion when the IDSA output not alerts, expressed by the term $TNPV(/A_O/I_E)$ with $/A_O$= no alert, $/I_E$= no intrusion incident.
- *False-Negative Prediction Value (FNPV)*: Probability of no output alert obtained, when threat event attack intrusion happen, expressed by the term $FNPV(/A_O|I_E)$ with $/A_O$= no alert, I_E= intrusion incident.

These terms are important from a perspective of detection rate usability. For an IDSA two cases are essential:

- $TPPV(A_O|I_E)$.: Output an alerts when an intrusion incident happen, which means that the IDSA has to take action.
- $TNPV(/A_O|I_E)$: Output not alert and no intrusion incident happen, which means the IDSA not take action.

Applying the Bayes' theorem [51] to calculate probabilities $TPPV(A_O|I_E)$ and $FNPV(/A_O|I_E)$ results in two equations for the Bayesian positive detection rate and the Bayesian negative detection rate:

$$TPPV((A_O|I_E)) = \frac{TPPV(I_E)TPPV((A_O|I_E))}{[TPPV(/I_E)TPPV(/A_O|/I_E)] + [TNPV(/I_E)FPPV(A_O|/I_E)]}$$

and

$$TNPV(/A_O|I_E) = \frac{TNPV(/I_E)TNPV(/A_O|I_E)}{[TNPV(/I_E)TPPV(A_O|I_E)] + [TNPV(/I_E)FNPV(/A_O|/I_E)]}$$

The Bayesian positive and negative detection rates are functions of the variables Base Rate B, False Positive Rate α, and False Negative Rate β. In case False Positive Rate α and False Negative Rate β decrease, *TPPV* and *TNPV* increase that obtain better intrusion detection results. As shown in [52] *FPPV* has a dominant impact on *TPPV* assuming B is very low and thus *TPPV* depends only on *FPPV*. Furthermore, *PPPV* is less sensitive against changes *FNPV* values, however, *CID* is. In this regard, Gu et al. [50] conclude that *PPPV* and *TNPV* are useful when evaluating an IDSA from a usability perspective, but more in general there exist no objective method to integrate these metrics in an easy way, which be part of further research work. However,

$$H(X|Y)$$

is expanded as

$$H(X|Y) = CID = -B(1-\beta)logTPPV - B(1-\beta)\,log(1-TNPV) - (1-B) \\ \times (1-\alpha)logTNPV - (1-B)alog(1-TPPV)$$

From the above equations, it is seen that the *IDC* metric has incorporated both *TPPV* and *TNPV* in measuring the intrusion detection capability *IDC*. Hence, *CID* unifies all existing commonly used metrics, i.e., *TPPV*, *FPPV*, *TNPV*, and *FNPV*. It also factors in *B*, a measure of the IDSA operational intrusion incident of the threat event attack to computer systems, networks, infrastructure resources, and others.

3.4   Intrusion Prevention

As described in Sect. 3.1 an IDS is a cybersecurity tool that, like other measures such as antivirus software, firewalls, and access control schemes, is intended to strengthen cybersecurity by detection of possible threat event attacks to computer systems, networks, infrastructure resources, and others. In contrast, an Intrusion Prevention System (IPS) responds to detected threat event attacks attempting to prevent it from succeeding. The IPS can change the content of threat event attacks or change the cybersecurity environment. It also could change the configuration of other security controls to disrupt a cyberattack incident like reconfiguring, e.g., a network device to block access from the threat event actor or altering a host-based firewall to a target to block incoming threat event attacks [51, 52]. Some IPS removes or replaces malicious portions of threat event attacks to make it benign. Due to the high false alarm rates of AID [53], IPS incorrectly identifies a legitimate nonintrusive normal activity as malicious event and responds to that detected activity inaccurately.

3.4.1   Intrusion Prevention System

An Intrusion Prevention System (IPS) continuously monitors computer systems, networks, infrastructure resources, and others to take action to detected malicious threat event attacks, which means responding to malicious cyberattacks. Thus, the IPS control malicious threat event attacks by intelligent algorithms to take further actions, e.g., closing access points or configure firewalls to prevent future malicious threat event attacks, and others. In contrast, the IDS is a passive type of security system that scans data sets or packet traffic to detect security problems and threat event attacks, and report back requesting for required further action(s). Against this background, the IPS takes measures to protect computer systems, networks, infrastructure resources, and others. For this purpose, the IPS is installed inline directly in the legitimate data or packet transmission path, and can block individual data set (s) or packet(s) or interrupt and reset the connection(s) of the alerted threat event attacks. In today's IPS these systems works directly together with a firewall and actively influences its rules. Since the IPS works in-line, data or packet analysis achieved in real time. Indeed, the IPS must not slow down the data or packet traffic stream or suspend the analysis of the data or packets due to high transmission speeds.

In order to detect anomalies or direct threat event attack signatures or patterns, IDS and IPS use the same methodological approaches. Known cyberattack signatures or patterns found in the analyzed data and packet traffic streams are compared with the ones stored in a database, which refers to the *Known-Knowns* cybersecurity risk level of information security, as shown in Table 2.10. The more extensive and up-to-date this database is, the more effective is this type of intrusion detection.

In comparison to signature or pattern-based detection, additional statistical and anomaly-based methods are event signatures or patterns, which refer to the *Known-Unknowns* cybersecurity risk level, as shown in Table 2.10. Furthermore, modern IPS use more advanced methods such as AI and work partly self-learning, for instance based and Deep Learning approaches.

Like IDS, IPS is divided into different types, like:

- *Host-based Intrusion Prevention System (HIPS)*: Directly installed at computer systems, networks, infrastructure resources, and others that be protected. It analyzes all received and sent data or packet traffic streams, and the data or packets provided by the system itself, such as logs. If an IPS respond to a threat event attack, it can intervene in the data or traffic stream to the information and communication systems;
- *Network-based Intrusion Prevention System (NIPS)*: Network-based intrusion prevention systems directly monitor network traffic and can be installed as a separate device or integrated into a firewall inline. Computer systems, networks, and infrastructure resources connected to networks be attacked via the network. The IPS examines the transmitted data or packet traffic stream at the protocol or application level. Furthermore, there are IPSs that incorporate the data or packet traffic streams transferred and deviations from the usual data or packet traffic streams between specific sources and targets in threat event attack detection;
- *Firewall Integrated Intrusion Prevention System (FIIPS)*: In firewall-based systems the IPS function is directly integrated. Such integrated systems are good to use for networks in small or medium-sized organizations. An advantage of such a solution lies in a simple management of the system whereby the IPS can directly influence the firewall rules on detected threat event attacks, without configuring complex connections and rules for the communication between IPS and firewall.

3.4.2 Intrusion Prevention System Architecture

The Intrusion Prevention System Architecture (IPSA) is based on the conceptual approach that the prevention component lies in the direct communication path between the attack surface and the mission critical and crucial systems. Therefore, the IPSA actively analyzing and taking actions to all data or packet traffic stream that enter the computer system, network, infrastructure resource, and others, as shown in Fig. 3.7.

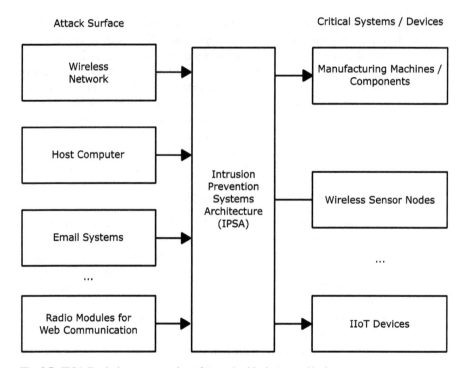

Fig. 3.7 IPSA lies in between attack surface and critical systems/devices

The work task of the IPSA is scanning data or packet traffic streams coming from the attack surface for different threat events, such as:

- Denial of Service (DoS) attack.
- Distributed Denial of Service (DDoS) attack.
- Various types of exploits.
- Worms.
- Viruses.

For this purpose, IPSA performs data or packet traffic stream inspection in real time, deeply inspecting every data or packet traffic stream that travels across the surface site. If any malicious or suspicious data or packets traffic is identified, IPSA carries out one of the following actions:

- Blocking data or traffic flows from the attack surface.
- Dropping malicious threat events protecting the critical computer systems, networks or infrastructure resources.
- Removes or replaces any malicious content that remains to the computer system, network, or infrastructure resources following a threat event attack. This is done by repackaging payloads, removing header information, and removing any infected attachments from file or email servers.
- Resetting the connection.

- Sending an alert to the cybersecurity team in charge.
- Terminates the Transmission Control Protocol (TCP) session that has been exploited and blocks offending source Internet Protocol (IP) address or user account from unauthorized accessing any application, target hosts or other computer systems, networks, infrastructure resources.

These are essential constraints when developing an IPSA. Beside these activities, the IPSA must also respond accurately, to eliminate threat event attacks and false-positive detection rates, to avoid legitimate data or traffic packets misread as threat event attacks. As inline cybersecurity prevention entity, IPSA works efficiently and effectively to avoid degrading computer systems, networks, or infrastructure resources performance, and reacts fast, because threat event atttacks may happen in near real time. Furthermore, IPSA solutions are used to identify issues with corporate security policies, deterring employees and network guests from violating rules these policies contain. In this regard, the main difference between IPSA and IDSA is finally the action they take when potential threat events are detected, which result in the following characteristics:

- *Intrusion Prevention System Architecture (IPSA)*: Control access to computer systems, networks, infrastructure resources, and others to protect them from abuse and threat events. IPS is designed to monitor intrusion data and take the necessary action to prevent threat event attacks from successful developing.
- *Intrusion Detection System Architecture (IDSA)*: Not designed to block threat event attacks. Designed to monitor the computer systems, networks, infrastructure resources, and others, it sends alerts to these components if a potential threat event attacks are detected.

Moreover, IPS typically configured to use a number of different approaches to protect computer systems, networks, infrastructure resources, and others from unauthorized access. These include [54]:

- *Signature-based IPS Architecture SiIPSA)*: This approach use predefined signatures of well-known threat event attacks, the *Known-Knowns* of threat risk level in Table 2.10. When a cyberattack is initiated that matches one of the *Known-Knowns* signatures or patterns, the SbIPSA respond with necessary action(s);
- *Anomaly-based IPS Architecture (AIPSA)*: This approach monitors for any abnormal or unexpected behavior to computer systems, networks, infrastructure resources, and others. If an anomaly is detected, the AbIPSA blocks access to the mission critical and crucial target host immediately;
- *Policy-based IPS Architecture (PIPSA)*: This approach requires cybersecurity teams or administrators to configure security policies according to organizational security policies due to computer systems, networks, infrastructure resources and business activities. When an activity occurs that violates the security policy, an alert created and sent to cybersecurity teams or administrators for further action(s).

In this context, IPS solutions offer proactive prevention against some of today's most notorious computer systems, networks, infrastructure resources exploits. When deployed correctly, the IPS prevents severe damage caused by malicious or unwanted threat event attacks. Several techniques used for IPS, divided into the following groups:

- *IPS Stops Intrusion Attack*: Examples of how this could be achieved are:
  - Block access to targeted critical systems and devices or possibly other likely targets from offending user account(s), IP address(es), or other threat event attack attributes.
  - Block all access to targeted system, service, application, or other resources.
  - Terminate network connection or user session that is being used for intrusion attacks.

- *IPS Changes Security Environment*: IPS could change configuration of other security controls to disrupt an intrusion attack. Common examples are:
  - Cause patches to be applied to a host if IPS detects that the system has vulnerabilities.
  - Reconfigure a network device, e.g., firewall, router, switch, to block access of the threat event actor, and alter a system-based firewall to a target to block incoming threat event attacks.

- *IPS Changes Intrusion Attack's Content*: Some IPS technologies can remove or replace malicious portions of threat event attacks to make it benign.
  - A simple example is an IPS that removes an infected file attachment from an e-mail and then permits the cleaned email to reach its recipient.
  - A more complex example is an IPS that act as a proxy and normalizes incoming requests, which means that the proxy repackages the payloads of the request(s), discarding header information. This might cause certain threat event attack intrusions discarded as part of the normalization process.

Therefore, the main task of intrusion prevention is defending computer systems, networks, infrastructure resources, and others detecting a threat event attack and possibly repelling it. Detecting hostile threat event attacks depends on the number and type of appropriate actions, which can be obtained from publicity available data, found in the National Vulnerability Database (NVD), the US Government Repository of Standards Vulnerability Management Data, or the CVE database, a dictionary of publicly known information security vulnerabilities and exposures [26]. Therefore, intrusion prevention requires well-selected investigations of threat event attacks because threat event actors seeking out and exploiting systems, networks, devices, and applications vulnerabilities to attack, causing serious problems to targeted computer systems, networks, infrastructure resources, and others.

Besides the foregoing mentioned advancements of Next Generation FireWalls (NGFWs), Intrusion Detection and Prevention (IDP) get closer responding to threat event attacks in near real-time, to protect the most critical and crucial data and business assets. In a more common sense, NGFW is a hardware- and software-based security solution used to detect and block sophisticated threat event attacks. It work based on security guidelines on the application layer on the protocol and port checking of classic firewalls and enable data analysis at the application level. Therefore, NGFW combine the functionalities of conventional firewalls, which include [52]:

- *Packet-Filtering Firewalls (PFF)*: Operate at network layer (Layer 3) of the Open System Interconnection (OSI) model. Processing decisions based on network addresses, ports or protocols. This firewall enables high performance filtering.
- *Quality of Service (QoS)*: Manage available network bandwidth to make sure that important network services are given priority over less important traffic.

And features that normally not available in firewalls, which includes intrusion prevention options:

- *Secure Socket Layer (SSL)*: Technology responsible for data authentication and encryption for Internet connections. It encrypts data being sent over the Internet between two systems so that it remains private.
- *Secure Shell (SSH)*: Protocol that is frequently permitted through firewalls. Unrestricted outbound SSH is very common, especially in smaller and more technical oriented organizations. Inbound SSH is usually restricted to one or very few servers.
- *Deep Packet Inspection (DPI)*: Information extraction or complete packet inspection is a network packet filtering type. It evaluates the data part and the header of a packet transmitted through an inspection point, weeding out any non-compliance to protocol, spam, viruses, intrusions, and other, define criteria to block the packet from passing through the inspection point.
- *Reputation-based Malware Protection (RMP)*: Leverages anonymous software patterns, which automatically identify new threat event attacks. Support vector machines employed on behavioral log to identify threat event attacks.
- *Malware Filtering (MWF)*: Aims to stop threat event attacks, based on a stochastic security game framework in which the cost function consists of two parts: i) network resource costs and ii) edegrades performance of Kalman filtering, indexing by the trace of the estimation error covariance matrix [55];
- *Application Awareness (AA)*: Represent the capacity of computer systems, networks, infrastructure resources, and others to maintain information about connected applications to optimize their operation that they run or control.

In this context, a NGFW represent a firewall generation that integrates intrusion detection and prevention, malware filtering, and many other security functions to enable a more advanced control of data traffic flow, as indicated in Fig. 3.8, showing the NGFW security services platform.

The blocks in Fig. 3.7 have the following meaning [52]:

- *Firewall*: Providing multi-layer and protocol inspection, network segmentation, and access control.

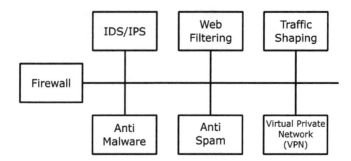

Fig. 3.8 NGFW security services platform

- *Intrusion Detection and Prevention System (IDPS)*: Featuring wide range of detection techniques, e.g., header-based, pattern matching, protocol-based, heuristic-based, anomaly-based and rich customization capabilities.
- *Anti-Malware*: Providing malware protection on all web, mail, and file transfer traffic.
- *Web Filtering*: Enforcing access to allowed web content and filtering high risk URLs such as anonymizers and known hostile addresses.
- *Anti-Spam*: Mitigating directory harvest attacks that is a technique used by spammers in an attempt to find valid/existent e-mail addresses at a domain by using brute force, spam, and enforcing email policy.
- *Traffic Shaping*: Apply Quality-of-Service (QoS) to various applications' traffic such as Instant Messaging (IM), web, streaming video and audio, or Peer to Peer (P2P) if allowed.
- *Virtual Private Network (VPN)*: Provide remote access and secure site-to-site interconnection over untrusted networks. Support protocols such as IPSec, SSL.

Hence, if any malicious or suspicious data set(s) or packet(s)s is detected, the NGFW carries out the following action: detection and prevention as well as reprogram or reconfigure the NGFW to prevent a similar threat event attack occurring in the future.

3.5 Intrusion Detection and Prevention System Architecture

A major challenge for organizations in digital transformation is in the one hand awareness about cybersecurity risks that requires securing mission critical and crucial entities, and in the other hand to meet the resulting cybersecurity practices to defend threat event attacks. In this regard, methods for logging data, detecting intrusions, preventing intrusions evolving for years, which are essential part of today's research [56]. Therefore, this section presents an architectural solution to enable intrusion detection and prevention of computer systems, networks, infrastructure resources, and others. The main Intrusion Detection and Prevention System Architecture (IDPSA) functionalities summarized in Fig. 3.9. The Intrusion Detection System (IDS) part in Fig. 3.9 based on the generic approach shown in Fig. 3.3, expanded by an ANN classifier. The IDS work inline to detect any suspicious activity with the help of the ANN. ANNs is machine learning algorithms, inspired by a mathematical model of the human central nervous system. The term Artificial Neural Network coined in 1943 when McCulloch and Pitts published the idea of a mathematical model based on a biological neuron [57]. Later, Hebb [58] and Rosenblatt [59] implemented the ANN by their unsupervised and supervised training algorithms. The unsupervised learning ANN classifies and visualizes system input data to separate normal behavior from abnormal or more specific intrusive ones. Most of these systems use the SOM learning algorithm, while the supervised learning algorithm refers to the perceptron approach. The first SOM application was to learn normal (regular) system activity characteristics and identify statistical variations from the normal (regular) trends, published by Fox et al. [60].

In [27] multiple SOMs are used for intrusion detection, where a collection of specialized maps are used to process network traffic for each protocol such as Transmission Control Protocol (TCP), User Datagram Protocol (UDP), and Internet Control Message Protocol (ICMP). Each neural net trained to recognize the normal (regular) activity of a single protocol. Today, these early approaches are more advanced intelligent ANN based IDS, using the backpropagation algorithm to train ANNs. The backpropagation algorithm train the ANN based on the chain rule method. This means that after each forward pass through an ANN, backpropagation performs a backward pass while adjusting the model's parameters (weights and biases). Therefore, backpropagation aims to minimize the cost function by adjusting ANN weights and biases, whereby the level of adjustment is determined by the gradients of the cost function to those parameters.

An ANN with only one or two hidden layer belongs to the category shallow learning, a type of machine learning to learn from data achieved by pre-defined features. However, there are unsupervised and supervised shallow system or network architectures. As stated in Sect. 8.1.1.1 Supervised Machine Learning uses labels to learn a task; Unsupervised Machine Learning (see Sect. 8.1.1.2) is performing a machine-learning task without labels. In shallow learning, feature extraction performing separately, not as a part of the ANN.

Executing this approach requires data gathering and pre-processing, which means that all incoming data is collected, transformed or normalized to standard forms. Thereafter, feature extraction of data is required, which means features are objects of data that could be used like performance evaluation for number of data or packets transferred between computer system or network entities, delay in transfer of data or packets, number of dropped data or packets, and others. The ANN type used in IDPSA is a Feedforward Artificial Neural Network (FFANN) that commonly is in its simplest form a single layer perceptron. In FFANN a series of inputs is connected with the layer and are multiplied by the weights. Summarizing each value is required to get the weighted output values. If the sum of values is above a specific threshold, e.g., set at zero (0), the value produced is often one (1), whereas if the sum falls

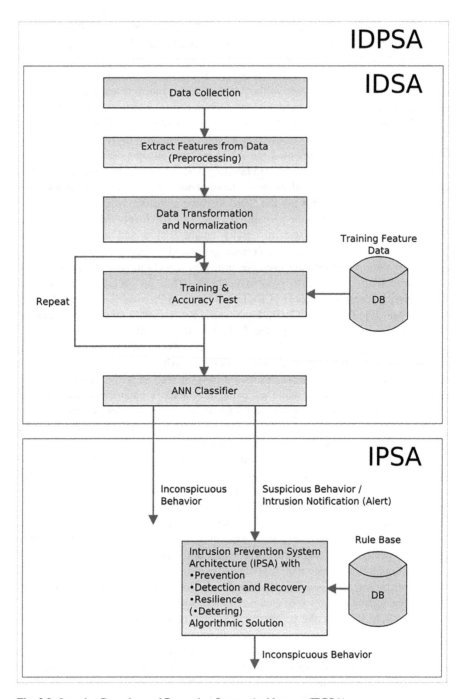

Fig. 3.9 Intrusion Detection and Prevention System Architecture (IDPSA)

below the threshold, the output value is minus one (-1). The single layer perceptron is an important model of FFANN, often used in classification tasks. The goal of the FFANN in the IDPSA is to approximate some function f^*. For example, a classifier

$$y = f^*(x)$$

maps an input x to a category y. Hence, a FFANN defines a mapping

$$y = f(x; \theta)$$

and learns the value of the parameters θ that result in the best function approximation.

The usage of the FFANN requires training, based on the features mentioned before. The next step after training the FFANN is to test it in place with the features assigned to normal (regular) and abnormal behavior, based on a performance metric, as shown in Table 3.5, which describe the accuracy of the detection rate and false alarm rate of the IDSA shown in Fig. 3.9.

The performance of the IDS, shown in the IDSA part in Fig. 3.8, can be calculated by the ratio of correct classification of total test data by adopting to a common model of performance measures as described in [3, 61, 62] with the terms

- *Accuracy (A_{cc})*: Refers to the overall effectiveness of the chosen algorithm in terms introduced in Table 3.5.
- *Detection Rate (DtR)*: Refers to the number of impersonation attack events detected divided by the total number of impersonation cyberattack events in the test data set(s).
- *Precision (P)*: Refers to the number of impersonation attack events detected among the total number of events classified as a cyberattack event.
- *False Alarm Rate (FAR)*: Refers to the number of normal events in the test data set(s);
- *False Negative Rate (FNR)*: Refers to the number of threat events that are unable to be detected.
- *Score (S_C)*: Refers to the harmonic mean of *P* and *DR*.
- *Correlation Coefficient (CC)*: Represents the correlation between detected and observed data.

Accuracy is the ratio of correct detections, indicated by a manifold of different rates, such as:

- *True Positive Rate (TPR)*: Number of threat event attack intrusions correctly classified.
- *True Negative Rate (TNR)*: Number of threat event intrusions correctly classified as a likelihood level rare or unlikely, divided by the sum of True Positive Rate *(TPR)*.
- *True Negative Rate (TNR) and False Negative Rate (FNR)*: Number of intrusions of threat event attacks, incorrectly classified as a likelihood rare or unlikely, as

well as False Positive Rate (*FPR*) as number of threat event intrusions incorrectly classified as a threat event attack.

Finally, *Accuracy* (*A_{CC}*) is the ratio of TPR + TNR divided by TPR + PNR + FPR + FNR as follows:

$$A_{CC} = \frac{TPR + TNR}{TPR + PNR + FPR + FNR}$$

The *Detection Rate (DtR)* is the ratio of the number of correct detection (*TPR*), divided by the sum of True Positive Rate (*TPR*), and False Negative Rate (*FNR*):

$$DtR = \frac{TPR}{TPR + FNR}$$

The *Precision (P)* is the ratio of the number of correct detection that represents the number of intrusions correctly classified as threat event attack intrusion, divided by the sum of True Positive Rate (*TPR*) and False Positive Rate (*FPR*):

$$P = \frac{TPR}{TPR + FPR}$$

The *False Alarm Rate (FAR):* Ratio of the number of False Positive Rate (*FPR*), representing threat event intrusion incorrectly classified as a threat event attack, which is divided by the sum of True Negative Rate (*TNR*) as the number of intrusions that are correctly classified as a likelihood level rare or unlikely, and the False Positive Rate (*FPR*), the number of intrusions that are incorrectly classified as a threat event attack:

$$FAR = \frac{FPR}{TNR + FPR}$$

The *False Negative Rate (FNR):* Ratio of the number of False Negative Rate (*FNR*) that represent the number of intrusions incorrectly classified as a likelihood rare or unlikely, divided by the sum of False Negative Rate (*FNR*) and True Positive Rate (*TPR*). *FNR* is the number of intrusions incorrectly classified as a likelihood rare or unlikely, and *TPR* representing the number of intrusions that are correctly classified as a threat event attack:

$$FNR = \frac{FNR}{FNR + TPR} (FNR).$$

The *Score (S_C):* Ratio of the number of True Positive Rate (*TPR*) representing the number of intrusion incidents that are correctly classified as a threat event attack with weight 2, divided by the sum of True Positive Rate (*TPR*) with weight 2 and False Positive Rate (*FPR*) and False Negative Rate (*FNR*). *TPR* represent the number of intrusion incidents correctly classified as a threat event attack, *FPR* as number of intrusion incidents incorrectly classified as a threat event, and *FNR* as number of intrusions incorrectly classified as a likelihood rare or unlikely:

$$S_C = \frac{2TPR}{2TPR + FPR + FNR}$$

The *Correlation Coefficient (CC)*: Introduced in [3] is the product of True Positive Rate (*TPR*), the number of correctly classified threat event attacks and (*TNR*) as number of correctly classified likelihood level rare or unlikely, subtracted by the product of False Positive Rate (*FPR*) as number of incorrectly classified threat event attacks and False Negative Rate (*FNR*) as number of incorrectly classified likelihood rare or unlikely. This term is divided by the square root of True Positive Rate (*TPR*) and False Positive Rate (*FPR*), True Positive Rate (*TPR*) and False Negative Rate, True Negative Rate (*TNR*) and False Positive Rate (*FPR*), True Negative Rate (*TNR*) and False Negative Rate (*FNR*):

Correlation Coefficient

$$= \frac{(TPR \times TNR) - (FPR \times FNR)}{\sqrt{(TPR + FPR)(TPR + FNR)(TNR + FPR)(TNR + FNR)}}$$

In this context of abnormal behavior be identified using a statistical threshold approach. Thus, the FFANN learns classifying any feature in normal (regular) or abnormal (irregular) classes through an iterative process, based on the profile created by FFANN for both classes. A stable trained FFANN recognize and classify the data, packets and control messages in the computer system, network or infrastructure resource under test in real-time, generating a decision for normal (regular) or abnormal (irregular) behavior. In abnormal behavior through threat event intrusions, the IDSA alerts, and the IPSA immediately respond to protect. In this context, both parts of the IDPSA immediately react. Therefore, IPSA is the key element protecting data or traffic flows between computer systems, networks, infrastructure resources, and others in threat event attacks through

- *Prevention*: Response onto threat event attack vulnerability alerts;
- *Response and Recovery*: Respond onto problems from threat event attack vulnerability to recover to a normal (regular) state;
- *Resilience*: Largely self-resilient in the face of threat event attack vulnerabilities;
- *Deterring*: Action taken to prevent threat event actors intruding vulnerabilities by making them afraid about strike-back attacks.

The performance metric to measure the effectiveness of an IDSA can be divided into three classes [62]:

- *Threshold Metric*: Include features such as Classification Rate (*CR*)—ratio of correctly classified threat event attacks and the total number of threat event

Table 3.9 Confusion matrix

Event	Predicted threat event attack	Predicted normal event
Threat event intrusion	*TPR:* Intrusions successfully detected by the IDS	*FNR***:** Intrusion missed by IDS classified as normal/ nonintrusive
Normal event	*FPR:* Normal/nonintrusive event, wrongly classified as intrusive by IDS	*TNR***:**Normal/nonintrusive event, successfully classified as normal/nonintrusive by the IDS

attacks, F-Measure (*FM*)—estimate how accurate a classifier work, Cost per Example (*CPE*), and others. This metric consider whether the prediction is below a threshold, whereby the threshold lies in between 0 and 1.

- *Ranking Metric*: Include False Positive Rate (*FPR*), Detection Rate (*DR*), Precision (*P*), Area under Curve (*AuC*)—curve used to visualize the relation between *DR* and *FPR* of a classifier, and to compare the accuracy of classifier(s). This measure is effective but has some limitations, depending on the ratio of threat events to normal data or packet traffic events.
- *Probability Metric*: Include Root Mean Square Error (*RMSE*) and lies in the range from 0 to 1. Metric is minimized when predicted value for each threat event class coincide with the true conditional probability of that class, being normal class.

These metrics used to calculate the performance of IDS classification results in the form of the confusion matrix. The confusion matrix refers to classification results. It represents *TPR*, *TNR*, *FPR*, and *FNR* (Table 3.9).

3.6   Intrusion Detection and Prevention Methods

Today's complex IDPSA require full security assess to enable stable and resilient system operation. For testing stability and resiliency, several methods used:

- *Function and Performance Test*: Validates security components under valid data and packet traffic due to threat event attack conditions.
- *Impairment Test*: Validates performance when communication is impaired. Typically used with delayed, dropped, or erroneous packets.
- *Resiliency Test*: Validates operation under degraded or failure conditions, such as sensor failure, actuator failure, and others.
- *Stress Test*: Validates system or components beyond normal (regular) operational capability to observe how system or components operate safe.

Beside these test methods other types of cybersecurity tests are:

- *Access Control Test*: Ensures computer systems, networks, infrastructure resources and applications can be accessed only by authorized and legitimate users. The test objective is to differentiate software components to ensure that the application implementation conforms to security policies to protect computer systems, networks, infrastructure resources and other from unauthorized access.

- *Ethical Hacking Test*: Person who cyberattack computer systems, networks, infrastructure resources and other, mimicking the behavior of cyber-hackers. Monitoring computer systems, networks, and other is done to identify expose security flaws and vulnerabilities, and to identify potential cyber threat attacks that malicious hackers might take advantage of.
- *Security Risk Assessment Test*: Involves the security risk(s) by reviewing and analyzing potential cyberattack risk(s). Type of cyberattack risk classified in high, medium and low categories, based on their severity level. Thereafter, the respective mitigation strategy follows, based on security posture of computer systems, networks, infrastructure resources, and other. Security audits conducted for access points, inter-network access, intra-network access, and data protection at this level.
- *Security Scanning Test*: Enhance the scope where testers conduct security scans to evaluate computer systems, networks, and infrastructure resources and applications weaknesses. Each scan sends malicious requests to systems and testers check for behavior that could indicate security vulnerabilities. SQL Injection, XPath Injection, XML Bomb, Malicious Attachment, Invalid Types, Malformed XML, Cross Site Scripting and other are scans executed to check for vulnerabilities, which are then analyzed at length to fix them.
- *Vulnerability Scanning Test*: Tests entire computer systems, networks and infrastructure resources to detect vulnerabilities, loopholes, and suspicious vulnerable signatures. This scan detects and classifies system weaknesses and predicts the effectiveness of countermeasures taken.

Another important test method is Security Penetration Test (SPT), a simulated test that mimics threat event attack(s) of a threat event actor. Test aims to gather information about computer systems, network and infrastructure resources, identifying entry points to these entities attempting a break to determine the cyber-security weakness. This type of text is like a white hat cyberattack. When implementing the SPT, the same techniques, tools and expert knowledge used as by real threat event actors. However, experienced penetration testers are required that use automated and manual test procedures to present realistic threat event attack scenarios.

In addition to technical analysis, social-level attacks also part of SPT to test the security awareness of employees of industrial, public and private organizations with the intension to disseminate information and the conscious or unconscious use of unauthorized information or applications. Depending on the targeted object, distinguished strategies for SPT described by TechTarget [63, 64]:

- *External Penetration Strategy*: External tests deal with threat event attacks. The methods used are carried out from outside the object to be attacked, i.e., through the Internet. This test is carried out with no or complete knowledge of the vulnerable technical environment. Typically, this penetration test begins with public available information about the object, subsequent network spanning, and other.

- *Internal Penetration Test Strategy*: Internal tests carried out within vulnerable technical environments. The penetration test simulates a threat event attack on the internal network. The focus is to understand what might happen if the network was successfully penetrated, or what an authorized user could do to capture specific information of the compromised network. One important threat event attack is sniffing, used to a considerable extent with internal penetration tests. The sniffer is direct connected to the network in promiscuous mode, which enables to collect a considerable amount of information. For sniffing, a variety of free and commercial tools are available, such as Wireshark (the former Ethereal), the Microsoft Message Analyzer (the successor to Netmon), or the Viavi Observer Analyzer.
- *Blind Test Strategy*: In blind tests, one tries to simulate actions and procedures of a real cyberattacker hacker). However, the test team has only limited or no information about the object before performing the penetration test. The penetration test team uses public available data to collect information about the targeted object to perform the penetration tests. These blind tests provide a lot of information about the targeted object(s) that otherwise would remain unknown. For example, this type of penetration test can raise problems such as additional Internet access points, directly connected networks, public available confidential/protected information, and other. However, blind tests are more time-consuming and expensive, because the necessary effort of the test team for target search is higher.
- *Double-Blind Test Strategy*: Double-blind test is an extension of the blind testing strategy. It is an important test approach due to the possibility to check security monitoring and identification of security incidents and the escalation and reaction procedures of the targeted object.
- *Targeted Testing Strategy*: In case of targeted or systematic tests, sometimes referred to as a lights-turned-on approach, the penetration test team is involved in the test. The test activities and the information regarding the target and network design generally known. Targeted penetration testing be more efficient and cost-effective if the goal of the test focus on the technical side or design of the network, rather than on incident response and other workflows of the targeted object. In contrast to blind tests, a systematic test carried out in less time and with less effort. The only difference is that this may not provide a complete picture of the targeted objects vulnerabilities and reactivity.

In addition to the foregoing methods, a large number of distributed computing resources required in the future in Industry 4.0 intelligent manufacturing ecosystems. Their connected intelligent systems and devices require advanced cybersecurity methods to protect the intelligent system-based infrastructure against possible threat events. However, the growing number of intrusions through threat event attacks stands against. This requires that intrusion detection and prevention methods have to be manifold too, adapting to specific threat event attack methods. Against this background, advanced intrusion detection and prevention methods introduce the mathematical methods of machine learning and Deep Learning, introduced in Chap. 8.

3.7 Exercises

3.7.1 Intrusion Detection

What is meant by the term *Intrusion?*
Describe the main characteristics of an Intrusion.
What is meant by the term *Intrusion Detection?*
Describe the main characteristics of Intrusion Detection.
What is meant by the term *Intrusion Detection System?*
Describe the main characteristics of Intrusion Detection System.
What is meant by the term *Host-based Intrusion Detection?*
Describe the main characteristics of Host-based Intrusion Detection.
What is meant by the term *Network-based Intrusion Detection?*
Describe the main characteristics of Network-based Intrusion Detection.
What is meant by the term *Specification-based Intrusion Detection?*
Describe the main characteristics of Specification-based Intrusion Detection.
What is meant by the term *Anomaly-based Intrusion Detection?*
Describe the main characteristics of Static and Dynamic Anomaly-based Intrusion
    Detection.
What is meant by the term *Misuse-based Intrusion Detection?*
Describe the main characteristics of Blacklist-based und Whitelist-based Misuse-
    based Intrusion Detection.
What is meant by the term *Intrusion Detection System Architecture?*
Describe the main characteristics and capabilities of an Intrusion Detection System
    Architecture.
What is meant by the term *Artificial Neural Network?*
Describe the main functional characteristics and capabilities of an Artificial Neural
    Network.
What is meant by the term *Feedforward Neural Network?*
Describe the main characteristics and capabilities of Feedforward Neural Networks.
What is meant by the term *Pre-processing in Intrusion Detection?*
Describe the main characteristics and capabilities of Pre-processing in Intrusion
    Detection.
What is meant by the term *Intrusion Detection Capability Metric?*
Describe the main characteristics and capabilities of the Intrusion Detection Capa-
    bility Metric.

3.7.2 Intrusion Prevention

What is meant by the term *Prevention?*
Describe the main characteristics of Prevention.
What is meant by the term *Intrusion Prevention?*

Describe the main characteristics and capabilities of Intrusion Prevention.

What is meant by the term *Intrusion Prevention System Architecture?*

Describe the main characteristics and capabilities of Intrusion Prevention System Architecture.

What is meant by the term *Signature-based Intrusion Prevention Architecture?*

Describe the main characteristics and capabilities of a Signature-based Intrusion Prevention System Architecture.

What is meant by the term *Anomaly-based Intrusion Prevention System Architecture?*

Describe the main characteristics and capabilities of an Anomaly-based Intrusion Prevention System Architecture.

What is meant by the term *Policy-based Intrusion Prevention System Architecture?*

Describe the main characteristics and capabilities of a Policy-based Intrusion Prevention System Architecture.

What is meant by the term *NGFW Security Services Platform?*

Describe the main components of NGFW Security Services Platform.

What is meant by the term *Performance Measures?*

Describe the main characteristics and capabilities of Performance Measures.

What is meant by the term *Confusion Matrix?*

Describe the main characteristics and capabilities of Confusion Matrix as a 2×2 Matrix.

What is meant by the term *Resilience Test?*

Describe the main characteristics and capabilities of the Resilience Test.

What is meant by the term *Security Scanning?*

Describe the main characteristics and capabilities of Security scanning.

What is meant by the term *Vulnerability Scanning?*

Describe the main characteristics and capabilities of Vulnerability Scanning.

References

1. Heady, R., Luger, G., Maccabe, A.B., Servilla, M.: The Architecture of a Network Level Intrusion Detection System. In: Technical Report 390-20, Department Computer Science, University of New Mexico, 1990
2. Anderson, J.: An Introduction to Neural Networks. MIT Press, 1995
3. Kim, K., Aminanto, M.E., Tanuwidjaja, H.C.: Network Intrusion Detection using Deep Learning – A Feature Learning Approach. Springer Nature, 2018
4. Tung, B.: A Graph Theory Approach to Combining Intrusion Detection. In: ISI Technical Report ISI-TR-2004-587 funded by NSF under award 0209046, 204
5. Tiwari, R., Kumar, R., Bharti, A., Kishan, J.: Intrusion Detection System. In: International Journal of Technical Research and Application, Vol. 5, pp. 38–44, 2017
6. Hay, A., Cid, D.: OSSEC Host Intrusion Detection Guide. Syngress Pub. 2008
7. Yeo, H., Che, X., Lakkaraju, S.: Understanding Modern Intrusion Detection Systems: A Survey. In: Cryptography and Security, pp.1–9, 2017, arXiv:1708.07174v2 (Accessed 12.2022)

8. Lichodzilewski, P. Network based Anomaly Detection using Self-Organizing Maps. PhD Thesis, Dalhousie University, Canada, 2002

9. Om, H., Sakar, T.K.: Designing Intrusion Detection System for Web Documents Using Neural Network. In: Communication and Network, Vol. 2, pp. 54–61, 2010

10. Ingham, K.: Protecting Network Servers. In: Technical Report, Department of Computer Science, University of New Mexico, 2003

11. Ghorbani, A.A., Lu, W., Tavallee, M.: Network Intrusion Detection and Prevention Concepts and Prevention. Springer Publ. 2010

12. Lee, W., Park, C.T., Stolfo, S.J.: Automated Intrusion Detection NFR: Methods and Experience. In: Proceeding of the Workshop on Intrusion Detection and Network Monitoring, pp. 1–7, 1999

13. Ranum, M.J.: Experiences Benchmarking Intrusion Detection Systems. In: NFR Security Technical Publications, 2001

14. Ko, C., Ruschitzka, M., Levitt, K.: Execution Monitoring of Security-Critical Programs in Distributed Systems: A Specification-based Approach. In. Proceedings IEEE Symposium Security and Privacy, pp. 175–189, 1997

15. Sekar, R., Uppuluri, P.: Synthesizing Fast Intrusion Prevention/Detection Systems from High-Level Specifications. In: USENIX Security Symposium Proceedings, pp. 63–78, 1999

16. Uppuluri, P., Sekar, R.: Experiences with Specification-based Intrusion Detection. In: Recent Advances in Intrusion Detection, pp. 172–189, Springer Publ. 2001

17. Raihan, M., Zulkernine, M.: Detecting Intrusions Specified in a Software Specification Language. In: Proceedings IEEE 29th Annual International Computer Software and Applications Conference (COMPSAC), pp. 143–148, 2005

18. Ko, C.: Logic Induction of Valid Behavior Specifications for Intrusion Detection. In: Proceedings in IEEE Symposium Security and Privacy, pp. 142–153, 2002

19. Helmer, G., Wong, J., Slagell, V., Honavrar, V., Miller, L., Wang, Y., Wang, X., Stakhanova, N.: Software Fault Tree and Colored Petrinet based Specification, Design, and Implementation of Agent based Intrusion Detection System. In: International Journal of Information and Computer Security, Vol. 1, pp.109–142, 2007

20. Balepin, I., Maltsev, S., Rowe, J., Levitt, K.: Using Specification-based Intrusion Detection for Automated Response's. In: Recent Advances in Intrusion Detection, pp.136–154, Springer Publ. 2003

21. Sekar, R., Gupta, A., Frullo, J., Shanbhag, T., Tiwari, A., Yang, A., Zhou, S.: Specification-based Anomaly Detection: A New Approach for Detecting Network Intrusions. In: Proceedings 9th ACM Conference on Computer and Communications Security, pp. 265–274, 2002

22. Stakhanova, N., Basu, S., Wong, J.: On the Symbiosis of Specification-based and Anomaly-based Detection. In: Computers and Security, Vol. 29, No. 2, pp. 253–268, 2010

23. Berthier, R., Sanders, W.H.: Specification-based Intrusion Detection for Advanced Metering Infrastructures. In: Proceedings IEEE Pacific Rim International Symposium on Dependable Computing, pp. 184–193, 2011

24. Kumar, S., Stafford, E.H.: A Pattern Matching Model for Misuse Intrusion Detection. In: Proceedings 17th National Computer Security Conference, pp. 11–21, 1994

25. Cappers, B.: Interactive Visualization of Event Log for Cybersecurity. PhD Thesis, TU Eindhoven, 2018

26. Möller, D.P.F., Haas, R.E.: Guide to Automotive Connectivity and Cybersecurity – Trends, Technologies, Innovations, and Applications. Springer Publ. 2019

27. Rhodes, B., Mahaffey, J., Cannady, J.: Multiple Self-Organization Maps for Intrusion Detection. In: Proceedings 23rd National Information Security Conference, pp. 32–42, 2000

28. Depren, O., Topallar, M., Anarim, E., Ciliz, K.M.K.: An Intelligent Intrusion Detection System (IDS) for Anomaly and Misuse Detection in Computer Networks. In: Expert Systems with Applications, Vol. 29, pp. 713–722, 2005

29. Kemmerer, R.R., Vigna, G.: Intrusion Detection: A Brief History and Overview. In: IEEE Security and Privacy Magazine, 2002

30. Veeramreddy, V., Rama Prasad, V.V., Munivara Prasad, K.: A Review of Anomaly based Intrusion Detection Systems. In: International Jopurnal of Computer Applications, Vol. 28, No. 7, pp. 26–35, 2011

31. Kohonen, T.: Self-Organizing-Map. Springer Publ., 2001

32. Modi, C.: A Survey of Intrusion Detection Techniques in Cloud. In: Journal of Network and Computer Applications, Vol. 36, No. 1, pp. 42–57, 2013

33. Mitchell, R., Chen, I.R.: A Survey of Intrusion Detection Techniques for Cyber-Physical Systems. In: ACM Computer Survey Vol. 46, No. 4, pp. 1–55, 2014

34. Shieh, S., Gligor, V.: A Pattern-Oriented Intrusion Detection Model and its Applications. In: Proceedings Symposium on Security and Privacy, pp. 327–342, 1991

35. Denning, A.: An Intrusion Detection Model. In: IEEE Transactions on Software Engineering, Vol. 13, pp. 222–232, 1967

36. Smaha, S.E.: Tools for Misuse Detection. In: Proceedings International Social Security Association, pp. 711–716, 1993

37. Lunt, T.F., Tamaru, A., Gilham, F., Jagannathan, R., Jalali, C., Neumann, P.G., Javitz, H.S., Valdes, A., Garvey, T.D.: A Real-Time Intrusion Detection Expert System (IDES). In: Final Technical Report SRI Project 6784, Contract No. N0003S89-C-0050, SRI Computer Science Laboratory, SRI International, 1992

38. Möller, D.P.F., Haas, R.E., Akhilesh, K.B.: Automotive Electronics, IT, and Cybersecurity. In: Proceedings IEEE/EIT Conference, pp. 575–580, 2017

39. Karim, E., Proha, V.V.: Cyber-Physical Systems Security. In: Applied Cyber-Physical Systems, pp. 75–84, Eds. S.S. Shuh, U. J. Tanik, J. N. Carbone, A., Springer Publ. 2014

40. Zimmer, C., Bhat, B., Mueller, F., Mohan, S.: Time-Based Intrusion Detection in Applied Cyber-Physical Systems. In: Proceed. 1st ACM/IEEE International Conference on Cyber-Physical Systems, pp. 100–118, 2010

41. Kubica, J., Moore, A.: Probabilistic Noise Identification and Data Cleaning. In: Research Report CMU-RT-TR-02-26 Robotic Institute Carnegie Mellon University. 2002

42. Xiong, H., Panday, G., Steinbach, M., Kumar, V.: Enhancing Data Analytics with Noise Removal. In: IEEE Transactions on Knowledge and Data Engineering, Vol. 18, No. 3, pp. 304–319, 2006

43. Struc, V., Zibert, J., Pavestc, N.: Histogram Remapping as a Preprocessing Step for Robust Eye Recognition. In: WSEAS Transactions on Information Science and Applications. Vol.6, No. 3, pp. 520–529, 2009

44. Sada, A., Kinoshita, Y., Shiota, S., Kiya, H.: Histogram-Based Image Processing for Machine Learning. In: Proceedings IEEE 7th Global Conference on Consumer Electronics (GCCE), pp. 272–275, 2018. https://doi.org/10.1109/GCCE.2018.8574654 (Accessed 12.2022)

45. Holzimger, A., Stocker, C., Peischl, B., Simonic, K.-M.: On Using Entropy for Enhancing Handwriting Preprocessing. In: Entropy, Vol. 14, pp. 2324–2350. 2012

46. Boulila, W.: A Top-Down Approach for Semantic Segmentation of Big Remot Sensing Images. In: Earth Science Informatics, 2019

47. Yang, H.: Data Mining Concepts and Techniques 1, Chapter 3 Course Script. Department of Informatics San Francisco State University, 2020

48. Davis, J.J., Clark, A.J.: Data Preprocessing for Anomaly Based Network Intrusion 'Detection': A Review. In Computers and Security, pp. 1–35, 2011. https://www.researchgate.net/profile/Andrew-Clark-42/publication/234130888_Post_review_version/links/0912f50f756d0 78993000000/Post-review-version.pdf (Accessed 12.2022)

49. Caruana, R., Niculescu-Mizil, A.: Data Mining in Metric Space: An Empirical Analysis of Supervised Learning Performance Criteria. In: Proceedings 10th ACM International Conference on Knowledge Discovery and Data Mining, pp. 69–78, 2004

50. Gu, G., Fogla, P., Dragon, D., Lee, W., Scoric, B.: Measuring Intrusion Detection Capability: An Information-Theoretic Approach. In: Proceedings ACM Symposium on Information, Computer and Communications Security, pp. 90–101, 2006

51. Axelsson, S.: The Base-rate Fallacy and its Implications for the Difficulty of Intrusion Detection. In: Proceedings of ACM Conference on Computer and Communication Security, pp. 1–7, 1999
52. Abdel-Aziz, A.: Intrusion Detection and Response – Leveraging Next Generation FireWall Technology. In: SANS Institute Reading Room Site Report, 2020
53. Al-Jarrah, O.Y., Alhussein, O., Yao, P.D., Muhaidat, S., Taha, K., Kim, K.: Data Randomization and Cluster-based Partitioning for Hotnet Intrusion Detection. In: IEEE Transactions on Cybernetics, Vol. 46, pp. 1796–1806, 2015
54. Scarfone, K., Mell, P.: Guide to Intrusion Detection and Prevention Systems. National Institute of Standards and Technology (NIST), Special Publication 800-94, 2007
55. Liu, S., Liu, X. P., El Saddik, A.: A Stochastic Security Game for Kalman Filtering in Networked Control Systems under DoS Attacks. In: Proceedings 3rd IFAC International Conference on Intelligent Control, pp. 106–111, 2013
56. Krenke, P.S., Pal, A., Colaco, A. (Eds,): Proceedings or the 3rd International Conference on Frontiers of Intelligent Computing: Theory and Applications. Springer Publ. 2014
57. MacCulloch, W.S., Pitts, W.: A Logical Calculus of the Ideas Immanent in Nervous Activity. In: Bulletin Mathematical Biophysics, Vol. 5, pp.115–133, 1943
58. Hebb, O.O.: The Organization of Behavior. John Wiley Publ., 1949
59. Rosenblatt, F.: The Perceptron: A Probabilistic Model for Information Storag5 and Organization in the Brain. In: Psychology Review, Vol. 65, pp. 386–408, 1958
60. Fox, K.L., Henning, R.R., Reed, J.H., Simonian, R.: A Neural Network Approach Towards Intrusion Detection. In: 13th National Computer Security Conference, pp. 125–134, 1990
61. Möller, D.P.F.: Cybersecurity in Digital Transformation: Scope ans applications. Springer Briefs on Cybersecurity Systems and Networks, 2020
62. Kumar, G.: Evaluation Metrics for Intrusion Detection Systems – A Study, In: International Journal of Computer Science and Mobile Applications, Vol. 2, No. 11, pp. 11–17, 2014
63. Yasar, K., Mehta, P.: Penetration Testing. https://www.techtarget.com/searchsecurity/definition/penetration-testing (Accessed 12.2022)
64. Bigelow, S.J.: Complete Guide to Penetration Testing Best Practices. https://www.techtarget.com/searchsoftwarequality/tip/Everything-you-need-to-know-about-software-penetration-testing (Accessed 12.2022)

Chapter 4
Cyberattacker Profiles, Cyberattack Models and Scenarios, and Cybersecurity Ontology

Abstract Cyberattacks often dominate today's headlines. However, what are the odds of industrial, public, and private organizations becoming the next victims of misusing organizations' computer systems, networks, infrastructure resources, and others? The number and the sophistication of cyberattacks on computer systems, networks, and infrastructure resources is on the rise, and cyberattackers sniffing around for vulnerabilities. Furthermore, cyberattackers easily try to overcome, for example, password authentication rules as one of the several cyberattack scenarios in the cyberattack space. The different types of cyberattack scenarios mainly cause resource disruption, for example, through a Denial of Service (DoS) attack that affects system operation, or disclosure of resources, e.g., by an eavesdropping attack, which alone cannot disrupt the system operation. Thus, cyberattacks cause serious problems because a safe and reliable operation of computer systems, networks, infrastructure resources, and others is a major concern in industrial, public, and private organizations. However, with an increasing understanding of how the industrial, public, and private organizations' systems work, operation manuals and technical data sheets are often available on the Internet. Thus, cyberattackers become easily skilled at determining the weaknesses of the system(s) they try to exploit to obtain unauthorized access. One international well-known cyberattack was achieved by Stuxnet malware, which supposedly infected industrial control systems and disrupted their operations (Falliere et al., W32 Stuxnet Dossier, 2011; Schenato, IEEE Trans Autom Control, 54(5):1093–1099, 2009). Another big cyberattack was WannaCry that crippled computer systems in more than 150 countries. Furthermore, cyberattackers also use patterns to intrude threat event attacks that are difficult to trace and identify. Therefore, methods and tools are necessary to monitor computer systems, network, infrastructure resources, and others to detect data breaches and respond immediately to the identified threat event attack. Most data breaches exploit well-known cybersecurity weaknesses in the targeted systems. One possible solution to this problem is to study the characteristics of threat event attacks and thereafter extrapolating their characteristics into future possibilities. However, intrusions of threat event attacks are not easy to understand and to define in terms of behavior and/or action(s), but more easily in terms of their effects. Nevertheless, there is intrusion by threat event attacks that can't be detected alone by statistical methods that must be watched specifically. Collecting essential knowledge about threat event

D. P. F. Möller, *Guide to Cybersecurity in Digital Transformation*, Advances in Information Security 103, https://doi.org/10.1007/978-3-031-26845-8_4

181

attacks and threat event actors is important to gain deeper understanding of cyber-security violations' impact. Against this background, this chapter introduces cyberattack models and scenarios. Section 4.1 focuses on threat event attacks, threat event actors, and threat event impacts in the context of how to defend threat event attacks, and the potential policy of threat event attack raised. This includes answering questions such as "What is the threat event actor's objective?" "What is the threat event actor's goal?" or "What may be the threat event actor's preferred attack method to achieve his attack goal?" and refers to cyberattackers' profile in Sect. 4.1.1. Section 4.2 introduces cyberattack models, including modeling formalisms in Sect. 4.2.1 and a generic cyberattack model in Sect. 4.2.2. Section 4.3 refers to cyberattackers' behavior modeling, divided in Sect. 4.3.1 that introduces into generic cyberattackers' behavior modeling, and Sect. 4.3.2 that refers to simulation of cyberattacks. Section 4.4 considers the topic cybersecurity ontology as a formal specification of a shared conceptualization within a knowledge model. It contains Sect. 4.4.1: introduction in ontology, and Sect. 4.4.2 cybersecurity ontology. Section 4.4.2 is divided into Sect. 4.4.2.1, generic cybersecurity data space ontology framework, and Sect. 4.4.2.2, cyberattack ontology model. Section 4.5 contains comprehensive questions from the chapter topics cyberattackers profile, cyberattack models, and cybersecurity ontology, followed by references for further reading.

4.1   Introduction

Cyberattacks represent a sensitive issue in the era of digital transformation. The impact and prevalence of cyberattacks (also termed threat event attacks) have grown and need action toward better protection from today's cybersecurity risk(s). These risk(s) rise in number and sophistication of cyberattacks on computer systems, networks, infrastructure resources, and others. Therefore, methods and tools are necessary to identify probably attacked systems, to detect break-in, and to respond immediately about identified malicious cyberattacks. Most cyberattack-based break-ins prevalent today exploit well-known cybersecurity weaknesses and types of threat event attacks. One solution to this problem is to study the characteristics of threat event attacks based on the results obtained and extrapolate their characteristics to gain an understanding of how threat event attacks will look like in the near future. In this regard, an intrusion is any set of action(s) that attempt to compromise the IT cybersecurity, based on the confidentiality, integrity, and availability (CIA) Triad (see Sect. 1.6.2), which refers to confidentiality, integrity, and availability. Confidentiality in the CIA Triad relates to the nondisclosure of data and/or business assets by unauthorized internal or external parties. Integrity in the CIA Triad refers to the trustworthiness of data or business assets, meaning that no-unauthorized change of data or business asset contents or properties occur. Availability in the CIA Triad ensures the timely access to data or business assets functionalities. Therefore, collecting knowledge about threat event attacks and threat event actors is essential

Table 4.1 Intrusion break-in types and detection approaches

Intrusion break-in type	Detection method
Attack	Specific formulation or execution of a plan to carry out a threat event attack
Attempted break-in	Atypical behavior profiles or violations of cybersecurity constraints
Denial of Service (DoS)	Atypical usage of system resources
Leakage	Atypical usage of I/O resources
Malicious use	Atypical behavior profiles, violations or cybersecurity constraints, or use or special privileges
Masquerade attack	Atypical behavior profiles or violations of cybersecurity constraints
Penetration	Successful cyberattack obtains (unauthorized) access to files and programs or the control state of a computer system
Risk	Accidental or unpredictable exposure of information, or violation of operations integrity due to malfunction of hardware or incomplete or incorrect software design
Threat event	Possibility of deliberate, unauthorized attempt to: Access information Manipulate information Render a system unreliable or unusable
Vulnerability	Known or suspected flaw in hardware or software design or operation of a system that exposes the system to penetration or its information to accidental disclosure.

in order to develop a cybersecurity strategy. A method to deal with this matter is the intrusion detection method with its two approaches in anomaly intrusion detection and misuse intrusion detection (see Chap. 3). Anomaly intrusion detection (see Sect. 3.1.1) refers to intrusions of threat event attacks that is detected based on irregular (anomalous) behavior. Let's look at two simple examples:

- An authorized user A only uses the organization's computer resources in his office between the regular working hours of 9:00 am and 5:00 pm. An activity in his account late at night is irregular (anomalous) and might be an intrusion.
- An authorized user B might always login his organizations account outside of regular working hours through the organization's server access. A late night login session from another host to his account would be considered unusual (anomalous) and might be considered an intrusion.

Table 4.1 shows some intrusion break-in types and detection approaches [3].

Due to the many possibilities to break-in, methods required observing and capturing actual activities and store the result(s). These results must regularly be updated to generate an output that indicate whether the actual system behavior is regular (normal) or irregular (anomalous). Let M_i be n measures.

$$M_1, M_2, \ldots, M_n$$

to determine whether an intrusion occurs or not. In this context, each measure M_i has two values, 1 and 0. 1 implies that the measure is irregular (anomalous) and 0 implies that the measure is regular (normal). Let H be the hypothesis that the object currently monitored is undergoing an intrusion through a threat event attack. The reliability and sensitivity of each irregular measure M_i is determined by probabilities [4]

$$P(M_i = 1 | H)$$

and

$$P(M_i = 1 | /H).$$

The combined belief in H is

$$P(H | M_1, \ldots, M_n) = P(M_{1,\ldots,}M_n | H) \times \frac{P(H)}{P(M_1, \ldots, M_n)}.$$

This requires the joint probability distribution of the set of measures conditioned in H and $/H$. Let this simplification be the calculation at the expense of accuracy, which relies on the assumption that each measure M_i depends only on H and is independent of the other measures $M_j, j \neq i$, which yield

$$P(M_{1,\ldots,}M_n | H) = \prod_{i=H}^{n} P(M_i | H)$$

and

$$P(M_{1,\ldots,}M_n | /H) = \prod_{i=H}^{n} P(M_i | /H)$$

which finally leads to

$$\frac{P(H | M_1, \ldots, M_n)}{P(/H | M_1, \ldots, M_n)} = \frac{P(H)}{P(/H)} \times \frac{\prod_{i=1}^{n} P((M_i | H)}{\prod_{i=1}^{n} P((M_i | /H)}$$

meaning determining the odds of an intrusion in regard to the values of various irregular (anomalous) measures, from the prior odds of the intrusion and the likelihood of each measure being irregular (anomalous) in the presence of an intrusion.

In [5] covariance matrices used to account for the interrelationships between measures $M_i, i = 1,\ldots,n$. If measures represented by the vector M, the compound irregular (anomalous) measure is determined by

$$M^T C^{-1} M$$

with C as covariance matrix representing the dependence between each pair of irregular (anomalous) measures M_i and M_j.

Other intrusion detection methods use Bayesian networks and other forms of belief networks. Applying Bayes law to the equations above result in the conditional probability

$$P(Intrusion|Event\ Pattern).$$

Besides the foregoing methods, expert systems, predictive pattern generation, neural networks, and model-based approaches are also used in intrusion detection applications. The model-based cyberattack scenario approach is based on measures, which are stored in a database, whereby each of which comprises a sequence of behaviors making up a cyberattack. Therefore, at any actual measure, the model-based approach considers a subset of cyberattack scenarios as likely ones, under which the observed object might currently undergo a cyberattack. Therefore, the model-based approach seeks to verify the current scenario with the scenarios stored in the database, seeking information to substantiate or refute an assumed cyberattack scenario. The outcome indicates whether a cyberattack happens or not, which requires further action or not. Thus, the mapping from behavior to activity must have a high likelihood of appearing in the behavior, expressed by the probability ratio [4].

$$\frac{P(Activity|Behavior)}{P(Activity|/Behavior)}.$$

Therefore, simple and meaningful cyberattack defense methods required trust-worthiness identifies the incident level(s) of cyberattack risks. Available methods are based on probabilistic approaches, machine learning, and Artificial Neural Network (ANN) approaches and others. These methods are often integrated in intrusion detection and prevention system architectures to defend against cyberattacks (see Chap. 3). In this regard, industrial, public, and private organizations provide enormous efforts to secure their data and business assets from cyberattacks. They use techniques, keeping their organizations daily business and assets cyber-secure, defending from cyberattacks, initiated by threat event actors. These cyberattacks try to breach an organization's digital infrastructure to infiltrate malicious software, to easily access valuable data and business assets. Therefore, knowledge is required to understand threat event attacks in order to provide deeper insight into possible cyberattack scenarios as well as the cyberattacker's profile. This knowledge enables organizations to understand the premises of threat event actors, when using regular data communication pathways to breach and steal data or manipulate valuable assets, and others. These exploits are directed through various conducts, e.g., from remote locations by mostly unknown individuals using the Internet. In this context, cyberattack scenarios originate in different types of cyberattackers (also termed adversaries or threat event actors), as shown in Table 4.2.

4.1.1 Cyberattacker Profiles

Cyberattacks against computer systems, networks, infrastructure resources, and others are linked to a cyberattacker and/or a cyberattacker group. A cyberattacker profile describes templates or classes of attackers, as listed in Table 4.2. The profiles are a generic description of settings and intuition, and not an exhaustive listing of possible actions, motivations, or capabilities of the attacker. To protect computer systems, networks, infrastructure resources, and others from cyberattackers' threat event attacks, it is necessary to create a cyber-secure cyber-barrier environment around the targeted objects. This requires, besides essential knowledge in cybersecurity, the awareness of possible threat event attacks that may occur, understanding the profile of cyber attackers, and development of cyberattack models for better understanding the scope of possible threat event attacks with their intrinsic cybersecurity risks. A cyberattacker model ideally characterizes the possible interactions between the cyberattacker and the object under attack, In particular, the cyberattacker model defines constraints for the cyberattacker, e.g., finite computational resources and others.

Table 4.2 shows a collection of cyberattacker profiles, often referred to in the literature as cyberattacker profile. Nevertheless, the boundary conditions between the different attacker profiles are not well defined, and sometimes it is hard to classify a specific real-life cyberattacker as one specific profile [12]. For example, a cyberattacker, sometimes called black hat hacker or structured hacker, is a cyberattacker with extensive security knowledge and skills. This category of cyberattackers takes advantage of known vulnerabilities, and potentially has the knowledge and intention of finding new vulnerabilities. In this regard, the profile belongs to the categories of *known-knowns* and *known-unknowns* (see Sect. 2.2.1). The attack goals of cyberattackers with the profile of cybercriminals can range from blackmailing to espionage or sabotage.

Let's assume a cyberattacker with the profile cracker is able to launch unique cyberattacks to computer systems, networks, infrastructure resources, and others, which means he has access to hardware, software, and Internet connectivity. A possible simple but powerful threat event attack is the resonance attack on load frequency in industrial digital control systems, whereby the threat event actor (cyberattacker) compromises the input of the load frequency, according to a resonance source that forces the targeted digital control system to oscillate at its resonant frequency, to take it out of order. Thus, it is essential to identify and categorize the types of executable cyberattacks for this caber-attacker profile in the digital transformation era, and to examine the possible consequences (risks), to respond on it. This requires using effective defense procedures, working nearly in real-time.

Table 4.2 Cyberattacker types and their characteristics and capabilities

Cyberattackers	Characteristics and capabilities
Basic user	Known as *script kiddie, unstructured hacker, hobbyist,* or even *crackers.* Someone who uses already established and potentially automated techniques to attack a system. This attacker has average access to hardware, software, and internet connectivity, similar to what an individual can obtain through purchase with personal funds or by theft from an employer [6, 7]
Cybercriminals	Individuals or groups use technology to commit malicious activities on computer systems, networks, digital infrastructure resources, and others with the intention of stealing sensitive information to generate a monetary or asset benefit. Cybercriminals sometimes named *black hat hacker* or *structured hacker,* are agents or facilitators of cyber-crime. They often access the cybercriminal underground markets found in the dark net to trade malicious code and services, such as hacking tools. Defending cybercriminal activities requires cybersecurity awareness, understanding scopes of possible threat event attacks and the risks that occur, and developing scenario models of threat event attacks for an effective defense strategy. Compared with threat event actors, cybercriminals are unlikely to focus on a single object, but conduct activities on broad masses of victims by similar platform types, programs used, or online behavior [6–10]
Cyberterrorists	Any premeditated, politically motivated attack against information systems, networks, programs, data, and others that threatens violence or results in violence. Often includes any cyberattack that intimidates or generates fear in the targeted object. Cybercriminal attackers often do this by damaging or disrupting critical infrastructure, which is the body of systems, networks, and assets essential for the continuous undisturbed operation of the target [7, 8, 10];
Disgruntled employees (insider)	People planning to leave their job are a danger committing a cybersecurity breach. They cause damage, steal data by sending files to personal email accounts, create cybersecurity failures by uploading confidential information to a third-party without permission, which puts the organization at a higher cybersecurity risk level [7–10]
Hacktivists	A portmanteau word, which combines hacker and activist, as defined in [11]. This class of attackers uses their hacking abilities to promote a political agenda. Often related to freedom of information (e.g., anonymous) [12]
Organized criminal groups	Also referred to as gang, group, mafia, syndicate, and other. In general, a networked subculture and community of criminals, involved in organized crime, also referred to as gangland or underworld
Nation states	Nation state actors work for a government and have a "license to hack" to disrupt or compromise targeted governments, organizations, or individuals to gain access to valuable data or secret assets. They can create incidents that have international significance [6, 7, 10, 13]

To better defend industrial control systems against threat event attacks, the US Government Accountability Office (GOA) has published the "Guide to Supervisory Control and Data Acquisition (SCADA) and Industrial Control System Security

(ICSS) draft," which provides a description of various threat event attack types, which are summarized and shown in Table 4.3 [14, 15].

As Table 4.3 indicates, no general cybersecurity strategy solution is available due to the different cyberattackers' profiles with their sophisticated and dangerous cyberattack intentions. Therefore, analyzing the criminological profile of an individual cyberattacker or cyberattacker groups is important, because cybercrimes are the fastest growing cybercriminal offense compared to others that become more complex by increasing the transnational character of crime [16, 17]. The data of the Global Cyber Risk Perception Survey 2018 showed that nearly two-thirds of respondents rated cyber-risks as one of the highest risks in their organizations [16], which determine cybercrimes' rapid development, implementation, and use of Information Technology (IT) for criminal purposes on the Internet environment. In this context, the rapid development of cloud computing and cloud services opens up a profitable environment for cybercriminals, and allows cyber-crimes to occur at an alarming rate globally, by making use of anonymity and today's technological offers. To combat cybercrimes, cyberattacker profiling by digital forensic (see Sects. 2.2.2 and 2.2.3) is an essential issue due to cyberattackers' complex cybercriminal offenses. Therefore, cyberattacker profiling tries to identify and catch the anonymous perpetrators of cyberattackers profiles and activities. In order to achieve objective profiling results, two strategies are common: the inductive and the deductive profiling methods. The inductive profiling method is based on statistical and comparative analysis that leads to applicable traits/characteristics shared by criminal cyberattackers that commit the same type of cyber-crimes [18]. The deductive profiling method creates a profile based on theories developed at the cyber-crime scene solely based on evidence.

As reported in [19], the best methodology in profiling cyberattackers is based on a six-element approach that consists of attack signature, attack method, motivation level, capability factor, cyberattack severity, and demographics [20]. As described in [19], the cyberattack signature and cyberattack method investigate what tools used for the cyberattack, along with the methods used by the threat event actor to commit the intrusion of social engineering, malware, Denial of Service (DoS) attacks and other. The motivation level is key to determining the complexity of the threat event attack along with the capability factor (see Sect. 3.3), which highlights the availability and use of specific hacking tools, techniques, and potential resources available to the cyberattackers' profile. The cyberattack severity defines the impact the threat event actor had on the targeted organization, and one of the most critical metrics in the development of the cyber attackers profile is the identification of the geographical location brought about through the demographic element of the methodological approach [19]. These methodologies all have the same goal in mind although their approaches in identifying significances vary.

Within the Internet environment, cyber-crime in the web is an actual risk problem that requires specific security solutions such as the formal model of web-personality, based on an algorithm of personal data verification. The creation of a person psychological profile of the web-personality is one of the most important stages of

Table 4.3 Threat event types and their impacts

Threat event attack types	Description
Bot-network operators	Instead of a break-in into computer systems or networks for bragging rights, they take over multiple systems in order to coordinate cyberattacks and to distribute phishing schemes, spam, and malware attacks.
Criminal groups	Seek to cyberattack systems to gain money. Specifically, organized crime groups are using spam, phishing, spyware, and malware to commit identity theft and online fraud. International corporate spies and organized crime organizations also pose threat event attacks to conduct industrial espionage; large-scale monetary theft, and to hire or develop hacker talent.
Foreign intelligence services	Using cyber tools as part of their information gathering and espionage activities. In addition, several nations are aggressively working to develop information warfare doctrine, programs, and capabilities. Such capabilities enable a single entity to have a significant and serious impact by disrupting the supply, communications, and economic infrastructures that support military power - impacts that could affect the daily lives of citizens.
Hackers	Break into computer systems or networks for the thrill of bragging rights in the hacker community. While remote cracking once required a fair amount of skill or computer knowledge, hackers can download cyberattack scripts and protocols from the internet and launch them against targeted sites. Cyberattack tools have become more sophisticated and have also become easier to use. The worldwide hacker community can pose relatively high threat event attack levels of an isolated disruption causing serious damage.
Insiders	A disgruntled insider in an organization is a principal source of computer crime. They may not need a great knowledge about computer intrusions because their knowledge of a targeted system often allows them to gain unrestricted access to cause damage or to steal valuable data or assets. Insider threat event attack also includes outsourcing vendors and employees who accidentally introduce malware into computer systems or networks.
Phishers	Individuals, or small groups, who execute phishing schemes in an attempt to steal identities or valuable information for monetary reason. Phishers may also use spam, spyware, and malware to accomplish their objectives.
Spammers	Individuals as well as dubious organizations who distribute unsolicited e-mails with hidden or false information in order to sell products, conduct phishing schemes, distribute spyware and malware, or cyberattack public and private organizations.
Spyware/malware authors	Individuals as well as dubious organizations with malicious intent carry out attacks against users by producing and distributing spyware and malware. Several destructive computer viruses and worms have harmed files and hard drives.
Terrorists	Seek to destroy, incapacitate, or exploit critical infrastructures in order to threaten national security, cause mass casualties, weaken the economy, damage public morale and confidence. Terrorists may use phishing schemes, spyware as well as malware in order to generate funds or gather sensitive information.

Table 4.4 Essential steps in computer forensics

Steps in computer forensic	General conditions
Identification/collecting	Purpose of investigation along how to perform the necessary tasks
Preservation	Requires integrity of data
Analysis/extraction	Requires to make sure that this step is done correctly and accurately
Documentation	Records all evidence found and how it was found
Presentation	Summarizes everything found along with explanations

the algorithm developed in [21]. The web-personality is a psychological profile description of the likely character, behavior, problems and interests of an individual (cyberattacker). The manifestations of user personality in website choice and the behavior on online social networks are investigated in [22]. Psychological profiling may be described as a method of suspect identification that seeks to identify an individual's mental, emotional, and personality characteristics and attitudes such as manifested in things done, studied from web-content and others [23].

As in traditional crimes, understanding the unknown cyber-crime is important, supported by answering questions such as:

- What motivates a cybercriminal to get involved in cyber-crime?
- What keeps a cybercriminal in his cybercriminal behavior?
- How a cybercriminal chooses his targets?

However, there is no simple answer possible in cyber-crime. In this regard, identifying an individual cyberattacker profile enables drafting a picture from the puzzles gathered due to the lack of reliable data, which often hinders efforts to create substantive cyberattacker profiles behind cyber-crimes. Therefore, systematic profiling of cyberattackers cyber-crime allows to identify the areas in which they act, which is summarized by the term computer forensic (see Sect. 2.2.2). The goal of computer forensic is "to perform a structured investigation and maintain a documented chain of evidence to find out exactly what happened on a computing device and who was responsible for it" [24]. Thus, computer forensics' aim is to identify, preserve, recover, analyze, and present facts and opinions about digital information. As a consequence of digitization in digital transformation, the process of computer forensic has become essential in solving today's growing cybercrimes landscape. Computer forensic techniques are based on different types of data retrieved. Computer forensics consists of the process steps shown in Table 4.4 [25].

Furthermore, the national and international data exchange of identified, categorized, and documented threat events attacks, stored cyber-secure in databases, accessible through secure keys, is an important initiative to identify cyberattackers profile, so that it can be blocked. Moreover, cyberattack models used to understand the mechanism of threat event attacks before and after they happen, to provide a higher level in cybersecurity. Thus, a proper utilization of modelling techniques enables further planning during an ongoing threat event attack. Integration of scenarios of threat event attacks in the cyberattack model analysis is an effective approach to deepen understanding of possible vulnerabilities. In this context, modeling cyberattack scenarios provides a significant methodology to understand ongoing cyberattack characteristics and intentions. This is required i) to identify cyberattacker(s) and their chosen target, and ii) to test and evaluate cybersecurity strategies and be alert of suspicious activity.

4.2   Cyberattack Models and Scenarios

Modeling cyberattack scenarios enables classifying threat event attacks, specifies cyberattack mechanisms, and verifies protection mechanisms, evaluating the resulting consequences and others, which make modeling cyberattack scenarios an essential approach to investigate how to detect cyberattacks and how to protect against them.

4.2.1   Modeling Formalisms

The formalizations of modeling are useful if they succeed in seizing the essential features of the dynamic system under test. However, two major factors are important when developing models of real-world systems:

1. A model is always a simplification of reality, but should never be so simple that its answers are not true.
2. A model has to be simple enough to be easy to work with and investigate.

Therefore, a suitable model is a compromise between mathematical difficulty, with regard to used equations, and the accuracy of the obtained result. The corresponding relationships are as shown in Fig. 4.1.

From Fig. 4.1, it can be seen that there is no reason to develop expensive models because the quality gained is less than the increase in cost. This is important because a mathematical model is a very compact way to describe complex real-world systems. A complex model not only describes the relations between the system inputs and outputs, it also provides detailed insight into the system's structure and internal relationships. This is because the main relations between the variables of the real-world system modeled mapped into appropriate mathematical expressions, e.g., the relationship between a system's input and output variables. This means they permit extrapolations that allow to generalize, mostly correctly, from past experience to future events to investigate how the real system can be manipulated for ones purposes, which is a kind of uncertainty.

Against this background, models used for many different purposes are to explain intrinsic behavior and data, to provide a representation of data, time behavior, and more. Therefore, mathematical model-building is a method supporting to solve complex problems, making use of properly chosen models. Various simulation models capture different aspects of a system enabling evaluation of complex

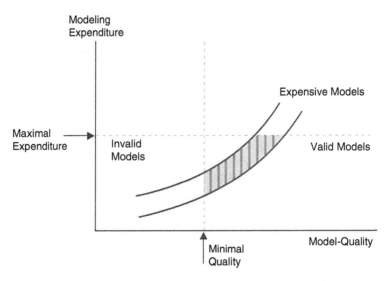

Fig. 4.1 Dependence of the modeling expenditure (costs) on the degree of accuracy (model quality) [26]

simulation scenarios, where each one represents a certain aspect of a system under test or a certain operational strategy. If a model structure is determined for the system under investigation, the next step is to describe the model formalisms in terms of mathematical equations. They describe the dynamics of the system under test at a mostly abstract level of detail. The mathematical notation is eligible to describe the systems behavior, functions, notations, and principles. In this context, a model has achieved its purpose when an optimal match is obtained between simulation results, based on the model, and datasets, gathered through measurements and experiments at the real physical system. Hence, a model helps to explain an object's behavior while investigating essential scenarios to predict the behavior. A model is an abstraction of the (real) system behavior under test, e.g., the behavior of a threat event actor attacking a system resource. The dynamics of a system resource is often modeled using state-space differential equations or differential algebraic equations, which capture the interactions among the components of the system resources. These differential equation models are often nonlinear or linearized approximations. Considering the system resource comprising n components labeled 1, ..., n, and each component is associated with a vector state $x_i(t)$, which describes the evolving properties of component i at time t. The state vector of each component is modeled by a linear differential equation, with the status update for component i influenced by a local input $u(t)$. For convenience, the system dynamic of the state vector $x\text{-}(t) = [x_1(t),\ldots,x_n(t)^T]$ and the input vector $u(t) = [u_1(t),\ldots,u_n(t)^T]$ is described in the following form:

$$x'(t) = A(t)\mathbf{x}(t) + B(t)\mathbf{u}(t)$$

where $A(t)$ and $B(t)$ are, respectively, $n \times n$ and $n \times p$ matrices with $A(t)$ as system resource component matrix and $B(t)$ as input matrix. Hence,

$$A(t) = \begin{bmatrix} A_{11}(t) & \cdots & A_{1n}(t) \\ \cdot & \cdot & \cdot \\ A_{n1}(t) & \cdots & A_{nm}(t) \end{bmatrix}$$

and

$$B(t) = \begin{bmatrix} B_1(t) & & \\ & \cdot & \\ & & B_n(t) \end{bmatrix}.$$

Model investigations capture properties of the (real) system, and keep it as simple as possible for investigation. Therefore, it is often necessary to use different concepts of models in the scientific sector, which depends on a priori knowledge based on [27]:

- Input variables and/or input values
- Output variables and/or output values
- System behavior that follow the concepts of probability and/or statistics
- For decision of the unknown

Applying system inquiries to the systems under investigation requires transforming the characteristics of the systems in such a way that a particular system notation found out of the set of possible system descriptions from the system's theory approach. In any case, a set of mathematical equations is obtained that describe the important variables of the system. Describing a system based on the translation of system knowledge into the language of a mathematical notation depends on the adequate form of its representation. This notation consists of symbols for representing operations, unspecified numbers, relations, and any other mathematical objects. In this context, there are different forms of descriptions that take into account the different circumstances of the application. For a time-invariant, continuous-time system, the abstract mathematical model is expressed by the notation MM_{TICTS}, based on ordinary differential equations as a set of dynamical equations as a structure of the form [27]:

$$MM_{TICTS} = <U, X, Y, f, g, T>$$

with $u \in U$ as set of inputs, $x \in X$ as set of states, $y \in Y$ as set of outputs, f as rate of change function, g as output function, T as time domain, with

$$x' = f(x, u)$$
$$y = g(x, u).$$

It can be convenient for some applications to transform these mathematical equations into a standard or normal form. The important point here is to decide on state variables, which are essential to characterize the system. Such model formalism is a specific set structure.

Let MM equal the sum of the forms of representation used: input, output, and state. Thus, the notation of the mathematical model of a system under test is called dynamic, if it can be defined as set structure Σ:

$$\sum = <X, U, v, T, a, b>$$

with state variable X, set of output values Y, set of input values U, set of admissible controls v, time domain T, state transition map a, and read-out map b.

In some cases, it can be necessary to specify unmeasurable and/or random inputs. These system disturbances described as impacts of uncontrollability and/or un-observability of the modeled dynamic system, mathematically described by stochastic, continuous-time models, as a structure in the form of:

$$MM_{SCT} = <U, V, W, X, T, f, g>$$

with

$$x' = f(x, u, w, t)$$
$$y = g(x, v, t).$$

The vectors v and w represent random model disturbances. In a case where v and w be random or stochastic vector processes, meaning that the stochastic properties of these vectors are not related to the model specification, then x and y will be the same process.

In many cases, the dynamic system thought built of a collection of events. Even the state variables change at specific time instants. A mathematical description based on the notation of a mathematical discrete-event model yields in a structure in the form of:

$$MM_{DEVS} = <V, S, Y, \delta, \lambda, \tau, T>$$

with V: set of external events, S: sequence of states, Y: set of outputs, δ: transition function, λ: output function, and τ: time function, and T as time domain.

Many dynamic systems have properties that vary continuously in space, described based on distributed models. The mathematical expression for distributed

models based on partial differential equations, which result in the following structure
in the form:

$$MM_{PDE} = \; <U, \Theta, S, Y, F, \delta, \lambda, \tau, T>$$

with

$$x = f\left(\Theta; \frac{\partial}{\partial t}\Theta t \partial z; u; z; t\right) z \in Z;$$
$$x = r(\Theta; z; t)z \; dom \; Z;$$
$$y = g(\Theta; z; t)z \in Z :$$

Apart from the independent variable t, the space coordinate z is introduced.
Vector Θ of the dependent variables can vary in space and time. The equation
(s) hold(s) in a spatial domain Z, while conditions given by r provided on the
boundary of the domain $domZ$. There is input u and output y.

Let's now assume X is a variable that can assume any of several possible values
over a range of such possible values, and assume that X is a variable in which the
range of possible values is finite or countably infinite, and then the probability mass
function of X is.

$$p(x_i) = P(X = x_i)$$
$$p(x_i) \geq 0$$
$$\sum_i p(x_i) = 1.$$

Let's assume X is a variable in which the range of possible values is the set of real
numbers; the probability mass function of X is

$$P\langle a \leq X \leq b\rangle \int_a^b f(x)dx$$

$$f(x) \geq 0 \; for \; all \; x \; in \; R$$

$$\int_{Rx} f(x) = 1.$$

Let $E(X)$ be the expected value of the random variable X. The expectation
function is given by

$$E(X) = \sum_i x_i p(x_i)$$

and

$$E(X) = \int\limits_{-\infty}^{\infty} xf(x)dx$$

if X is discrete. The expected value, also called the mean, is μ. Defining the n-th moment of X results in the variance of the random variable X

$$V(X) = E\left[(X - E(X))^2 \right] = \left[(X - \mu)^2 \right].$$

4.2.2 Generic Cyberattack Models

Accessibility and connectivity are key issues in the digital transformation era, but this comes with a number of unprecedented potential risks like stolen valuable data, loose privacy and/or identity, get infected by malware, and others. Therefore, computer systems, networks, and infrastructure resources can get infected in the digital transformation cyber-space. Hence, research on the analysis of threat event attacks is required to gain knowledge about the nature of threat event attacks and cyberattackers' profiles, their motivation and behavior as well as security weakness to mitigate threat event attacks. Therefore, developing cyberattack models of potential threat event attacks are an essential knowledge base for designing intrusion detection and prevention system models. Based on these models, threat event attacks and cyberattackers' profiles can be identified to enhance cybersecurity of potentially targeted objects.

A threat event actor is termed a Cyber-Threat Attacker (CTA), who attacks a target after analyzing the attack capability against the target. After analyzing the capability of the target, the adversary may find that he has more capability to attack than the targeted object can defend. This approach is essential when dealing with advanced CTAs who have already gained some control of targeted computer systems and/or networks.

Let's assume the cyberattacker (also termed adversary) has also analyzed the infrastructure of the targeted resources to take over Command and Control (C2) of any targeted object. Against this background, identifying technical and behavioral patterns and attack goals of cyberattackers are used to make assumptions about the profile of a cyberattacker. In this context, cyberattacker models and scenarios help in profiling cyberattackers' specific scope, but profiling needs collecting knowledge about potential cyberattackers' attack scope, which includes questions such as:

- What is the objective of threat event actors?
- What is the scope of threat event actors?
- What may be the threat event actors' preferred attack method?

In this context, profiling a cyberattacker depicts the attack potential or the threat event attack risk as a measure of a minimum effort to be expended in a threat event attack to become successful, defined by threat event attackers' disclosure, knowledge, and resources. Thus, the outcome of profiling is an expectation function about the successfulness of the threat event attack. Applying modeling techniques to threat event attacks are important, but must be understood, explored, and validated for the respective cybersecurity risk(s) in the cyber-space of the digital transformation era [28].

Today, a number of cyberattack modeling techniques exist and are used to analyze cyberattack risks. These cyberattack model types are the Attack Graph or Attack Tree Models [29, 30], the Attack Vector [31] and the Attack Surface Models [32], the Bayesian Net Model [33], the Diamond Model [34], the Open Web Application Security Project (OWASP) Cyber Threat Model [35], the Kill Chain [36], the Petri Net Model [37], as shown in Table 4.5.

In Table 4.6, advantages and disadvantages of cyberattack model types are shown.

The cyberattack modeling techniques Diamond Model [34], Kill Chain [36], and Attack Graph [29, 30] for threat event attack modeling are discussed in detail in [38]. The OWASP Automated Threat Handbook [39] currently describes 20 cyber-threat incidents. OWASP currently works on a Top 10 publication list describing the 10 most significant classes of application vulnerabilities [40]. In this list, each vulnerability includes two threat event attack modeling constructs: threat event agents—types of threat event actors which could exploit the vulnerability—and cyberattack vectors—descriptions of how the vulnerability could be exploited—in effect, descriptions of either threat event attacks or fragments of threat event attack scenarios.

Cyberattack (threat event attack) modeling is the process of developing a representation of adversarial threat event attacks with regard to possible symptoms of used threat event attacks, scenarios, and specific incidents in the cyber-space of digital transformation, sources, targeted sectors, and others. Hence, the development of adequate adversary cyberattack models is to create substantive profiles of adversaries behind cyber-criminal attacks and their intended attack scenarios with specific values such as adversary expertise, adversary resources, adversary motivation, patterns of threat event attacks, incidents of threat event attacks, and others. This requires specifying an Adversary Attack Behavior Model (AABM), interpreted as hint that threat event attackers may use distinct paths or alternative approaches to reach their threat event attack goals.

As mentioned in [41], the adversary profile depicts the attack potential as a measure of the minimum effort expended in a cyberattack to be successful. In ISO/IEC 15408:2009, the cyberattack potential is defined as a "measure of the effort to be expected in attacking a Target of Evaluation (TOE), expressed in terms of an adversary's expertise, resources and motivation." Besides this, ISO/IEC15408-1: 2009 gives guidelines for specification of Security Targets (ST) and provides a description of the arrangement of components throughout the model. The standard replaced by ISO/IEC AWI 15408-1. In this context, adversaries' cyberattack

Table 4.5 Attack model types and their characteristics and capabilities

Cyberattack model type	Characteristics and capability
Attack graph model	Automatically generate three types of input: Attack templates, configuration file, and attacker profile. Attack templates represent generic (known or hypothesized) attacks, including conditions like operating system version for an attack to be possible. Configuration file gives information about a specific system to analyze, including network topology and configuration of particular network elements (workstations, printers, or routers). Attacker profile contains information about assumed attacker's capabilities like possession of automated toolkit or sniffer and skill level. Models customization of generic attack templates to attacker profile and network are specified in the configuration file. Attack templates represent pieces of known attacks or hypothesized methods moving from one state to another. Their combinations lead to descriptions of new attacks that is any path in the attack graph that represents an attack, which could be cobbled together from many known attacks.
Attack tree model	Represent attacks and their countermeasures as a tree structure. Root node is goal of attack. Availability of several root nodes results in different attack goals. Leaf nodes represent attacks. Once a tree is created, different values are assigned to leaf nodes. Simplest of these values are Boolean, e.g., possible vs. impossible.
Attack surface model	Resource is part of an attack surface, if attacker can use the resource to attack. Entry points and exit points are relevant resources. A resource's contribution to the attack surface reflects the resource's likelihood being used in attacks, e.g., a method running with root privilege is more likely to be used in attacks than a method running with no root privilege.
Attack vector model	Pre-defined attack detection method can be placed over an abuser case that penetration tests can be performed with a flow without confusion (doesn't mean can penetrate easily). Attack vector models neither promise security flaws nor guaranteed vulnerabilities. Attack vectors are manifold from privilege escalation: Session attacks, brute force, default login, and others, to injection like SQL injection, XSS, cross site request forgery (CSRF) to unrestricted file upload like reverse shell, DDOS, malware, and others, to vulnerabilities, which differ based on functionality.
Bayesian net model	Models decision problems containing uncertainty based on directed acyclic graph, where each node represents a discrete random variable of interest. Each node contains states of random variable that it represents and a conditional probability table (CPT) gives conditional probabilities of this variable such as realization of other connected variables based upon Bayes rules. Model learns dataset containing signatures of Normal connections and signatures of several types of known attacks. Bayesian process begins with classifying connections of learning dataset into two classes: Normal/intrusion by using association rules, accelerate learn process from dataset and deduction process about new connection (normal or intrusion).
Diamond model	Consists of four basic elements: Adversary, infrastructure, capability, and victim. Adversary is a threat event actor or a set of actors who attack a victim after analyzing their capability against the victim. Initially the adversary starts with no knowledge of victim's capability. After analyzing the capability of a victim, the adversary may find that they have more capability as victim to attack or not. Model helps when dealing with more advanced

(continued)

Table 4.5 (continued)

Cyberattack model type	Characteristics and capability
	attackers, such as those who have already gained some control over the network. The adversary also analyzes infrastructure of technical and logical ability to command and control any of victim's network.

Kill chain model	Modeling techniques define attack as action chain. Structured attack, since attacker progresses the attack in ordered chain according to plan. Kill chain technique is described by US DoD to attack target. Kill chain stages: Find, fix, track, target, engage, assess. Kill chain is applied in several areas, e.g., cybersecurity to describe attack steps within a counter measure framework. Model consists of seven attack steps:
	    Reconnaissance
	    Weaponization
	    Delivery
	    Exploitation
	    Installation
	    Command and control (C2)
	    Action on objectives

Open web model	Modeling tool to create threat model diagrams as part of secure lifecycle development follows values and principles of threat event modeling manifesto, used to record possible threat event attacks and decide on their mitigation given visual indication of threat event model components and threat surfaces. Runs either as web application or desktop application [https://owasp.org/www-project-threat-dragon/].
Petri net model	Support detection of malicious attacks by identification of suspicious threat event attacks. Identification process based on a set of rules describing suspicious activities using the same semantics as colored petri nets like color sets, places, and appropriate markings.

corresponds to the effort required creating and carrying out the targeted attack objectives. Therefore, adversaries' cyberattack corresponds to the effort required creating and carrying out the initiated attack objectives. Knowledge about the cyberattack is based on knowledge of the goals of the cyberattacker (threat event actor), which is essential in analyzing cyberattack scenarios by cyberattack models. Scenario analysis requires, on the one hand, the identification of the threat event actors' intention with regard to the goals and tasks of the possibly intended attack(s), and, on the other, the identification of possible countermeasures by an Intrusion Detection and Prevention System (IDPS). Furthermore, the attack space is based on a-priority knowledge of the target, available to the cyberattacker, disclosure resources that enable the cyberattacker to obtain target information during the cyberattack, and disruption resources to affect the targeted operation by the cyberattack, as reported in [42]. However, this requires that a scenario analysis of security attacks have enough information about the targeted system(s) to allow validation of the systems' security requirements with respect to particular cyberattacks. In [43], a security attack scenario refers to an attack situation describing the targeted computer system or network with their security capabilities and possible cyberattack with their goals, to identify the security capabilities of the

Table 4.6 Cyberattack model types, advantages and disadvantages

Cyberattack model type	Advantages	Disadvantages
Attack graph model	Comprehensive overview of system cybersecurity	Reveals only known vulnerabilities, becomes soon inappropriate used in larger systems.
Attack tree model	Modular model construction, construction speed, intuitiveness	High abstract level, difficult to display operations adapted to target, difficult to capture coordinated operation.
Attack surface model	Continuous discovery of assets, connections, and vulnerabilities	Integrations with other tools can suffer from technological limitations, e.g., lack of APIs.
Attack vector model	Enables to exploit system vulnerabilities, including human element	Probabilistic attack modeling on network is a complex issue. Skilled attacker is more likely to be ready for unexpected exploits.
Bayesian net model	Suitable for real-time security analysis	Not tangible enough to develop models based on graphical display. Deals with need for statistical analysis in information security.
Diamond model	Model provides opportunities leverage real-time intelligence for network defense, correlation across intrusions, threat events classification, prediction of adversary operations, and planning mitigation strategies	Most difficult cyberattack type to defend against is zero-day attack.
Kill chain model	Defense model for identification and prevention of cyber intrusions activity. Model identifies what adversaries must complete to achieve their objective.	Security gaps exist because model has not been modified since its creation.
Open web model	Structured representation of all information that affects security of an application	Updating threat models is advisable after events such as: New feature is released, security incident occurs, architectural or infrastructure changes.
Petri net model	Improved capture of coordinated actions during attack, identifying model elements	Model quickly becomes impractically large and difficult to create.

targeted computer system or network to prevent cyberattackers achieving their goals. This requires identifying to intentions of possible cyberattacks and identifying possible countermeasures, taking into account the cyberattackers' capabilities. Also high-level descriptions of scenarios of cyberattacks used to describe cyberattacker's behavior or to model some other specific behaviors [44, 45].

4.2.3 Generic Intent-Based Cyberattacker Models

In organization three types of actors active, mostly the external cyberattackers CAE_1 ..., CAE_n, the internal cyberattacker CAI, and the internal cyberattack defender D. Without loss of generality, it is considered that each external cyberattack be launched by a different external cyberattacker.

Cyberattackers are afraid of detection, so they try to make optimal attacking decisions based on their knowledge of the potential expected defensive mechanisms of the targets' object(s). That results in varying the strategy by the attack(s) launched. Therefore, a generic intent-based cyberattacker model is described that depends on the decision whether or not to launch the cyberattack, depending randomly between two choices according to a probability function with the utility function of a cyberattacker as follows:

$$
u_{CA} = \begin{cases} 1, & \textit{launching an undetected cyber-attack} \\ -\beta, & \textit{launching a detected cyber-attack,} \\ 0, & \textit{abstains.} \end{cases}
$$

where β_{CA} is a predefined cyberattacker preference parameter. If the cyberattacker is afraid to be detected, then the decision refers to b > 0. In this regard, the objective of a cyberattacker is to choose the strategy that maximizes its expected utility.

Furthermore, the defender model must satisfy two requirements defending cyberattacks: (1) detect intrusion of cyberattack(s) and (2) reduce the number of false positive rates (see Sect. 3.1.1) to make a proper tradeoff between the detection rate and the false positive rate. For each cyberattacker

$$
CA_i(1 \leq i \leq m)
$$

let

$$
I_{CA}(i)\epsilon[0, 1]
$$

be the loss of the defender associated with CA_i [46]:

$$
I_{CA}(i) = \begin{cases} 1, & \textit{CA}_i \textit{ launching an undetected cyber-attack} \\ b, & \textit{CA}_i \textit{ launching a detected cyber-attack,} \\ 0, & \textit{CA}_i \textit{ abstains.} \end{cases}
$$

Let's assume b $\geq$ = 0 for a detected cyberattack. This case corresponds to cost for the repair of the damage issued by the launched cyberattack.

In [46], a proper tradeoff is defined for the defender between the detection rate and the false positive rate such that the tradeoff parameter

$$\gamma \in [0, 1]$$

that considers the higher the value of γ, the smaller is the false positive rate and hence the smaller is the detection rate [46]. 1f CA_i chooses to attack, then the expected loss of the defender is

$$E[I_{CA}(i)] = \gamma + (1 - \gamma) \cdot b.$$

If CA_i chooses to abstain, $I_{CA}(i) = 0$.

The loss of the defender from false alarm is a labor-intensive process of manually handling false alarms, which is one of the most important challenges of intrusion detection systems. The more false positives the intrusion detection system generates, the more is the resource, e.g., labor, the defender has to spend to distinguish between false positives and real cyberattacks, thus punishing only real cyberattackers [46].

Let

$$I_F(\gamma) \in [0, 1]$$

be the loss from false positives in one time slot with the tradeoff parameter γ, which yields in

$$I_F(0) = 1,$$
$$I_F(1) = 0,$$

and

$$I_F(.)$$

monotonically decreasing with γ [46].

4.3   Cyberattacker Behavior Modeling

Modeling cyberattackers' behavior is an integral concept in cyber-crime defense. A cyberattacker behavior model represents a formalization of a cyberattacker behavior to measure the effects of his intention (see Sect. 4.2.3) to attack computer systems or networks to intrude through cybersecurity flaws. However, such a model must contain an analysis of cyberattacker behavior and incorporate knowledge to defend the targeted computer system or network. The formalization of the cyberattacker behavior model based on algorithms or simply on a series of statements with regard to potential capabilities and goals of cyberattacker(s). Therefore, cybersecurity methods integrate cyberattacker behavior models to verify data flows, data packets, protocols, and others to identify malicious incidents. Another important scope in cybersecurity is digital forensic (see Sect. 2.2.2), a method that benefits from the use of cyberattack models to prove that a detection process is forensically sound [47]. McKemmish [48] defined forensic soundness in the context of digital evidence as the combination of four criteria: meaning, errors, transparency, and experience. Hence, forensic soundness is integral to the admissibility of evidence, and is analogous to the aim of maintaining data and business assets security. Thus, cyberattacker behavior models are an approach modeling possible cyberattacks based on the perspective of either a defender of threat event attacks, using an asset-centric model of cyberattack(s), or a model of a cyberattacker, based on the intention as part of an intention-centric threat model attack [49, 50].

Cyberattack-centric models enable in detail distinct cyberattacks, e.g., spoofing attacks. These attacks are modeled in detail to allow developers of defense systems to reinforce the defense from such cyberattacks. Thus, cyberattacker behavior models represent a complete cyberattack scenario of assumptions, capabilities, and goals. Hence, cyberattacker behavior models represent a more general approach to model attacks on computer systems and consider the application of threat event attack models and cyberattack behavior models to be distinct [44, 45]. Network attack models may also be modeled in various levels of detail, e.g., from complex packet-level descriptions to less detailed network descriptions [51, 52].

The topic of cyberattack behavior models should also take into account employee behavior, described by a model with four dimensions, which are knowledge about cybersecurity risk(s), context, motivation, and behavior. Knowledge refers to clear insights into sustainable save habits, meaning if the human factor is neglected in cybersecurity awareness, this promotes in terms of cybersecurity, unintentionally careless employee behavior. This requires motivating employees to memorize cybersecurity knowledge in a sustainable manner. Context stands for common nomenclature of individual cybersecurity risk factors, which encourages efficient risk behavior. Motivation means intrinsic employees' willingness to respond correctly to respond to threats and to enhance knowledge. Safe behavior is the central element of an organization's strong cybersecurity culture. Protecting an organization also occurs by safe habits of the activities of their employees, e.g., lock screen when leaving the desk, check email scan for suspicious activity, or inform the IT department about risks and incidents in a timely manner or to evaluate to what extent the cybersecurity awareness behavior is influenced, to continuously adjust and effectively minimize cybersecurity risks.

4.3.1 Generic Cyberattacker Behavior Modeling

Threat event modeling, risk assessment, and cyberattack modeling are commonly used in cybersecurity risk management. Thus, threat event modeling deals with the probability of a threat event attack. The risk score describes how likely it is that this threat event attack will cause damage, and cyberattack modeling shows how existing vulnerabilities can be leveraged to cause that damage. In practice, threat event attack and cyberattack modeling are sometimes equated, and thus security solutions actually integrate both approaches. A certain amount of cybersecurity incidents depends on human behavior. The term humans include system administrators, software developers, end users, personnel responsible for securing organizations systems, and finally the cyberattackers themselves who attack organizations' systems. This means that a variety of behavioral models have to be taken into account, which increases the complexity of behavior modeling. Thus, a restriction to attacker behavior model is made.

Cyberattacks with their risk score and cyberattack models are general elements to describe cyberattackers' behavior. Let's assume a cyberattack model based on an a-priory knowledge of the targeted computer system or network is available to the cyberattacker, representing the core knowledge of a Cyberattack Behavior Model (*CABM*). A general goal of the *CABM* is to support cyberattack scenarios to be successful. Let's assume the cyberattacker has gained the most important a priori Knowledge of the Targeted Computer System he tries to Attack ($KTCS_A$), consisting of overall Knowledge of the Targeted Computer System Network (KTCSN), and Data Available at Attack Time I (DAT_i). Thus, cyberattackers' total a-priory knowledge about the targeted computer system yields

$$KTCS_{ap} = (KTCSN, DAT_i)$$

Let's assume that the resulting Cyberattackers' Attack Policy (*CAAP*) to be described by

$$CAAP = \left(KTCS_{ap}, PDCTCS_A \right)$$

with cyberattackers' total a-priority Knowledge of the Targeted Computer System ($KTCS_{ap}$) as, and Probability of Data Corruption to the Targeted Computer System ($PDCTCS_A$). In this context, profiling a cyberattacker is essential to depict the cyberattack potential and the cybersecurity risk through a threat event attack as a measure of a minimum effort expended in a Cyberattack Being Successful (*CABM*). Based on these assumptions, a *CABM* making use of essential characteristics in a cyberattacker model creation is based on:

- Cyberattack Defenders' Knowledge of Cyberattackers' Capabilities (*CADKAC*), has to be taken into account for classification of cyberattackers' capabilities such as Cyberattackers' Capability Skills (*CACS*), Cyberattackers' Goal(s) Intentions (*CAGI*), and assumptions with regard to Possible Cyberattackers' Profile to Attack (*PCAPA*) a computer system, which ultimately gives insight into Cyberattackers' Attack Policy (*CAAP*).
- Threat event Attack Defenders' Probability of Countermeasures (*ADPCs*) of identified attacks to computer system infrastructure.

which result in a scenario-based Cyberattacker Defending Model (*CADM*)

$$CADM = (CADKAC, ADPC)$$

with

$$CADKAC = (CACS, PCAPA)$$

and

$$CADM = (CACS, PCAPA, ADPC)$$

with *CADM* as cyberattack defense model. This finally results in the balance of power of cyberattackers and cyber-defenders as final outcome

$$CAAP \Leftrightarrow CADM.$$

Based on these generic assumptions, a methodological approach with regard to the balance of power of cyberattackers and cyber-defenders has been derived based on *CAAP*, and the respective *CADM*. The specific information and data for modeling, and finally simulating the model, can be obtained from known security standards, e.g., ISO/IEC 15408:2009, ISO/IEC 18045, ISO/IEC 27000: 2012, ISO/IEC 17799:2005, NIST SP-800:30, and others. Furthermore, security dictionaries, e.g., Common Vulnerabilities and Exposures (CVE), Common Attack Pattern Enumeration and Classification (CAPEC™), Open Web Application Security Project (OWASP), Comprehensive Lightweight Application Security Process (CLASP), and others, are references for information and data. In this context, ISO/IEC 15408:2009 defines the attack potential as "measure of the effort to be expected in attacking a Target of Evaluation (*TOE*), expressed in terms of an adversary's (cyberattackers) expertise, resources and motivation." ISO/IEC15408-1:2009 gives guidelines for specification of Security Targets (*ST*) and provides a description of the components and architecture throughout the model. However, the ISO/IEC15408-1: 2009 standard was replaced by ISO/IEC AWI 15408-1. Furthermore, the US governmental report in [53] mentions that a more fundamentally cyber-secure ecosystem can help tip the balance toward those protecting networks and away from malicious threat event actors. The PhD thesis in [54] investigates the obligation for cyber health and the capability approach to well-being, a base developing cyber adversary models.

4.3.2   Cyberattacker Simulation Model

The many benefits of simulation and analysis are described in Sect. 1.2.17. In general, modeling and simulation of complex objects provide analysis and insight on how to optimize systems and thus their behavior. However, in cyberattacker behavior modeling, there is no widely accepted modeling methodology that allows for building an accurate model with a reasonable degree of accuracy. Nevertheless, a model base found taking into account the results in Sect. 4.2.1. Let's assume that a cyberattacker simulation model is procedural in character and expressed in discrete event system specification, a theoretically well-grounded means of expressing a modular Discrete Event Simulation (DEVS) model is a structure in the form of

$$MM_{DEVSCASM} = < U, S, Y, \delta int, \delta ext, \lambda, t_a >$$

with U as a set of input event types, S is the sequential state set, Y is the set of external event types generated as output, δint (δext) is the internal (external) transition function dictating state transitions due to internal (external input) events, λ is the output function generating external events as the output, and ta is the time advanced function.

With $MM_{DEVSCASM}$, the abstract model base for the system entity structure of a cyberattacker simulation model is given. The attributes of entities classified into subject and target whereby subject is an active entity that can take action on other entities, e.g., a threat event, and target is a passive entity that may be affected by a subject. In this context, behavioral attributes are intention and target which denote the characteristic and the target of the threat event, respectively. The system entity structure also contains the representation of the attacked system, e.g., network connection and others.

Furthermore, $MM_{DEVSCASM}$ refers to the probabilities of cyberattackers' attacks and targets by the following attributes:

- Statistical data processing
- Vulnerability analysis
- Computer system or network structure components
- Attackers' knowledge-based model
- Attack scenario generation based

and should not neglect

- Random generation capability
- Cyberattackers' learning capability

which finally make the cyberattacker simulation model complex. For $MM_{DEVSCASM}$, the Discrete Events System (DEVS) representation of the cyberattacker model is shown in Fig. 4.2 [55]:

The attacker model outputs a sequence of attacking commands according to its attacking scenario. The basic mechanism that produces this behavior is the next scenario command and hold-in active attacking-time phrase in the external transition function, as shown in Fig. 4.2. This phrase returns the model to the same phase, active after each external transition and schedules it to undergo a next transition in a time given by attacking-time. Just before the internal transition takes place, the output of next command is proceeded on [55].

Fig. 4.2 DEVS
representation of the
cyberattacker model

state variable

   scenario-type, target-host

external transition function

   case input-port

     in: case phase

       passive: next command := scenario-table

         hold-in active attacking-time

      active : continue

     else: continue

internal transition function

   case phase

     active: passive

output function

   case phase

     active: send packet(command) to port out

Let's now assume cyberattackers' attack is applied based on the knowledge gained of the targeted computer system or network at a given time, which corresponds to the Cyberattackers' Assumptions, Capabilities, and Goals (*ACG*), resulting in the cyberattackers' attack action *A*. In this case, *A* be defined as triple

$$a = (s, t, e)$$

where $s \in S$ is a source node, $t \in T$ is a target node, and $e \in E$ is an exposure such that S takes action on t by exploiting exposure e. Preferences of cyberattacker be known, which determines the probabilities for selection $a \in A$, which results in cyberattack paths in the output of the simulation model. Furthermore, cyberattackers' current intent due to his knowledge K at a given point in time results in the set of exposures E. However, cyberattackers knowledge is a complex data structure, which represents the attributes of the targeted object, uncovered by cyberattackers throughout the cyberattack. A discussion of knowledge constructs of cyberattackers is described in the PhD thesis in [56]. Cyberattackers' preferences can be determined by $a \in A$, which results in the sequence of attack actions

$$\{a_1, a_2, \ldots, a_n\}$$

taken by cyberattackers attack during a cyberattack scenario, representing an attack path. The opportunities for a cyberattack at a given time depend on the intent and accumulated knowledge of the targeted object, and the Kill Chain used. This describes the stages over which a scenario of a cyberattack could transpire, to assess the opportunities, which is the possible attack actions based on the cyberattackers intent and accumulated knowledge. The notion of a Kill Chain integrated in the

CABM is a set of reduction functions describing the types of cyberattack actions that match the objective of a particular Kill Chain state. The type of cyberattack, e.g., DDoS, data extraction, and others or the set of Kill Chain states may vary [56]. The Kill Chain originated from a military attack model called F2T2EA, which stands for:

- Find cyberattackers targets suitable for engagement.
- Fix their location.
- Track and observe.
- Target with suitable weapon or asset to create desired effects.
- Engage cyberattackers.
- Assess effects.

The method was developed to provide organizations with a guide that describes how cyberattacks can be identified, prevented, or rendered harmless before they can possibly cause irreversible damage. The cyberattack model represents a complete Kill Chain, i.e., a cyberattack chain of a successful cyberattack(s). The substeps of the Kill Chain model (see Sect. 4.2) are also referred to as Indicators of Compromise (IoC).

In [57], the selection process of the Kill Chain based on fuzzy logic captures a balance between rule-based behavior models and probabilistic models. The fuzzy logic-based parameters depend on gained knowledge of the targeted object by cyberattackers, and the outcomes of past actions. This approach allows the description of cyberattackers' behavior by controlling the membership functions of each Kill Chain as well as the definition of the Kill Chain outside the Minimum Viable Kill Chain (*MVKC*). To determine the membership for a particular Kill Chain, a set of attack stimuli is generated based on input data. The set of attack stimuli used to define the linguistic variables in the fuzzy rules are currently represented in the *CABM*. This set is separated in three categories: (1) cumulative targeted objects discovered; (2) newly discovered targeted objects; and (3) past successes and failures. These stimuli influence the definition of fuzzy rules for each of the cyberattack stages, enabling a representation of an array of cyberattack types. The membership function defined for each stage of the Kill Chain to describe the process by which the cyberattacker chooses cyberattack types. The following example shows a fuzzy inference rule set for a cyberattacker type using a rule base for *MVKC* in the form:

R₁: **IF** *scanned ratio is* **low** **OR** *newly compromised targeted object is* **high** **THEN** *state Kill Chain is* recon.

R₂: **IF** *newly scanned ratio is* **high** **AND** *newly compromised targeted object is* **low THEN** *state Kill Chain is* **breach-in.**

R₃: **IF** *targeted object with intent is* **high THEN** *state Kill Chain is* **exfiltration.**

with scanned ratio, newly compromised targeted object, targeted object with intent as input quantities, state Kill chain as output quantity, and recon, breach in, exfiltration as linguistic terms of the output quantity.

The result of fuzzy inference is initially a resulting fuzzy set for the output variable. In order to get a sharp output quantity, the resulting output fuzzy set has to be defuzzified. For defuzzification, the linguistic variables are accumulated in each of the membership functions, denoted by μ, to create a set of membership values for each of the Kill Chains (kc). Thus, defuzzification is an inverse transformation, which maps the output of the aggregated fuzzy inference rule set domain back into the crisp (number oriented) output. Defuzzification realized by decision-making algorithms that select the best crisp value based on a fuzzy inference rule set. There are several defuzzification methods, including

- *Center of Gravity (COG)*: Most prevalent and physically appealing defuzzification method. The basic principle is to find the point x where a vertical line would slice an area into two equal areas. Let A_i and x_i denote the area and center of gravity of *i-th* sub-region, which results in

$$x = \frac{\sum_{i=1}^{n} A_i x_i}{\sum_{i=1}^{n} A_i};$$

  with A as linguistic terms of input quantity, x as input quantity, and x as crisp output value.
- *Mean of Maximum (MOM)*: Calculate the most plausible result. *MOM* uses the typical value of the consequent term of the most valid rule as the crisp output value.

The *COG* method returns the value of the center of area under the curve and the *MOM* approach is the point where balance is obtained on a curve.

The Kill Chain is prioritized in the order from least importance, **THEN** clause *reconnaissance*, to the most important, **THEN** clause *exfiltration*. If defuzzification is based on the maximum method, it only considers active rules with the highest degree of fulfillment. Hence, the maximum of the associated output fuzzy quantity determines the sharp output size. Thus, the membership to one of the Kill Chains is determined by the maximum value of the fuzzy quantities based on the Basic Quantity G. Then

$$\mu_1 \cup \mu_2 : G \rightarrow [0, 1] \; with \; (\mu_1 \cup \mu_2)(u) = MAX \; (\mu_1(u), \mu_2(u))$$

which is the union of fuzzy sets μ_1 and μ_2, represented by the Maximum operator.

The reduction function from the Kill Chain returned from the defuzzification process that then is applied to A. Set A now represents the available a with the contribution of the cyberattacker's intent and the opportunities [57].

To demonstrate the impacts that specific types of cyberattackers will have on cybersecurity of a targeted object and how the configuration of the targeted object affects the success of the cyberattackers' attacks, simulations will be executed to calculate targeted object configurations with regard to cyberattacker types and behaviors. There are different tools available for this purpose such as the Cyber Attack Scenario and Network Defense Simulator, the Network Security Simulation, XM Cyberattack Path Management Platform that is a SaaS-based platform, Threatcare, and others.

4.4 Cybersecurity Ontology

4.4.1 Introduction to Ontology

The rapid growth in data through today's digital technologies expands the importance of cybersecurity with regard to the increase of cybersecurity threats, because data are the most important business value in the digital transformation era. However, public and private organizations are currently coping with cybersecurity issues without collaboration to solve this problem. Nevertheless, some public and private organizations possess some forms of standards trying to solve this problem based on these standards. This requires developing ontologies for cybersecurity, which provides a common understanding of cybersecurity domains. The word ontology comes from the Greek term onto, which means existence or being real, and logia, which means science, or study. To this extent, the term ontology specifies some sort of shared understanding. Hence, in a more formal sense, the term ontology also represents some kind of description logic. Furthermore, ontology may indicate that certain object types are subsets of another, and indicate what can be said about the objects in the respective domains. As an outcome, the ontology also can specify which properties each object has, and what value or range of values each property can take. In this regard, the ontology defines the discourse about that object. Therefore, ontology is a description of what exists specifically in a specific domain, e.g., every component that exists in an information system. This includes the relationship and hierarchy between these components. In this regard, the focus of ontology is not primarily discussing whether these components are the true essence and/or core of the computer system or not. Furthermore, it is important to note that the ontology does not describe whether the components within the computer system are more real compared to the process that takes place within the computer system. Rather, they are naming components and processes and grouping similar ones together into categories. The purpose of the ontology is to understand and describe the underlying structures that affect the domain-specific components or systems. In this context, ontology of a domain specifies the domain-specific objects, concepts, and relations in that domain, which may be assumed as a generally structured description of items. Thus, ontology also indicates that certain object types are subtypes of another, and specify, which properties each object has, and what value or range of values each property can take. Therefore, the ontology of a domain defines the discourse about the domain, and if an item does not appear in ontology, then no statement can be given about that item. In this context, the ontology specifies some sort of shared understanding of a domain. In other words, the term ontology assumed analogous to description logic. Some of the major characteristics of ontologies are that they ensure a common understanding of information and that they make explicit domain assumptions. As a result, the interconnectedness and interoperability of the model make it invaluable for addressing the challenges of accessing and querying data.

Assuming ontology represents a formal explicit description of concepts in a domain of discourse (classes), properties of each concept describing the various features and attributes of the concept (slot), and restrictions on slots (facets). Therefore, ontology, together with a set of individual instances of classes, constitutes a knowledge base of a domain. Some common components of ontologies include:

- *Individuals*: Represent instances or objects.
- *Classes*: Represent sets, collections, concepts, classes in programming, types of objects, or things.
- *Attributes*: Represent aspects, properties, features, characteristics or parameters that objects and classes can have.
- *Relations*: Represent ways in which classes and individuals can be related to one another.

Classes are the focus of most ontologies. *Classes* describe concepts in the respective domain, e.g., a class of smartphones represents all smartphones. Specific smartphones are instances of this class. A smartphone with a 5G feature is an instance of the class of 5G smartphones. A class can have subclasses that represent concepts that are more specific than the superclass, e.g., the class of smartphones divided in 5G, LTE, and others. Alternatively, the class of all smartphones is divided into iOS, Android, and other operating systems-based smartphones.

Slots describe properties of classes and instances, e.g., the 5G smartphone Mate 30 Pro, the future cellular standard and successor to LTE-Advanced, produced by Huawei. Two *slots* may describe the smartphone in this example, the *slot body* with the value 5G and the *slot marker* with value Huawei. At class level, instances of the class smartphone slots describe their technical specs such as possibilities of photography and videography, chipset, power, speed, touchless gesture control, energy efficiency, and others.

Instances of the class smartphone, and its subclass 5G, have a *slot maker*, the value of which is an instance of the class Huawei, as shown in Fig. 4.3. All *instances* of the class Huawei have a *slot produces* that refers to all smartphones, instances of the class smartphone and its subclasses Huawei produces.

4.4.1.1 Ontology Types

Ontology is a formal, explicit specification of a shared conceptualization in which the knowledge model is built upon the following types:

- *Entity*: Represents an object or thing, e.g., person, smartphone manufacturer, smartphone user, and many others.

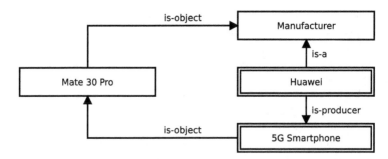

Fig. 4.3 Classes, instances, and relations in ontology domain smartphone. Double bordered boxes represent classes and normal bordered boxes used for instances. Direct links represent slots and internal links such as instance-of and subclass-of

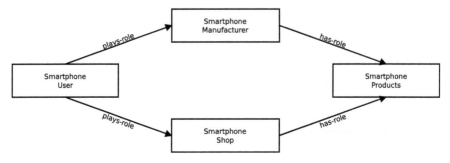

Fig. 4.4 Ontology knowledge model

- *Relation*: Represents relationships between entities, e.g., smartphone manufacturer and smartphone user customer relationship.
- *Role*: Describes the participation of entities in a relation, e.g., in business roles of manufacturer and users built a relation.
- *Resource*: Represents the properties associated with an *entity* or a *relation*, for example, a name or date, and others. Resources consist of primitive types and values, such as strings or integers.

The ontology for the foregoing types of the knowledge model is shown in Fig. 4.4, which has one sub-type or category of an object, which is the user. The user can take one or two different roles in a product's relation with another user entity, a smartphone manufacturer, or a smartphone shop role. The products relation has a single property associated with it, e.g., products type, while the user has a number of properties, such as name, age, gender, profession.

Therefore, ontology specifies the objects, concepts, and relations within the respective application domain, and be stated as structured list of items. In this regard, it is a formal naming and definition of types, properties, and interrelationships of items that really or fundamentally exist for a particular domain of discourse. Furthermore, ontology indicates that certain object types are subtypes of another type. Hence, ontology of a domain defines the discourse about the domain. However, if an item does not appear in ontology, that item cannot reasoned. In this regard, ontology of a domain refers to a specific knowledge, e.g., knowledge of the data in the domain of interest. This includes a vocabulary of terms, definitions of these terms, and a specification of the terms and concepts' interrelations. To this extent, ontology specifies some sort of shared understanding of a domain [58]. Hence, ontologies defined for particular purposes and in particular contents and the form ontology takes, partially influenced by those purposes and contexts [59]. Thus, understanding appropriate domain ontology is a great aid to knowledge acquisition, which means ontology designed with different levels of [60].

4.4.2 Cybersecurity Ontology

In recent years, there has been a need for using ontologies in cybersecurity supporting resolution of the cybersecurity problem, e.g., by making use of Semantic Web Languages and Ontologies (SWLO) for cybersecurity awareness [61]. Nevertheless, ontologies for cybersecurity go back to the early days of Semantic Web. In [62], the use of the Defense Advanced Research Projects Agency Agent Markup Language (DARPAAML), the precursor of the Web Ontology Language (OWL) for representing ontology for intrusion detection issues, is discussed. It compares DARPAAML against XML and discusses the inadequacy of the latter. In this study, ontology includes 23 classes and 190 properties/attributes. OWL is a semantic web computational logic-based language, designed to represent rich and complex knowledge about things and the relations between them [63]. It also provides detailed, consistent, and meaningful distinctions between classes, properties, and relationships. By specifying both object classes and relationship properties as well as their hierarchical order, OWL enriches ontology modeling in semantic graph databases, also known as Resource Description Framework (RDF). RDF is a model for data publishing and interchange on the Web standardized by the World Wide Web Consortium (W3C). In this regard, RDF triplestore is a type of graph database that stores semantic facts. OWL, used together with the OWL reasoner in RDF triplestores, enables consistency checks to find any logical inconsistencies, and ensures satisfiability checks to find whether there are classes that cannot have instances. The data in an RDF triplestore is stored in three linked data pieces, called a triple. Triples are also referred to as statements or RDF statements [64]. OWL is equipped with means for defining equivalence and difference between instances, classes, and properties. These relationships supports users match concepts even if various data sources describe these concepts somewhat differently. They also ensure the disambiguation between different instances that share the same names or descriptions [63].

Against this background, the development of cyberattack ontology is essential to enable data integration across disparate data sources, to raise cybersecurity and thus, minimizing cybersecurity risks. Cyberattack ontology is a knowledge base that contains type, mode, consequences, and other information of cyberattacks.

Developing a cyberattack ontology makes use of known security standards such as ISO/IEC 15408:2009, ISO/IEC 18045, ISO/IEC 27000: 2012, ISO/IEC 17799: 2005, NIST SP-800:30, and others as well as security dictionaries such as Common Vulnerabilities and Exposures (CVE), CAPEC™, OWASP, Comprehensive Light-weight Application Security Process (CLASP), and others. Therefore, the security ontology uses the foregoing standards like the Web Ontology Language (OWL) [63]. OWL is a semantic web computational logic-based language, designed to represent rich and complex knowledge about objects/things and the relations between them. It also provides detailed, consistent, and meaningful distinctions between classes, properties, and relationships. Thus, OWL-based ontology describes a domain in terms of classes, instances, and relations, including descriptions of characteristics of those objects/things, with regard to slots and internal links such as *instance-of* and *subclass-of*. Based on these conceptual aspects about cyberattack models and cyberattack scenarios as well as the available security standards, the cyberattack ontology approach illustrates Fig. 4.5.

In semantics, it is possible to execute precise searches and complex queries. Initially, this effort focuses on malware subjects initiated through threat event attacks, because malware is one of the most prevalent threat event attacks to cybersecurity risks. For this reason, the MITRE Corporation has developed the Malware Attribute Enumeration and Characterization (MAEC) language [65], which is a structured language for encoding and sharing high-fidelity information about malware incidents, based upon attributes such as behaviors, artifacts, and relationships between malware subjects. As described in [65], MAEC focuses on characterizing the most common malware types, including Trojans, worms, rootkits, and today's more advanced malware types. MAEC's core components include a

- *Vocabulary*: Represented by dictionaries defining three distinct levels of malware subjects, high-level mechanisms, mid-level behaviors, and low-level actions
- *Grammar*: Represented by schemas, a syntax for the vocabulary of actions, behaviors, and taxonomies, and an interchange format for structured information about these elements

and

- *Standardized Output Format (SOF)*: Used for particular use cases, including the description of a malware instance, malware intrusion set, or malware families in terms of MAEC's dictionaries and schemas

and provide a standard means of communicating information about malware attributes, as shown in Fig. 4.6.

Malware is responsible for a variety of malicious activities, ranging from spam email distribution via botnets to the theft of sensitive information via targeted social engineering attacks and others. Therefore, the protection of computer systems and networks from malware subjects is a primary cybersecurity concern for all organizations, as even a single instance of an uncaught malware subject can result in damaged computer systems and compromised data.

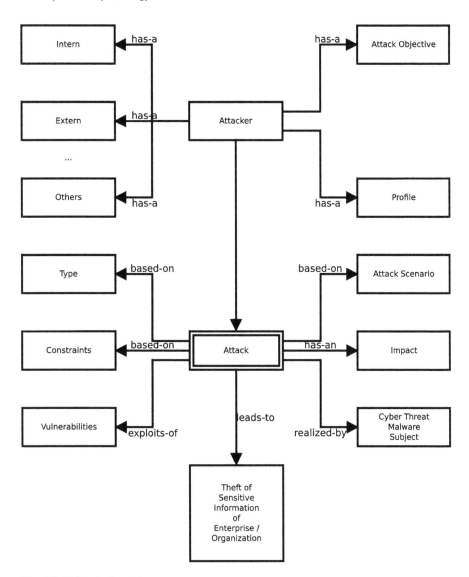

Fig. 4.5 Cyberattack ontology

Methods for detecting and combating malware subjects often rely on the characterization of malware attributes and their behavior, discovered through static and dynamic analysis techniques. The combination of the two allows for an en-compassing profile of malware subjects to be constructed based upon the mal-ware's disassembled binary and observed run-time behavior.

Before MAEC, the lack of an accepted standard for unambiguously characterizing malware subjects meant there was no clear method for communicating the specific malware attributes detected in malware analyses, or for enumerating its

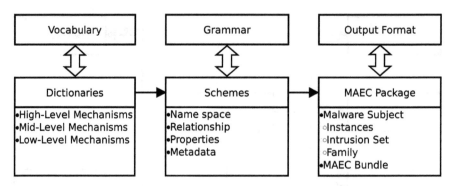

Fig. 4.6 MAEC's core components: Vocabulary, grammar, and forms of standardized output after [65]

fundamental makeup. The results included non-interoperable and disparate mal-ware reporting between public and private organizations, disjointed or inaccurate malware attribution, the duplication of malware analysis efforts, increased difficulties in determining the severity of a malware threat, and a greater delay between malware infection and detection as well as responses [66]. Therefore, the key to ontology development is to understand the cyber-domain that drives the kinds of entities, properties, relationships, and rules needed in the ontology. With regard to the complexity of cybersecurity analysis, the ontology development better consists of modular sub-ontologies, rather than a single, monolithic ontology [67]. Ontologies are grouped into three categories: Upper Level Ontology (ULO), Mid-Level Ontology (MLO), and Domain Level Ontology (DLO), according to their specific levels of abstraction of the architecture of the cybersecurity ontology concept to be developed. For more details, see references [67, 68]:

- *Upper Level Ontologies*: High-level, domain-independent ontologies providing common knowledge bases from which more domain-specific ontologies are derived. Standard upper ontologies are also referred to as foundational or universal ontologies.
- *Mid-Level Ontologies*: Less abstract, make assertions that span multiple domain ontologies. These ontologies provide concrete representations of abstract concepts found in the upper ontology. There is no clear demarcation point between upper and mid-level ontologies. Mid-level ontologies also encompass the set of ontologies that represent commonly used concepts, such as time and location. These commonly used ontologies are sometimes referred to as utility ontologies [69].
- *Domain Level Ontologies*: Specify concepts particular to a domain of interest and represent those concepts and their relationships from a domain-specific perspective. Domain-level ontologies may be composed by importing mid-level ontologies. They also extend concepts defined in mid-level or upper ontologies.

Developing the detailed architecture of the cybersecurity ontology requires, dependent on the category of interest, specific descriptions to abstract major categories, domain-specific concepts, and ontologies that span multiple concept categories. The descriptions of the major categories, the basis for cybersecurity ontology taxonomy, are

- *Entities*: Describe foundational incidents, collections, and others
- *Relations*: Describe relationships of detection and defense actions, organizational locations, and others
- *Role*: Describes cyber threat attackers and cyber-threat defenders
- *Resources*: Describe capability, infrastructure, behavior, malware subjects, and others

Malware resources' published attempts systematically categorize malware subject's ontology, are reported in [70], and descriptive languages implemented in Extensible Markup Language (XML) published in [71, 72]. The described ontology enables data exchange between security algorithms. The taxonomy of malware classes is illustrated in Fig. 4.7. Worthy to mention is an attempt at categorizing malware subject traits [73].

Furthermore, as discussed in [73], semantic Web technologies provide representation languages to build a common framework that allows data to be shared, integrated, and reused across applications, organizations, and community boundaries. The Web Ontology Language (OWL) and Resource Description Framework (RDF) represent the semantics of an entity as a set of objects or concepts, rather than strings of words. OWL and RDF provide constructs to represent machine readable and understandable information, enabling facilitating semantic integration and sharing of information from heterogeneous sources, which is essential in cybersecurity ontology development. Cybersecurity ontology facilitates data sharing across different formats and standards and allows reasoning to infer new information. This development finally ends up in the Unified Cybersecurity Ontology (UCO) framework, described in [73]. UCO helps to evolve cybersecurity standards from a syntactic representation to a semantic representation, showing several contributions in cybersecurity ontology. In this regard, UCO is an extension to an Intrusion Detection System Ontology [71, 72] to describe cyberattack incidents related to cybersecurity risks. Several projects that focus on individual components of a UCO framework analyze different data streams and assert facts in triplestore approach, as reported in [62]. In this context, UCO is essential for unifying information from heterogeneous sources and supporting reasoning and rule writing. Furthermore, UCO supports reasoning and inferring new information from existing information, and supports capturing specialized knowledge of cybersecurity analytics, using ontology classes, terms, and rules. The rules use terms from UCO to connect information within the framework with external information available on the web. The rule states that if the Web text description consists of some vulnerability terms, e.g., some security exploits, a certain product and some processes executed, which in turn logged through an opening up of an outbound port. Moreover, there is a possibility of a cyberattack on the host system [73]. Developing cybersecurity ontology of the respective cybersecurity domain enables data integration across disparate data sources. Formal defined semantics make it possible to execute precise

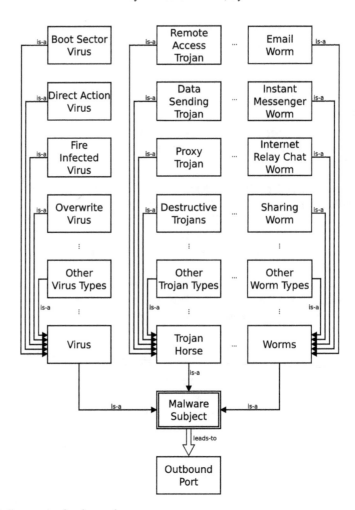

Fig. 4.7 Taxonomy of malware classes

searches and complex queries. Initially, this effort focuses on malware subjects, because malware is one of the most prevalent cyberattacks to cybersecurity risks.

Besides the OWL language, the MITRE Corporation has launched the Malware Attribute Enumeration and Characterization (MAEC) language [65, 66], a structured language for encoding and sharing high-fidelity information about malware subjects, based upon attributes such as behavior, artifact, and the relationship between malware subjects. Malware is responsible for a variety of malicious activities, ranging from spam email distribution via botnets to the theft of sensitive information via targeted cyberattacks. Therefore, the protection of computer systems and networks from malware is a primary cybersecurity concern for all organizations, as even a single instance of uncaught malware can result in damaged computer systems and compromised data. However, the key to ontology development means understanding the respective cyber-domain, which drives the kinds of entities, properties, relationships, and potentially rules essential to the cybersecurity ontology. With regard to the complexity of cybersecurity analysis, ontology development consists of modular sub-ontologies, rather than a single, monolithic ontology. Ontologies are grouped into three broad categories of upper-level, mid-level, and domain-level ontologies, according to their level of abstraction, as illustrated in Fig. 4.5. Several use cases support using mappings between UCO and general ontologies that can't be supported by individual ontologies alone, as described in [74] in more detail.

Let's assume datasets in the material data domain are ontology-based. Therefore, identification and defense of possible cyberattack incident intrusion must also be ontology-based. In this context, ontology is the abstract representation of objects of the real world, e.g., material data. Specifically, in the domain of material data research, this means developing a domain-specific ontology for cybersecurity.

With regard to cyberattack intrusion incidents on datasets, it is assumed that a cyberattack intrusion incident maps to the category unknown (see Table 2.10), pointing to unpredictable and unexpected cyberattacks. This represents a dynamically changing cybersecurity risk for the data infrastructure, which requires an adequate solution for unpredictable effects to make the data infrastructure cyber-secure. This means that a domain-specific semantic of unknowns is represented as a kind of uncertainty by their ontology. Such ontologies must be able to suggest suitable security services that may or may not be required, set at initial time of the dataset, and is customized and activated by the media dataset used. Therefore, developing architecture of a cybersecurity ontology framework is required to have the dataset infrastructure cyber-secured against cyberattacks. Such a framework requires the development of three key elements:

- Processes of business models
- Middleware of communication infrastructure
- Integration

as a domain-specific sematic representation of the cyberattack ontology.

Against this background, the cyberattack ontology framework includes the cybersecurity domain-specific ontology and data integration for different data sources in a common knowledge base. The integration is the required interaction between the dataset infrastructure and the cyberattack ontology layer that provides the requirements for cybersecurity to

- Detect cyberattack incidents
- Prevent cyberattack incidents
- Avert cyberattack incidents

for developing a cybersecurity domain-specific ontology.

The derivable development process is of multi-level, which, after evaluation, (feedback) initiates the next development step around the domain-specific context forming the ontologies, here cybersecurity relations of the essential attributes and relations, a critical step for the transformation of cybersecurity in the dataset infrastructure. In addition to the cybersecurity core ontology, the ontology for secure cyber-operations is developed.

Let an entity be a cyberattack incident that targets a dataset protected through the cybersecurity domain ontology. The cyberattack initiated by a cyberattacker who spied on vulnerabilities of the dataset infrastructure and is using this weakness to cyberattack the dataset infrastructure. The cybersecurity core ontologies thus form, in a certain sense, a strengths and weaknesses profile that maps the security requirements to the possible entities. Thus, the ontology aims at secure dataset infrastructure operation to reduce or eliminate cyberattack incidents that may arise when monitoring cyberattack incidents, as an important result for trusting the cybersecurity ontology.

4.4.2.1   Generic Cybersecurity Data Space Ontology Framework

Modeling threat event attacks is important to cybersecurity, but it is a complex and resource-demanding task. The aim of effective modeling of threat event attacks is to simplify model creation using data that are already available. However, the collected data often lack context, which means models less precise in terms of domain knowledge. The lack of domain knowledge can be addressed with ontologies.

Let's assume that a cyberattack map to the category *unknown*, pointing to unpredictable and unexpected cyberattack incidents, which represents a cybersecurity risk for the datasets. This requires domain-specific semantics of unknowns represented as a kind of uncertainty by their ontologies. Such ontologies must be able to suggest suitable cybersecurity services. Hence, the generic framework architecture of the cybersecurity ontology contains the components shown in the model in Fig. 4.8.

The generic cybersecurity ontology framework in Fig. 4.8 shows the essential system components, including the cybersecurity domain-specific ontology and data integration for different data sources in a common knowledge base, e.g., meta-datasets. This enables data integration and padding from ontology information and access to various datasets required. Furthermore, security services related to organizations' business process models, network devices, and the requirements ultimately required providing security against cyberattack incidents.

Datasets integration in the middleware layer provides the requirements for cybersecurity, for instance, cyberattack intrusion incidents and other vulnerabilities that create the security framework for using domain-specific cybersecurity ontologies.

4.4.2.2   Cyberattack Ontology Model

Let an entity be a cyberattack that targets a dataset protected by cybersecurity domain ontology. The cyberattack initiated by a cyberattacker who has spied on the vulnerabilities of data space and is using this weakness to attack the dataset. Let's assume that >100.000 known weaknesses of vulnerabilities exist which may result

Organizations Business Process Models	Middleware of Communication Infrastructure
Domain Data Space Service Model • Domain System Structure • Domain System Behavior • Domain Data Flow and Control • …. Technology Specific Modules • Information and Communication Technology • Domain Data Sets • Interaction with Organization • ….	Network and Process Observation • Threat • Vulnerability • Incident • Data Theft • Virus • Trojan Horse • Worm • ….
Integration	

Authentication			Cybersecurity Domain Ontology
Authorization	Standard Query Language	Standard Query Language	Cybersecurity Data Structure • Data Populations • Data Integration • Data Performance • ……
Encoding			
Integrity			

Fig. 4.8 Generic cybersecurity data space ontology framework

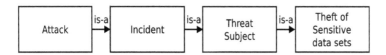

Fig. 4.9 Cyberattack model

in total in millions of them that have to be defended in an organization, taking into account the high number of hardware and software assets. Hence, vulnerability management is required for efficient periodization of incidents. However, this requires availability of time and resources for patching.

The cybersecurity core ontologies form, in a certain sense, strengths and weaknesses profiles that map the security requirements to the possible entities. The ontology for cyber-secure operations aims to reduce potential false positives in detecting potential cyberattacks that may arise when monitoring cyber-vulnerabilities. Thus, cybersecurity ontology represents a domain-specific model that defines the essential domain concepts, their properties, and the relationships between them and represents them as essential knowledge base to cyber-secure the respective application. The respective generic cyberattack model is shown in Fig. 4.9.

Table 4.7 Ontology-based security standard concepts

Security standard	Characteristics
Incident reporting	Structured threat information eXpression (STIX) makes use of OWL. STIX captures concepts and provides a high-level framework to hold the various cyber intelligence components together. STIX standard helps to glue together the lower level concepts such as events, devices, and the very other MITRE standards [75]. STIX is sponsored by the office of cybersecurity and communications at the US Department of Homeland Security.
Threat information	Malware attribute enumeration and characterization (MAEC) makes use of OWL. Common attack pattern enumeration and classification (CAPEC™) makes use of OWL. Is international in use and free for public use. It is a publicly available community list of common attack patterns along with a comprehensive schema and classification taxonomy. Attack patterns are descriptions of common methods for exploiting software systems. CAPEC™ [76] is co-sponsored by the MITRE corporation and the office of cybersecurity and communications at the US Department of Homeland Security
Risk information	Common Weakness Enumeration (CWE™) makes use of OWL. Is a community-developed list of common security weakness types. It serves as a common language, a measuring for software security tools, and a baseline for weakness identification, mitigation, and prevention efforts. CWE™ is co-sponsored by the MITRE Corporation [77].
Assets, target information	Common weakness scoring system (CWSS™) makes use of OWL. It provides a mechanism for prioritizing software weaknesses in a consistent, flexible, and open manner. It is a collaborative, community-based effort that is addressing the needs of its stakeholders across government, academia, and industry. CWSS™ is organized into three metric groups: Base finding, attack surface, and environmental: *Base finding metric group*: Captures the inherent risk of the weakness, confidence in the accuracy of the finding, and strength of controls *Attack surface metric group:* The barriers that an attacker must overcome in order to exploit the weakness *Environmental metric group:* Characteristics of the weakness that are specific to a particular environment or operational context Each group contains multiple metrics—also known as *factors*—that are used to compute a CWSS score for a weakness [78]. CWSS™ is co-sponsored by the MITRE corporation.

As shown in Fig. 4.9, cyberattack analysis is a security field that needs a scientific basis for sharing information among cyber-defending teams. One option is building OWL-based malware analysis ontology to provide more scientific approaches based on a malware analysis dictionary and taxonomy, and combining those in a competence model with the goal of creating an ontology-based cybersecurity framework. Meanwhile, several security standards developed, taking into account OWL, representing ontology-based security concepts, as shown in Table 4.7.

In this context, an Ontology (*O*) organizes Domain Knowledge (*DK*) in terms of Concepts (*C*), Properties (*P*), and Relations (*R*). In other words, an Ontology (*O*) at an elementary level is a triplet where *C* is a Set of Concepts (*SC*) essential for the domain, P is a Set of Concept Properties (*SCP*) essential for the domain, and R is a Set of Binary Semantic Relations (SBSR) defined between concepts in O [79].

A set of basic relations can be defined as $R_b = \{\Delta, \uparrow, \nabla\}$ with the following interpretations [80, 81]:

- For any two ontological concepts, $c_i, c_j \in C$, Δ denotes the equivalent relation, meaning $c_i \, \Delta \, c_j$. If two concepts, c_i and c_j, are declared equivalent in ontology, the instances of concept c_i can also be inferred as instances of c_j and vice-versa, which ultimately leads to the decision of what *unknown* represents and that the *unknown* is not clarified, and can be used for this purpose or in other words *unknown* can ultimately be transformed into *known*.
- $\uparrow$ is the corresponding generalization notation. In cases where the ontology specifies $c_i \uparrow c_j$, then c_j inhibits all property descriptors associated with c_i; and these need not be repeated for c_j while specifying the ontology, and *unknown* becomes known unknown.
- $c_i \, \nabla \, c_j$ means c_i has part c_j. If a cyberattack is given in the ontology as an aggregation of other concepts, it can be expressed with ∇ and *unknown* becomes *known*. In this context, the context of the ontology describes ways in which concepts can be divided and grouped together, making it easier to sort them into hierarchies of quantities in order to ultimately be able to decide whether there is a threat event attack and what category it belongs to prevent it.

The semantic of *Unknown-Unknowns* marks a type of uncertainties captured by their ontologies [82], or events not been thought of. In this regard, ontology defined as an abstract representation of real-world objects, which means that the ontology constitutes a domain-specific model defining the essential domain concepts, their properties, and the relationships between them, represented as a knowledge base.

4.5   Exercises

What is meant by the term *Cyber Attack Model?*
Describe the main characteristics and capabilities of a Cyber Attack Model.
What is meant by the term *Denial of Service Attack?*
Describe the main characteristics and capabilities of a Denial of Service Attack.
What is meant by the term *Cybersecurity Weakness?*
Describe the main characteristics and capabilities of Cybersecurity Weakness.
What is meant by the term *Penetration?*
Describe the main characteristics of Penetration and give a practical example.
What is meant by the term *Measures?*
Describe the main characteristics of Measures.
What is meant by the term Joint Probability Distribution?
Describe the main characteristics and capabilities of a Joint Probability Distribution.
What is meant by the term *Covariance Matrix?*
Describe the main characteristics and capabilities of the Covariance Matrix.

What is meant by the term *Bayesian Network?*
Describe the main characteristics and capabilities of a Bayesian Network.
What is meant by the term *Probability Ratio?*
Describe the characteristics and capabilities of the Probability Ratio.
What is meant by the term *Cybercriminals?*
Describe the main characteristics and Capabilities of a Cybercriminal.
What is meant by the term *Cyberterrorists?*
Describe the main characteristics and capabilities of Cyberterrorists.
What is meant by the term *Disgruntled Employee?*
Describe the main characteristics and capabilities of a Disgruntled Employee.
What is meant by the term *Foreign Intelligence Service?*
Describe the main characteristics and capabilities of Foreign Intelligence Service.
What is meant by the term *Cybersecurity Level?*
Describe the main characteristic steps of a Cybersecurity Level.
What is meant by the term *General Modeling Formalisms?*
Describe the main characteristics and capabilities of the General Modeling
    Formalisms.
What is meant by the term *Probability Mass Function?*
Describe the main characteristic and capabilities of the Probability Mass Function.
What is meant by the term *Random Variable?*
Describe the main characteristics of a Random Variable.
What is meant by the term *Attack Graph Model?*
Describe the main characteristic and capabilities of an Attack Graph Model.
What is meant by the term *Attack Tree Model?*
Describe the main characteristics and capabilities of the Attack Tree Model.
What is meant by the term *Attack Surface Model?*
Describe the main characteristics and capabilities of an Attack Surface Model.
What is meant by the term *Kill Chain Model?*
Describe the main characteristics and capabilities of the Kill Chain Model.
What is meant by the term *Cyberattacker Behavior Modeling?*
Describe **the** characteristically Process of Cyber-Attacker Behavior Modeling.
What is meant by the term *Cyberattack Simulation Model?*
Describe the main characteristics and capabilities of the Cyberattack Simulation
    Model.
What is meant be the term *Center of Gravity?*
Describe the main characteristics of the Center of Gravity.
What is meant by the term *Mean* Maximum?
Describe the characteristics of Mean Maximum.
What is meant by the term *Ontology?*
Describe the main characteristics of Ontology.
What is meant by the term *Cyberattack* Ontology?
Describe the main characteristics of the Cyberattack Ontology Approach in an
    example.
What is meant by the term *MAEC Language?*
Describe the main characteristics and capabilities of the MAEC Core Components.

What is meant by the term *Vocabulary?*
Describe the main characteristics of a Vocabulary.
What is meant by the term *Grammar?*
Describe the main characteristics of a Grammar.
What is meant by the term *Mid-Level* Ontology*?*
Describe the main characteristics and capabilities of the Mid-Level Ontology.
What is meant by the term *Domain Level* Ontology*?*
Describe the main characteristics and capabilities of the Domain Level Ontology.
What is meant by the term *Ontology Taxonomy?*
Describe the main characteristics and capabilities of the Ontology Taxonomy.
What is meant by the term *Taxonomy of Malware* Classes*?*
Describe the main characteristics by using an example for Taxonomy of Malware
Classes.
What is meant by the term *Ontology Framework?*
Describe the main components of a Cybersecurity Ontology Framework by using an
example.
What is meant by the term *Cyber Attack Model Ontology?*
Describe the main characteristics by using an example for the Cyber Attack Model
Ontology.

References

1. Falliere, N., Murchu, L., Chien, E.: W32 Stuxnet Dossier, 2011
2. Schenato, L.: To Zero or to Hold Control Inputs with Lossy Links?. In: IEEE Transaction on Automatic Control, Vol. 54, No. 5, pp. 1093–1099, 2009
3. McHugh, J.: Intrusion and Intrusion Detection. Int. J. Info. Syst. Vol. 1, pp. 14–35, 2001. DOI 10.1007/s102070100001 (accessed 12.2022)
4. Kumar, S., Spalfrd, E.H.: An Application of Pattern Matching in Intrusion Detection. In: Computer Science Technical Report, Paper 126, Purdue University, 1994
5. Gordon, S., Ford, R.: On the Definition and Classification of Cybercrime. In. Journal in Computer Virology, Vol. 2, No. 1, pp. 13–20, 2006
6. Corman, J., Etue, D.: Adversary ROI: Evaluating Security from the Threat Actor's Perspective, 2012
7. Heckman, R.: Attacker Classification to aid Targeting Critical Systems for Threat Modelling and Security Review, 2005. www.rockyh.net/papers/AttackerClassification.pdf. (Accessed 12.2022)
8. Cardenas, A.A., Amin, S.M., Sinopoli, B., Giani, A., Perrig, A., Sastry, S.S.: Challenges for Securing Cyber Physical Systems. In: Workshop on Future Directions in Cyber-physical Systems Security. DHS, 2009
9. Cardenas, A.A., Roosta, T., Sastry, S.: Rethinking Security Properties, Threat Models, and the Design Space in Sensor Networks: A Case Study in SCADA Systems. In: Ad Hoc Networks, Vol. 7, No. 8, pp. 1434–1447, 2009
10. LeMay, F., Ford, M.D., Keefe, K., Sanders, W.H., Muehrcke, C.: Model-based Security Metrics using Adversary View Security Evaluation (ADVISE). In: Proceedings of Conference on Quantitative Evaluation of Systems, QEST, 2011

11. Denning, D.E.: Activism, Hacktivism, and Cyberterrorism: The Internet as a Tool for Influencing Foreign Policy. In: Networks and Netwars: The Future of Terror, Crime, and Militancy. RAND Corporation, 2001
12. Rocchetto, M., Tippenhauer, N.O.: On Attacker Models and Profiles for Cyber-Physical Systems. In: Lecture Notes in Computer Science, Vol. 9879, pp. 467–469, Springer Publ. 2016
13. Ottis, R.: Theoretical Model for Creating a Nation-State Level Offensive Cyber Capability. In: European Conference on Information Warfare and Security, 2009
14. Department of Homeland Security's Role in Critical Infrastructure Protection Cybersecurity, GOA-05-434, 2005
15. https://ics-cert.us-cert.gov/content/cyber-threat-source-descriptions (Accessed 12.2022)
16. Jaishankar, K.; Cyber Criminology: Explorih Internet Crimes and Criminal Behavior. CRC Press, 2022
17. Sabillon, R., Cano, J., Cavaller, V., Serra, J.: Cybercrime and Cybercriminals: A Comprehensive Study. In: International Journal of Computer Networks and Communications Security, Vol. 4, No., pp. 165–176, 2016
18. Garcia, N.: The Use of Criminal Profiling in Cybercrime Investigations. In: ProcQuest, pp. 1–47, 2018
19. The Art of Cybercriminal Profiling. In: UK Essays, 2018. https://www.ukessays.com/essays/criminology/the-art-of-c<ybercriminal-profiling-7922.php?vrref=1 (Accessed 12.2022)
20. Wariko, A.: Proposed Methodology for Cyber Criminal Profiling. In: Information Security Journal: A Global Perspective, Vol. 23, No. 4–6, pp. 172–178, 2014. doi:10.1080/19393555.2014.931491 (Accessed 12.2022)
21. Fedushko, S., Bardyn, N.: Algorithm of the Cyber Criminals Identification. In: Global Journal of Engineering, Design and Technology, Vol. 2, No. 4, pp. 56–62, 2013
22. Kocsis, R.N.: Applied Criminal Psychology: A Guide to Forensic Behavioral Siences. In: Charles C. Thomas Publ., 2018
23. Rouse, M.: What is Computer Forensic (Cyber Forensic)? Definition from Whats.com. https://www.techtarget.com/searchsecurity/definition/computer-forensic (Accessed 12.2022)
24. McKlusky, Q.R., Chowdhury, M., Latif, S., Kambhampaty, K.: Computer Forensics: Complementing Cyer Security. In: Proceedings IEEE-IET 2022 International Conference, pp. 507–512, 2022
25. https://www.crime-scene-investigator.net/computer-forensics-digital-forensic-analysis-methodology.html (Accessed 12.2022)
26. Möller, D.P.F.: Mathematical and Computational Modeling and Simulation: Fundamentals and Case Studies. Springer Publ., 2005
27. Möller D.P.F.: Introduction to Transportation Analysis. Modeling and Simulation: Computational Foundations and Multimodal Applications. Springer Publ. 2014
28. Lin, X., Zavarsky, P., Ruhl, R., Lindskog, D.: Threat Modeling for Cross Site Request Forgery (CRSF) attacks. In: Proceedings IEEE 16th International Conference of Computational Science and Engineering, Vol. 13, pp. 486–491, 2009
29. Phillips, C., Swier, L.P.: A Graph-based System for Network-Vulnerability Analysis. In: Proceedings Workshop on New Security Paradigms, pp. 71–79, 1998; http://doi.acm.org/10.1145/310889.310919 (Accessed 12.2022)
30. Schneier, B.C.: Attack Trees. In: Dr. Doobs Journal, Vol. 24, No. 12, pp.21–29, 1999
31. Mulazzani, M., Schrittwieser, S., Leithner, M., Huber, M., Weippl, E.R.: Dark Clouds on the Horizon: Using Cloud Storage as Attack Vector and Online Stack Space. In: UNISiX Security Symposium, pp. 65–76, 2011
32. Mandhala, K.P., Wing, J.M.: An Attack Surface Metric. In: IEEE Transactions on Software Engineering, Vol. 37, No. 3, pp.371–386, 2011
33. Jemili, F., Zaghdoud, M., Ahmed, M.B.: A Framework for Adaptive Intrusion Detection System Using Bayesian Networks. 2007. https://www.researchgate.net/publication/4256770 (Accessed 12.2022)

34. Callagirone, S., Pendergast, A.: Betz, C.: The Diamond Model of Intrusion Analysis", DTIC Document, Technical Report, 2013
35. Joint Tactics, Techniques, and Procedures for Joint Intelligence Preparation of the Battlefield, U. S. Joint Chiefs of Staff, 2000
36. Hutchins, E.M., Cloppert, M.J., Amin, R.M.J.: Intelligence-driven Computer Network Defense Informed by Analysis of Adversary Campaigns and Intrusion Kill Chains. In: Leading Issues in Information Warfare and Security Research, Vol. 1, pp. 80 ff, 2011
37. Jasiul, B., Szypyrka, M., Sliw, J.: Detection and Modeling of Cyber Attacks with Petri Nets. In: Entropy 2014, Vol. 16, pp. 6602–6623; doi: 10.3390/e16126602 (Accessed 12.2022)
38. Al-Mohannadi, H., Mrza, Q., Namanaya, A., Awan, I., Cullen, A., Disso, J.: Cyber-Attack Modeling Analysis Techniques: An Overview. In: Proceedings 4th International Conference on Future Internet of Things and Cloud Workshops, pp. 69–76, 2016
39. Bodeau, D.J., McCollum, C.D., Fox, D.B.: Cyber Threat Modeling Survey: Assessment and Representative Framework. Homeland Security Systems Engineering and Development Institute, 2018
40. OWASP Top 10 Application Security Risks-2017, 2017. https://www.owasp.org/index.php/Top_10-2017_Top_10 (Accessed 12.2022)
41. Idrees,S., Roudier, Y., Apvrille, L.: Model the System from Adversary Viewpoint: Threats Identification Modeling. In: J. Garcia-Alfana, G. Gür (Eds.) Intrusion and Prevention Workshop, pp. 45–57, 2014
42. Texeira, A., Perez, D., Sandberg, H., Johansson, K.H.: Attack Models and Scenarios for Networked Control Systems. In: Proceedings ACM HiCoNss, pp. 55–63, 2012
43. Mouratidis, H. Giorgini, P., Manson, G.: Using Security Attacks Scenarios to Analyze Security during Information Systems Design. http://dit.unitn.it/~pgiorgio/papers/ICEIS04.pdf (Accessed 12.2022)
44. Kotenko, I., Doynikova, D.: The CAPEC based Generator of Attack Scenarios for Network Security Evaluation, In: Proceedings IEEE 8th International Conference on Intelligent Data Acquisition and Advanced Computing Systems, pp. 436–441, IEEE Publ., 2015.
45. Wang, B., Chai, J., Zhang, S.: A Network Security Assessment Model-based based on Attack Defense Game Theory. In: Proceedings IEEE International Conference on Computer Application and System Modeling, pp, 634–639, IEEE Publ. 2010
46. Jin, X., Dan, M., Zhang, N., Yu, W., Fu, X., Das, S.: Game Theory for Infrastructure Security: The Power of Intent-Based Adversary Model. In: Handbook on Securing Cyber-Physical Critical Infrastructure: Foundations and Challenges, S.K. Das, K. Kant., N. Zhang (Eds.), pp. 31–53, Elsevier Publ., 2012
47. Do, Q., Martini, B., Choo, K.-K.R.: The Role of the Adversary Model in Applied Security Research. In: Computers and Security, pp. 156–181, 2018. https://eprint.iacr.org/2018/1189.pdf (Accessed 12.2022)
48. McKemmish, R.: When is Digital Evidence Forensically Sound? In: Advances in Digital Forensics IV, I. Ray, S. Shenoi (Eds.), pp. 3–15, Springer Publ., 2008
49. Liu, P., Zhang, W., Yu, M.: Incentive-based Modeling and Inference of Attacker Intent, Objectives, and Strategies. In: ACM Transactions on Information and System Security, Vol. 8, No. 1, pp. 78–118, 2005
50. Myagmar, S., Lee, A. J., Yurcik, W.: Threat Modeling as a Basis for Security Requirements. In: Symposium ion Requirements Engineering for Information Security, pp. 1–8, 2005
51. Grunewald, D., Lützenberger, M., Chinnow, J.: Agent-based Network Security Simulation. In: Proceedings 10th International Conference on Autonomous Agents and Multiagent Systems, pp. 1325–1326, (Ed.:) International Foundation for Autonomous Agents and Multiagent System, 2011
52. Moskal, S., Wheeler, B., Kreider, D.: Context Model Fusion for Multistage Network Attack Simulation. In: Proceedings IEEE Military Communications Conference, pp.158–163, IEEE Publ., 2014
53. U.S. Department of Homeland Security – Cybersecurity Strategy, 2018

54. Darwin, J.T.: Cyber health and Informal Wellbeing. PhD Thesis at University of Darwin, 2019
55. Chi, S.-D., Park, J.S., Jung, K.-C., Lee, J.-S.: Network Security Modeling and Cyber Attack Simulation Methodology. In: Information Security and Privacy, Varadharvaran, V., Mu, Y. (Eds.), pp. 320–333, Lecture Notes in Computer Science, Vol. 2119, Springer Publ., 2001
56. Moskal, S.F.: Knowledge-based Decision Making for Simulation Cyber Attack Behaviors. PhD Thesis at Rochester Institute of Technology, 2016
57. Moskal, S.F., Yang, S.J., Kuhl, M.H.: Cyber Threat Assessment via Attack Scenario Simulation using an Integrated Adversary and Network Modeling Approach, In: Journal of Defense Modeling and Simulation, pp. 13–29, 2017
58. Uschold, M.. Knowledge Level Modeling: Concepts and Terminology. In: The Knowledge Engineering Review, Vol. 13, pp. 5–29, 1998
59. Chandrasekaran, B., Josephson, J.R., Benjamins, V.R.: The Ontology of Tasks and Methods. In: Proceedings 11th Banff Knowledge Acquisition for Knowledge for Knowledge-based System Workshop, 1998
60. Sadbolt, N., Hara, K.O., Cottam, C.: The Use of Ontologies for Knowledge Acquisition. In: Knowledge Engineering and Agent Technology, J. Cuena, Y. Demazeau, A.G. Serrano, J. Treut (Eds.), pp.19–42, IOS Press, 2004
61. Sheth, A.: Can Semantic Web Techniques empower Comprehension and Projection in Cyber Situational Awareness. ARO Workshop, 2007
62. Undercoffer, J., Pinkston, J., Joshi, A., Finn, T.: A Target-centric Ontology for Intrusion Detection. In 18th International Joint Conference on Artificial Intelligence, pp. 9–15, 2004
63. Bechhofer, S.: OWL: Web Ontology Language. In: Encyclopedia of Database Systems, L. Liu, M. T. Özsu (Eds.), Springer Publ., 2009. doi: 10.1007/978-0-387-39940-9_1073
64. https://www.ontotext.com/knowledgehub/fundamentals/what-are-ontologies/   (Accessed 12.2022)
65. MAEC - Malware Attribute Enumeration and Characterization. http://maec.mitre.org/ (Accessed 12.2022)
66. http://maecproject.github.io/about-maec/ (Accessed 12.2022)
67. Obrst, L., Chase, P., Markeloff, R.: Developing an Ontology of the Cyber Security Domain. http://ceur-ws.org/Vol-966/STIDS2012_T06_ObrstEtAl_CyberOntology.pdf   (Accessed 12.2022)
68. Obrst, L.: Ontological Architectures. In: Chapter 2, Part One: Ontology as Technology, in the book TAO - Theory and Applications of Ontology, Volume 2, J. Seibt, A. Kameas, R. Poli (Eds.), Springer Publ. 2010
69. Semy, S., Pulvermacher, M., Obrst, L.: Toward the Use of an Upper Ontology for U.-S. Government and U.S. Military Domains: An Evaluation. In: MITRE Technical Report, MTR 04B0000063, 2005
70. Swimmer, M.: Towards an Ontology of Malware Classes. http://www.scribd.com/doc/24058261/Towards-an-Ontology-of-Malware-Classes (Accessed 12.2022)
71. IEEE-SA – Industry Connections. http://standards.ieee.org/develop/indconn/icsg/malware.html (Accessed 12.2022)
72. MANDIANT: Intelligent Information Security. http://www.mandiant.com (Accessed 12.2022)
73. Zeltser, L.: Categories of Common Malware Traits. In: Internet Storm Center Handler's Diary, 2009. http://isc.sans.edu/diary.html?storyid=7186 (Accessed 12.2022)
74. More, S., Matthews, M., Joshi, A., Finn, T.: A Knowledge-based Approach to Intrusion Detection Modeling. In: Proceedings IEEE Symposium on Security and Privacy Workshops, pp. 75–81, 2012
75. https://attack.mitre.org (Accessed 12.2022)
76. Kotenko, I., Doynikova, D.: The CAPEC based Generator of Attack Scenarios for Network Security Evaluation. In: Proceedings IEEE 8th International Conference on Intelligent Data Acquisition and Advanced Computing Systems, pp. 436–441, IEEE Publ., 2015
77. https://cwe.mitre.org/cwss/cwss_v1.0.1.html (Accessed 12.2022)
78. https://dl.acm.org (Accessed 12.2022)

79. Kokkimakis, D.: Semantic Relations of Binary Compounds annotated with SBOMED CT. In: Studies in Health Technology and Informatics. Vol. 180, pp. 169–173. 2012. doi: 10.3233/978-1-61499-101-4-169 (Accessed 12.2022)
80. Möller, D.P.F.: Guide to Computing Fundamentals in Cyber-Physical Systems: Concepts, Design Methods, and Application. Springer Publ. 2016
81. Zhai, J., Zhon, Z., Shi, Z., Shen, L.: An Integrated Information Platform for Transportation Systems based on Ontology. In: IFIP Vol. 254, Research and Practical Issues on Enterprise Information Systems, pp. 787–796, I. Xu,A. Toja, S. Chaudhary, Springer Publ.2007
82. Möller, D.P.F.: Cybersecurity in Digital Transformation: Scopes and Applications. Springer Nature, 2020

Chapter 5
NIST Cybersecurity Framework and MITRE Cybersecurity Criteria

Abstract Today cyberattacks continue to evolve and are highly complex. They are also very expensive by the average cost of a breach-in cyberattack. The top ten most common cyberattack intrusion incidents for industrial, public, and private organizations are phishing attacks, negligent and malicious insiders, advanced persistent threats, zero day attacks, denial of service attacks, software vulnerabilities, social engineering attacks, and brute force attacks. Therefore, cybersecurity becomes an essential issue that generally focuses on the measures to protect valuable data, information, and business assets from malicious threat events that affect confidentiality, integrity, and availability of information. In this regard, it is vitally important that computer systems, networks and network-connected devices, infrastructure resources, and others stay up-to-date with current software operating systems, patches, and releases. Therefore, organizations need to institute policies and procedures that enforce the way their user's access information and interact with network or system resources. Here the NIST Cybersecurity Framework and the MITRE Cybersecurity Criteria come into play. The NIST Cybersecurity Framework is a set of best practices, standards, and recommendations that support organizations to improve their cybersecurity measures. It focusses on using business drivers to guide cybersecurity activities and considering cybersecurity risks as part of the organizations cybersecurity risk management. In this regard, the framework provides a common organizing structure for multiple cybersecurity approaches by assembling standards, guidelines, and practices that are working effectively today. The MITRE Cybersecurity Criteria enable a collective response against cybersecurity threat events, worked out in conjunction with industry and government authorities. It describes the common tactics, techniques, and procedures of advanced persistent threats against organizations' computer systems and networks and was later expanded to industrial control systems. In this regard, the MITRE Cybersecurity Criteria are fully committed to defending and securing cyber-ecosystems. NIST's and MITRE's goal is to develop cyber resiliency approaches and controls to mitigate malicious cyberattacks. Cyber resiliency enables anticipating, withstanding, recovering from and adapting to adverse conditions, stresses, cyberattacks, or compromises on computer systems, networks, infrastructure resources, and others. Against this background, this chapter introduces in Sect. 5.1 the NIST Cybersecurity Framework (NIST CSF) with their manifold possible uses and their great impact improving industrial, public, and private organizations' cybersecurity needs. Therefore, Sect. 5.1 introduces the process of cybersecurity risk management. Since NIST CSF is one of the most relevant cybersecurity frameworks, Sect. 5.1 introduces the NIST Cybersecurity Framework. Section 5.1.1 introduces CIS Critical Security Controls, Sect. 5.1.2 ISA/IEC 62443 Cybersecurity Standard, Sect. 5.1.3 MITRE Adversarial Tactics, Techniques, and Common Knowledge, Sect. 5.1.4 NIST 800-653, and in Sect. 5.1.5, the NIST Cybersecurity Framework. Section 5.2 focuses on the NIST Cybersecurity Framework for Critical Infrastructure and focuses in Sect. 5.2.1 on a NIST CSF Critical Infrastructure best practice use case, making use of a model approach in cybersecurity maturity. Against this background, Sect. 5.3 focusses on the MITRE Cybersecurity Criteria that provides a common taxonomy of Tactics, Techniques, and Procedures, applicable to defend cyberattacks, to withstand cyberattackers activities like unauthorized interaction with organizations' computer systems, networks, and infrastructure resources, to recover from potential malicious cyberattacks. Section 5.4 introduce the MITRE Cybersecurity Taxonomy, which refers to cyberattack possibilities and how to conquer them. Section 5.5 contains comprehensive questions on the topics of NIST Cybersecurity Framework and MITRE Cybersecurity Criteria. Finally, "References" refers to the used references for further reading.

5.1   Cybersecurity Frameworks

The National Institute of Standards and Technology (NIST), a division of the US Department of Commerce, developed the NIST Cybersecurity Framework (NIST CSF). NIST promotes and maintains measurement standards with active programs to advance innovation and cyber-secure industries. In this regard, NIST CSF is one of the several cybersecurity frameworks, along with CIS CSC, ISA/IEC 62443, MITRE Cybersecurity Criteria, and NIST 800-53, used in the cybersecurity domain to set maturity standards for cybersecurity.

5.1.1   CIS Critical Security Controls

The Center for Internet Security Critical Security Controls (CIS CSC) for Effective Cyber Defense, now called CIS Critical Security Controls (in short CIS Controls), published best-practice guidelines for computer security. This publication was initiated early in 2008, responding to experienced extreme data losses, developed by the SANS Institute, a private US for-profit company, founded in 1989. The SANS Institute specializes in information security, cybersecurity training, and certificates. However, ownership was transferred in 2013 to the Council on Cyber Security, and thereafter, in 2015, to the Center for Internet Security, a non-profit organization. Originally known as the Consensus Audit Guidelines, it is also known as the CIS

CSC, CIS20, CCS CSC, SANS Top 20, or CAG 20 [1]. CIS Controls or CIS 20 is a framework developed to tackle growing cybersecurity risks. The framework outlines 20 security controls that help organizations to safeguard their valuable data and assets from known attack vectors. The CIS security controls are a number of prioritized, well-vetted and supported security actions that organizations can take to assess and improve their current security state. Nevertheless, it is not a one-size-fits-all solution, in either content or priority. The new CIS CSC release, CIS CSC v8, includes cloud and mobile technologies [2]. Although the CIS Controls are not a replacement for any existing compliance scheme, the controls map to several compliance frameworks, e.g., NIST CFS, and regulations, e.g., PCI DSS and HIPAA. The 20 CIS CSC controls, shown in Table 5.1, are based on the latest information about common attacks. They reflect the combined knowledge of commercial forensic experts, individual penetration testers, and contributors from the US government agencies [3].

Getting value from the CIS CSC does not necessarily mean implementing all 20 CIS Controls at once. A more pragmatic approach to implementing the CIS Controls includes the following steps [3]:

- Discover the organization's information assets and estimate their value. Perform risk assessment and think through potential cyberattacks against the organization's systems and data (including initial entry points, spread, and damage). Prioritize the CIS Controls around the organization's riskiest assets.
- Compare the organization' current security controls to the CIS Controls. Make note of each area where no security capabilities exist or where additional work is needed.
- Develop a plan for adopting the most valuable new security controls and improving the operational effectiveness of the organization's existing controls.
- Obtain management buy-in for the plan and form line-of-business commitments for necessary financial and personnel support.
- Implement the controls. Keep an eye on trends that could introduce new risks to the organization. Measure progress and risk reduction, and communicate findings.

For the CIS CSC, there is no CIS Security Certification process or legal requirements for an organization to apply them. However, if organizations follow them, cybersecurity will be strengthened. Therefore, CIS organized them into three tiers by order of importance:

- Basic: CIS Controls 1–6
- Foundational: CIS Controls 7–16
- Organizational: CIS Controls 17–20

The CIS CSC Controls differentiates itself from other frameworks by offering securely configured settings for an extensive list of operating systems (OS) and devices, known as CIS benchmarks. The CIS benchmarks are the only consensus-developed security configuration recommendations both created and trusted by a global community of IT security professionals from academia, government, and

Table 5.1 CIC CSC controls

Control	Control topic	Control issues	What is critical?
1	Inventory of authorized and unauthorized devices	Organizations must actively manage all hardware devices on the network, that only authorized devices are given access and unauthorized devices can be quickly identified and disconnected before they inflict any harm	Cyberattackers continuously scanning the address space of organizations, waiting for new and unprotected systems attached to the network. This control is especially critical for organizations that allow Bring Your Own Device (BOYD), since hackers are specifically looking for devices that come and go off the enterprise's network
2	Inventory of authorized and unauthorized software	Organizations must actively manage all software on the network and ensure that only authorized software is installed. Security measures like application whitelisting can enable organizations quickly find unauthorized software before it has been installed	Cyberattackers look for vulnerable versions of software that they can remotely exploit. They can distribute hostile web pages, media files, and other content, or use zero-day exploits that take advantage of unknown vulnerabilities. Thus proper knowledge of what software has been deployed in the organization is essential for data security and privacy
4	Continuous vulnerability assessment and remediation	Organizations need to continuously acquire, assess and take action on new information, e.g., software updates, patches, security advisories, and threat event bulletins to identify and remediate vulnerabilities attackers could otherwise use to penetrate their networks	As soon as researchers report new vulnerabilities, a race starts among all relevant parties: Culprits strive to use the vulnerability for a cyberattack, vendors deploy patches or updates, and defenders start performing risk assessments or regression testing. Attackers have access to the same information as everyone else, and can take advantage of gaps between the appearance of new knowledge and remediation
5	Controlled use of administrative privileges	This control requires organizations to use automated tools to monitor user behavior and keep track of how administrative privileges are assigned and used to prevent unauthorized access to critical systems	The misuse of administrative privileges is a primary method for cyberattackers to spread inside an organization. To gain administrative credentials, they use phishing techniques, crack or guess the password for an administrative user, or elevate the privileges of a normal user account

(continued)

Table 5.1 (continued)

Control	Control topic	Control issues	What is critical?
			into an administrative account. If organizations do not have resources to monitor what's going on in their IT environments, it is easier for attackers to gain full control of their systems
6	Maintenance, monitoring, and analysis of audit logs	Organizations need to collect, manage, and analyze event logs to detect aberrant activities and investigate security incidents	Lack of security logging and analysis enables cyberattackers to hide their location and activities in the network. Even if the targeted organization knows which systems are compromised, without complete logging records, it will be difficult for them to understand what a cyberattacker has done so far and respond effectively to the security incident
7	Email and web browser protection, and analysis of audit logs	Organizations need to ensure that only fully supported web browsers and email clients are used in the organization in order to minimize their cyberattack surface	Web browsers and email clients are very common points of entry for hackers because of their high technical complexity and flexibility. They can create content and spoof users into taking actions that can introduce malicious code and lead to loss of valuable data and business assets
8	Malware defenses	Organizations need to make sure they can control the installation and execution of malicious code at multiple points in the organization. This control recommends using automated tools to continuously monitor workstations, servers, and mobile devices with anti-virus, anti-spyware, personal firewalls, and host-based intrusion prevention system functionality	Modern malware can be fast-moving and fast-changing, and it can enter through any number of points. Therefore, malware defenses must be able to operate in this dynamic environment through large-scale automation, updating, and integration with processes like incident response
9	Limitation and control of network ports, protocols, and services	Organizations must track and manage the use of ports, protocols, and services on network devices to minimize the windows of vulnerability available to attackers	Cyberattackers search for remotely accessible network services that are vulnerable for exploitation. Common examples include poorly configured web servers, mail

(continued)

Table 5.1 (continued)

Control	Control topic	Control issues	What is critical?
			servers, file and print services, and domain name system (DNS) servers that are installed by default on a variety of devices. Therefore, it is critical to make sure that only ports, protocols, and services with a validated business need are running on each system
10	Data recovery capability	Companies need to ensure that critical systems and data are properly backed up on at least a weekly basis. They also need to have a proven methodology for timely data recovery	Cyberattackers often make significant changes to data, configurations and software. Without reliable backup and recovery, it is difficult for organizations to recover from an cyberattack
11	Secure configurations for network devices	Organizations must establish, implement, and actively manage the security configuration of network infrastructure devices, such as routers, firewalls, and switches	Just as with operating systems and applications, the default configurations for network infrastructure devices are geared for ease of deployment, not security. In addition, network devices become often less securely configured over time. Attackers exploit these configuration flaws to gain access to networks or use a compromised machine to pose as a trusted system
12	Boundary defense	Organizations need to detect and correct the flow of information between networks of different trust levels, with a focus on data that could damage security. The best defense is technologies that provide deep visibility and control over data flow across the environment, such as intrusion detection and intrusion prevention systems	Culprits often use configuration and architectural weaknesses on perimeter systems, network devices, and internet-accessing client machines to gain initial access into an organization's network
13	Data protection	Organizations must use appropriate processes and tools to mitigate the risk of data exfiltration and ensure the integrity of sensitive information. Data protection is best achieved through the combination of encryption, integrity	While many data leaks are deliberate theft, other instances of data loss or damage are the result of poor security practices or human errors. To minimize these risks, organizations need to implement solutions that can

(continued)

Table 5.1 (continued)

Control	Control topic	Control issues	What is critical?
		protection, and data loss prevention techniques	help detect data exfiltration and mitigate the effects of data compromise
14	Controlled access based on need to know	Organizations need to be able to track, control, and secure access to their critical assets, and easily determine which people, computers, or applications have a right to access these assets	Some organizations do not carefully identify and separate their most critical assets from less sensitive data, and users have access to more sensitive data than they need to do their jobs. As a result, it is easier for a malicious insider or a cyberattacker or malware that takes over their account to steal important information or disrupt operations
15	Wireless access control	Organizations need to have processes and tools in place to track and control the use of wireless local area networks (LANs), access points, and wireless client systems. They need to conduct network vulnerability scanning tools and ensure that all wireless devices connected to the network match an authorized configuration and security profile	Wireless devices are a convenient vector for attackers to maintain long-term access into the IT environment, since they do not require direct physical connection. For example, wireless clients used by employees as they travel are infected on a regular basis and later used as back doors when they reconnected to the organization's network
16	Account monitoring and control	It is critical for organizations to actively manage the lifecycle of user accounts (creation, use, and deletion) to minimize opportunities for attackers to leverage them. All system accounts need to be regularly reviewed, and accounts of former contractors and employees should be disabled as soon as the person leaves the company	Attackers frequently exploit inactive user accounts to gain legitimate access to an organization's systems and data, which makes detection of the attack more difficult
17	Security skills assessment and appropriate training to fill gaps	It is critical for organizations to actively manage the lifecycle of user accounts (creation, use, and deletion) to minimize opportunities for cyberattackers to leverage them. All system accounts regularly need be reviewed, and accounts of former contractors and employees should	It is tempting to think of cyber defense as primarily a technical challenge. However, employee actions are also critical to the success of a security program. Cyberattackers often use the human factor to plan exploitations, for example, by carefully crafting phishing messages that look like normal emails,

(continued)

Table 5.1 (continued)

Control	Control topic	Control issues	What is critical?
		be disabled as soon as the person leaves the company	or working within the time window of patching or log review
18	Application software security	Organizations must manage the security lifecycle of all software they use in order to detect and correct security weaknesses. In particular, they must regularly check that they use only the most current versions of each application and that all the relevant patches are installed promptly	Cyberattackers often take advantage of vulnerabilities in web-based applications and other software. They can inject specific exploits, including buffer overflows, SQL injection attacks, cross-site scripting, and click-jacking of code, to gain control over vulnerable machines
19	Incident response and management	Organizations need to develop and implement proper incident response, which includes plans, defined roles, training, management oversight, and other measures that will help them discover attacks and contain damage more effectively	Security incidents are now a normal part of the daily life. Even large and well-funded organizations struggle to keep up with the evolving cyber threat landscape. Sadly, in most cases, the chance of a successful cyberattack is not "if" but "when." Without an incident response plan, an organization may not discover a cyberattack until it inflicts serious harm, or be able to eradicate the cyberattacker's presence and restore the integrity of the network and systems
20	Penetration tests and red team exercises	The final control requires organizations to assess the overall strength of their defenses (the technology, the processes, and the people) by conducting regular external and internal penetration tests. This will enable them to identify vulnerabilities and attack vectors that can be used to exploit systems	Cyberattackers can exploit the gap between good defensive intentions and their implementation, such as the time window between the announcement of vulnerability, the availability of a vendor patch, and patch installation. In a complex environment, where technology is constantly evolving, organizations should periodically test their defenses to identify gaps and fix them before an attack occurs

industry [4]. There are over 12,000 professionals in the CIS benchmarks communities, creating CIS benchmark recommendations.

5.1.2 ISA/IEC 62443-Cybersecurity Standard

For a long time, the areas of information technology (IT) and operational technology (OT) were considered separately. In industrial organizations, these areas are mostly assigned to different departments. However, the increasing degree of automation in industrial production and the increasing digitization of all areas of business through the digital transformation ensure that these two areas are growing closer together. The connecting element between the two areas is cybersecurity. Nevertheless, networking in the field of sensors and actuators has been indispensable in industry for years, and industrial systems have been isolated and separated from the office networks. But this has changed completely in recent years: industrial networks have been opened up and connected to other IT components. There is a trend toward connecting cloud infrastructures in order to use them to open up new business models. Indeed, networking leads to openness and entails the risk of misuse by unauthorized persons. In addition to a digitization strategy, every industrial organization, therefore, also needs a strategy to secure their production processes and their own know-how to secure their business model. Therefore, the IEC 62443 standard for industrial security was defined. The standard contains an established, implementable process model for an industrial cybersecurity strategy and is therefore clearly superior to uncoordinated individual activities. The main parts of the standard, published since the beginning of 2019, enable users of the standard sufficient security planning. In the case of industrial security, it must always be remembered that the availability of an OT system has top priority, directly followed by the integrity of the respective production processes. In contrast to classic IT, confidentiality in OT is often of lesser importance and only becomes relevant when intellectual property rights are considered. In this regard, the International Standard Organization (ISO) standard ISO 27000 series of standards define the protection goals for IT security, with the focus on the confidentiality of the information described by the CIA Triad. For the OT area, the main concern is the availability of the embedded automation systems, described by the inverse notation of the CIA Triad as AIC Triad. Against this background, the International Society for Automation (ISA) developed consensus-based industrial standards on automation. Thus, the ISA-99 standards development committee brings together industrial cybersecurity experts from across the globe to develop standards on industrial automation and control systems (IACS) security, following the AIC Triad paradigm with the primary concern of availability of the embedded automation control systems. This original and ongoing work on ISA-99 being adopted by the International Electrotechnical Commission (IEC) in creating the ISA/IEC 62443 standard, a non-regulated standard, is a consensus-based wide-ranging collection of multi-industry standards for the secure development of Industrial Automation and Control Systems (IACS). This standard describes the relevant OT requirements to avoid security risks for operators, integrators, and device manufacturers by defining a set of cybersecurity protection methods and techniques to defend industrial systems and networks against cybersecurity threat events. For this purpose, it categorizes these techniques to apply to all stakeholders, including manufacturers, asset owners, and suppliers. In this regard, the ISA/IEC 26443 standard addresses cybersecurity issues unique to industrial automation and control systems (IACS) and OT. The standard body regularly consults IACS security experts to maintain relevant guidance that applies to multi-industries and critical infrastructure. In this regard, ISA/IEC 62443 provides a series of requirements and methods to manage cybersecurity challenges in IACS and industrial environment, including, but not limited to,

- The relative criticality of data availability of industrial systems, operations, or functions
- Potential dangers to personnel, the environment, and society in the event of cyber-physical systems failures
- The increased need for compensating controls to protect legacy IACS/OT systems
- The relative difficulty of applying common IT security techniques, described in the CIA Triad, without severe systems modifications
- Prospects for financial loss due to a cyberattack incident-related drop-down in productivity
- Unique approaches to ensuring computer technological systems availability, reliability, and integrity in industrial environments

In this context, the ISA/IEC 62443 standard is a family of documents that illustrates a defense-in-depth model via zones and conduits, describing how to build a cybersecurity management system, and offers instructions for performing risk assessments in IACS/OT environments. Therefore, the IEC 62443 supports organizations defining IACS security maturity and posture and offers selection criteria of security products, programs, and service providers, routinely supplemented with technical reports, covering specific technological situations and solutions. The ISA/IEC 62443 standard consists of four categories, each of which is numbered consecutively from ISA/IEC 62443-1 to ISA/IEC 62443-4:

- *First Category*: Explains the commonly used terms, abbreviations, vocabularies, concepts, and example use cases.
- *Second Category*: Describes the safety management with the important program requirements, e.g., patch management, implementation guidance, and others. This part mainly affects plant operators and, to a certain extent, service providers.
- *Third Category*: Represents the functional requirements for a system and explains the underlying technologies, e.g., assessment approaches, authentication procedures, security requirements levels and technologies. In addition, it is defined here how the safety of a system can be assessed. This part is particularly relevant for plant construction, but also for component manufacturers.

- *Fourth Category*: Describes requirements for components and their development that refer to secure development of components products used in the make-up IACS. Examples of devices that might be compliant include PLCs, HMI displays, three-term controllers, and components of Distributed control systems (DCS), as well as more obvious networking components such as Ethernet switches.

The key practices identified by ISA/IEC 62443 are

- Security management
- Specification of security requirements
- Design by security
- Secure implementation
- Security verification and validation testing
- Management of security-related issues
- Security update management
- Security guidelines

Furthermore, the ISA/IEC 62443 standard classifies the technical requirements for products developed in accordance with the standard and the systems they contribute to assigned Security Levels (SL) according to their resistance against different cyberattack types. ISA/IEC 62443 states these levels assigned to each requirement, rather than to complete products. In this regard, the security levels are [5]:

- *Security Level 0*: No special protection
- *Security Level 1*: Protection against unintentional misuse
- *Security Level 2*: Protection against unsophisticated intentional misuse by aggressors with few resources, general skills, and little motivation
- *Security Level 3*: Protection against sophisticated intentional misuse by aggressors with moderate resources, some with specialist IACS knowledge, and moderate motivation
- *Security Level 4*: Protection against highly sophisticated intentional misuse by aggressors with extensive resources, considerable IACS-specific knowledge, and high motivation

Security requirements and security levels are strongly related. The more demanding the security levels, the more demanding the security requirements will be.

Furthermore, the ISA/IEC 62443 maturity levels for processes based on those from the Capability Maturity Model Integration (CMMI) framework, which supports organizations, streamlines process improvement, encouraging a productive, efficient culture that decreases risks in software, product, and service development. The CMMI also has the capability levels to appraise an organization's performance and process improvement as it applies to an individual practice area outlined in the CMMI model. The CMMI model supports to bring structure to process and performance improvement, whereby each level builds on the previous level. The capability levels are:

- Capability Level 0 – Incomplete: Inconsistent performance and an incomplete approach to meeting the intent of the practice area.
- Capability Level 1 – Initial: The phase where organizations start to address performance issues in a specific practice area, but there is not a complete set of practices in place.
- Capability Level 2 – Managed: Progress is starting to show and there is a full set of practices in place that specifically address improvement in the practice area.
- Capability Level 3 – Defined: There is a focus on achieving project and organizational performance objectives and there are clear organizational standards in place for addressing projects in that practice area.

In the context of the ISA/IEC 62443 maturity level certification, all process-related requirements practiced in their entirety during product development or integration. Thus, the maturity levels are as follows [5]:

- *Maturity Level 1 – Initial level*: Ad hoc and not fully documented.
- *Maturity Level 2 – Managed*: Development managed using documented, repeatable processes. Personnel are sufficiently expert, trained, and/or follow written procedures.
- *Maturity Level 3 – Defined/Practiced*: The process is repeatable throughout the supplier's organization. Personnel are demonstrably well *practiced*.
- *Maturity Level 4 – Improving*: Appropriate process metrics to monitor the effectiveness and performance of the process and demonstrate continuous improvement in these areas.
- *Maturity Level 5 – Optimizing*: Organizations processes are stable and flexible. At this final stage, an organization will be in constant state of improving and responding to changes or other opportunities. The organization is stable, which allows for more "agility and innovation" in a predictable environment.

The ISA/IEC 62443 standards do not directly supersede nor replace the ISA-99 and Purdue models. The Purdue reference model, as adopted by ISA-99, is a model for industrial control system network (ICS) segmentation that defines six layers within these networks, the components found in the layers, and logical network boundary controls for securing these networks [6]. The ISA/IEC 62443 standards leverage previous concepts, and divide cybersecurity and management of cybersecurity risk into several areas. These cover not only cybersecurity reference architectures, such as goal-based and the threat event-based approaches [7, 8]. The goal- based approach defines the security goals, and then selects the countermeasures to reach these goals [9]. The threat event-based approach analyzes the threat events and vulnerabilities of the system to be secured and then selects countermeasures mitigating the threat events and vulnerabilities [10], but also guidance for cybersecurity processes, requirements, technology, controls, cybersecurity acceptance with regard to factory testing, product development, security lifecycles, and a cybersecurity management system (CSMS).

The 62443 standards reach beyond ISA-99 in terms of coverage, cybersecurity and modern concepts, but ISA95 and the Purdue models may still have value for organizations that have specific cybersecurity requirements, e.g., when Industrial Internet of Things (IIoT) devices connected directly to the Internet or the cloud.

5.1.3 MITRE Adversarial Tactics, Techniques, and Common Knowledge

The Adversarial Tactics, Techniques & Common Knowledge (ATT&CK)1 project by MITRE is an initiative started in 2015 with the goal of providing a "globally-accessible knowledge base of adversary tactics and techniques based on real-world observations." ATT&CK provides a key capability that many organizations have struggled with in the past. A way to develop, organize, and use a threat-informed defensive strategy that may be communicated in a standardized way across partner organizations, industries, vendors, and products [11].

Tactics and Techniques are the key objectives the ATT&CK project centers around. A technique is a unique method that MITRE or the information security community has identified as being used by attackers to achieve some specific higher-level intrusion goal, or tactic, as shown at the top of each column in Table 5.2, following [11]:

Some examples of tactics include persistence, command and control, and defense evasion. All tactics include an organized list of techniques, which are specific ways of achieving that higher-level objective.

Examples of techniques enumerated with the Persistence tactic include Bootkit, Logon Scripts, and New Service. Each technique has a set of structured data that is associated with it and includes:

- A unique four-digit identifier in the form of T****, such as T1037 for Logon Scripts
- A tactic or tactics with which the technique is associated. A technique can be listed under more than one tactic
- Platforms the technique is applicable to, such as Windows or Linux
- System or permission requirements for cyberattackers to use that technique
- Defense strategies bypassed, such as whitelisting
- Data sources that can identify the use of the technique

Table 5.2 Tactics and techniques layout in ATT&CK after [11]

Initial access	Execution	Persistence	Privilege escalation	Tactics
Drive-by compromise	AppleScript	.bash_profile and .bashrc	Access token Manipulation	Techniques
Exploit public-facing application	CMSTP	Accessibility features	External remote services	
External remote services	Command-line interface	Account manipulation	AppCertDLLs	

Table 5.3 Execution tactic with multiple techniques and sub-techniques

Execution techniques	
Command and scripting interpreter	PowerShell; AppleScript; Windows Command Shell; Bash; VBScript; Python
Exploitation for client execution	
Inter-process communication	Component object model; dynamic data exchange

- Mitigations and detection methods for preventing or identifying the technique a cyberattacker is using

Table 5.3 shows an example of the metadata associated with the Scheduled Task technique, following [11].

Section 5.3 describes the MIITRE Cybersecurity Criteria in detail.

5.1.4   NIST 800-53

The NIST 800-53 provides a catalog of security and privacy control for information systems and organizations to protect organizational operations and assets and others from a diverse set of threat events and cybersecurity risks, including hostile cyberattacks, human errors, structural failures, foreign intelligence entities, privacy risks, and natural disasters. The controls are flexible and customizable and implemented as part of an organization-wide process to manage cybersecurity risks. The controls address diverse requirements derived from mission and business needs, directives, executive orders, guidelines, laws, policies, regulations, and standards. Finally, the consolidated control catalog addresses cybersecurity and privacy from a functionality perspective, i.e., the strength of functions and mechanisms provided by the controls, and from an assurance perspective, i.e., the measure of confidence in the cybersecurity or privacy capability provided by the controls). Addressing functionality and assurance helps to ensure that information technology products, and the systems that rely on those products, are sufficiently trustworthy [12]. Modern information technologies and systems can include a variety of computing platforms, e.g., industrial control systems, general purpose computing systems, cyber-physical systems, super computers, weapons systems, communications systems, environmental control systems, medical devices, embedded devices, sensors, and mobile devices such as smart phones and tablets. These platforms all share a common foundation: computer systems with complex hardware, software, and firmware providing a capability that supports the essential mission and business functions of organizations. Therefore, cybersecurity controls are the safeguards or countermeasures employed within an information technology and system or an organization to protect the confidentiality, integrity, and availability of the information technology and system and its information, and to manage information cybersecurity risk. Privacy controls are the administrative, technical, and physical safeguards employed within a system or an organization to manage privacy risks and to ensure compliance with applicable privacy requirements. Security and privacy controls are selected and implemented to satisfy cybersecurity security and privacy requirements level on a system or organization. Security and privacy requirements are derived from applicable laws, executive orders, directives, regulations, policies, standards, and mission needs to ensure the confidentiality, integrity, and availability of information processed, stored, or transmitted and to manage risks to individual privacy [12].

With an emphasis on the use of trustworthy, cyber-secure information systems and supply chain cybersecurity, it is essential that organizations express their cybersecurity and privacy requirements with clarity and specificity in order to obtain the systems, components, and services necessary for mission and business success. Accordingly, the NIST 800-53 publication provides controls in the System and Services Acquisition (SA) and Supply Chain Risk Management (SR) families that are directed at developers. The scope of the controls in those families includes information system, system component, and system service development and the associated developers, whether the development is conducted internally by organizations, or externally, through the contracting and acquisition processes. The affected controls in the control catalog include SA-8, SA-10, SA-11, SA-15, SA-16, SA-17, SA-20, SA-21, SR-3, SR-4, SR-5, SR-6, SR-7, SR-8, SR-9, and SR-11 [12].

The organization of this special publication is as follows:

- *Chapter one*: Describes the purpose and applicability, the target audience, the organizational responsibilities, the relationships to other publications, revisions, and extensions, and the publication organization
- *Chapter two*: Describes the fundamental concepts associated with security and privacy controls, including:

  - The structure of the controls
  - How the controls are organized in the consolidated catalog
  - Control implementation approaches
  - The relationship between security and privacy controls, and trustworthiness and assurance

- *Chapter three*: Provides a consolidated catalog of security and privacy controls, including a discussion section to explain the purpose of each control and to provide useful information regarding control implementation and assessment, a list of related controls to show the relationships and dependencies among controls, and a list of references to supporting publications that may be helpful to organizations
- *References, glossary, acronyms, and control summaries*: Provide additional information on the use of security and privacy controls

5.1.5 NIST Cybersecurity Framework

The NIST CSF is the cybersecurity framework (CSF) created by the National Institute of Standards and Technology (NIST), a division of the US Department of Commerce, was established to [13]

- Identify security standards and guidelines applicable across sectors of critical infrastructure
- Provide a prioritized, flexible, repeatable, performance-based, and cost-effective approach
- Help owners and operators of critical infrastructure identify, assess, and manage cybersecurity risks
- Enable technical innovation and account for organizational differences
- Provide guidance that is technology-neutral and enables critical infrastructure sectors to benefit from a competitive market for products and services
- Include guidance for measuring the performance of implementing the NIST CSF
- Identify areas for improvement that should be addressed through future collaboration with particular sectors and standards-developing organizations

The NIST CSF focuses on using business drivers to guide cybersecurity activities and considering cybersecurity risks as part of the organization's risk management processes. Therefore, NIST CSF consists of three parts [13]:

- *NIST CSF Core*: Refers to a set of cybersecurity activities, outcomes, and information references that are common across sectors and critical infrastructures. Elements of the NIST CSF Core provide detailed guidance for developing individual organizational profiles.
- *NIST CSF Profiles*: Vary for each organization, chosen and optimized depending on the organizations' unique challenges, needs, and opportunities to address different core objectives, supporting an organization to align and prioritize its cybersecurity activities with its business/mission requirements, risk tolerances, and resources.
- *NIST CSF Implementation Tiers*: Specify to what level an organization addresses each of the NIST CSF elements. The Tiers provide a mechanism for organizations to view and understand the characteristics of their approach to managing cybersecurity risk, which will help in prioritizing and achieving cybersecurity objectives.

In this context, the NIST CSF provides a set of control guidelines more targeted at industrial automation and control systems (IACS), rather than pure information systems (IT). Therefore, the NIST CSF provides a common structure for multiple approaches to cybersecurity by assembling standards, guidelines, and practices that are working effectively today. Furthermore, the NISR CSF offers a flexible way to address cybersecurity, including cybersecurity's effect on physical, cyber, and people dimensions. It is applicable to organizations relying on technology, whether their cybersecurity focus is primarily on information technology (IT), industrial

Table 5.4 NIST CSF functions (topmost line) and categories

Identify	Protect	Detect	Respond	Recover
Asset management	Access control	Anomalies and events	Response planning	Recovery planning
Business environment	Awareness and training	Security continuous monitoring	Analysis	Improvements
Governance	Data security	Detection processes	Mitigation	Communication
Risk assessment	Info protection process and procedures	./.	Improvements	./.
Risk management strategy	Maintenance	./.		
./.	Protective technology			

automation and control systems (IACS), cyber-physical systems (CPS), or connected devices more generally, including the Internet of Things (IoT) and the Industrial Internet of Things (IIoT). Moreover, the NIST CSF can assist organization in addressing cybersecurity as it affects the privacy of customers, employees, and other parties. Additionally, the NIST CSF outcomes serve as targets for workforce development and evolution activities [13]. The NIST CSF Core provides a set of activities to achieve specific cybersecurity outcomes, and references examples of guidance to achieve those outcomes. However, the NIST CSF Core is not a checklist of actions to perform. It presents key cybersecurity outcomes identified by stakeholders as supportive in managing cybersecurity risks. The NIST CSF Core comprises four elements: Functions, Categories, Sub-Categories, and Information references [13].

The NIST CSF Functions organize basic cybersecurity activities at the highest level. These Functions are: Identify, Protect, Detect, Respond, and Recover, as shown in Table 5.4. They support an organization in expressing its management of cybersecurity risks by organizing information, enabling risk management decisions, addressing threat events, and improving by learning from previous activities. The function also aligns with existing methodologies for incident management and helps show the impact of investments in cybersecurity, e.g., investments in planning and exercises support timely response and recovery actions, resulting in reduced impact to the delivery of services [13].

The NIST CSF categories, as shown in Table 5.4, are subdivisions of Functions into groups of cybersecurity outcomes closely tied to programmatic needs and particular activities, e.g., asset management, access control, and others. In

Table 5.5 Sub-activities corresponding to the NIST CSF categories

Identify	Protect	Detect	Respond	Recover
Asset management	Patch	Monitor device performance	Incident troubleshooting	Backup and restore all systems
Asset control	Standard secure configuration	Monitor account behavior	Remote software, malware, and others	Remediate software malware
Software inventory	Antivirus	Monitor and manage configuration changes	Kill Chain analysis to root cause threat event	Recovery procedures and processes
Configuration baseline	Application and device whitelisting	Monitor device logs	Communications	./.
Network connectivity and rules	Network segmentation	Monitor network traffic: flow and packets	./.	./.
Vulnerability assessment	Identity management and authentication controls	Analyze anomalies: single and correlations	./.	./.
./.	Change control procedures	./.	./.	./.

Table 5.5, the different potential sub-activities, corresponding to the NIST CSF categories, are illustrated [14].

The NIST CSF also enables to describe the maturity level of achieved cybersecurity procedures, policies, and clearly defined roles and responsibilities, as shown in Table 5.6.

From Table 5.5, several key findings emerged:

- Maturity scores were relatively low as IACS had not been subject to the cybersecurity advances that IT security has.
- Maturity scores were particularly low in asset management, protective technology and processes, detection of threat events, and recovery.
- Gaps exist in process and technology.
- Increasing maturity scores need better information on organizations' assets and potential vulnerabilities to generate momentum for continuous improvement.

More in general, the NIST CSF is a living document and will continue to be updated and improved as industry provides feedback on implementation. NIST will continue coordinating with the private sector and government agencies at all levels. As the NIST CSF is put into greater practice, additional lessons learned will be

Table 5.6 NIST CSF maturity levels of achieved cybersecurity profile

Maturity level	Technology	Process	People
#1 Initial	Despite security issues, no controls exist	No formal security program in place	Activities unstuffed or uncoordinated
#2 Developing	Some controls in development with limited documentation	Basis governance and risk management process, policies	Infosec leadership established, informal communication
#3 Defined	More controls developed and documented, but over-reliant on individual efforts	Organization-wide processes and policies in place, but minimal verification	Some roles and responsibilities established
#4 Managed	Controls monitored, measured for compliance, but uneven levels of automation	Formal infosec committees, verification and measurement processes	Increased resources and awareness, clearly defined roles and responsibilities
#5 Optimized	Controls more comprehensively implemented, automated, and subject of continuous improvement	Process more comprehensively implemented, risk-based and quantitatively understood	Culture supports continuous improvement to security skills, process, and technology

integrated into future versions. This will ensure the NIST CSF is meeting the needs of critical infrastructure owners and operators in a dynamic and challenging environment of new threat events, risks, and solutions [13].

5.2 NIST Cybersecurity Framework Critical Infrastructure

The critical infrastructure community includes industrial, public, and private organizations and other entities with a strategic mission in securing a nation's critical infrastructure. Organizations, active in the critical infrastructure sector, perform functions that are supported by the broad category of technology, including Information Technology (IT), Industrial Automation and Control Systems (IACS), Cyber-Physical Systems (CPS), Cyber-Physical Production Systems (CPPS), and others as well as connected devices including the Internet of Things (IoT), and the Industrial Internet of Things (IIoT). Against this background, the reliance on technology, communication, and interconnectivity has changed and expanded the potential vulnerabilities and increased possible cybersecurity risks to the complex systems operations. In this regard, the embedded sensors, actuators, and emerging technologies in IACS or CPPS produce and process enormous data volumes, increasingly used to deliver critical services and support business/mission decisions, that has potential impacts of a cybersecurity incident on an organization, the environment, communities, and the broader economy and society, which has to be considered.

To manage cybersecurity risks, a clear understanding of the respective organizations' business drivers and security considerations specific to its use of technology is required. Because each organizations cybersecurity risks, priorities, and systems are unique, the tools and methods used to achieve the outcomes described by the NIST Cybersecurity Framework (NIST CSF) will vary [15].

Recognizing the mission that the protection of privacy and civil liberties plays in creating greater public trust, the NIST CSF for Improving Critical Infrastructure Cybersecurity includes a methodology to protect individual privacy and civil liberties when critical infrastructure organizations conduct cybersecurity activities. Many organizations already have processes for addressing privacy and civil liberties. The methodology is designed to complement such processes and provide guidance to facilitate private cybersecurity risk management, consistent with an organization's approach to cybersecurity risk management. Integrating privacy and cybersecurity can benefit organizations by increasing customer confidence; enabling more standardized sharing of information, and simplifying operations across legal regimes [15].

The NIST CSF for Improving Critical Infrastructure Cybersecurity remains effective and supports technical innovation, because it is technology-neutral, while also referencing a variety of existing standards, guidelines, and practices, which evolve with technology. By relying on those global standards, guidelines, and practices developed, managed, and updated by industry, the tools and methods available to achieve the NIST CSF for Improving Critical Infrastructure Cybersecurity outcomes will scale across borders, acknowledge the global nature of cybersecurity risks, and evolve with technological advances and business requirements. The use of existing and emerging standards will enable economies of scale and drive the development of effective products, services, and practices that meet identified market needs. Market competition also promotes faster diffusion of these technologies and practices and realization of many benefits by the stakeholders in these sectors.

Building from those standards, guidelines, and practices, the NIST CSF for Improving Critical Infrastructure Cybersecurity provides a common taxonomy and mechanism for organizations to

- Describe their current cybersecurity posture
- Describe their target state for cybersecurity
- Identify and prioritize opportunities for improvement within the context of a continuous and repeatable process
- Assess progress toward the target state
- Communicate among internal and external stakeholders about cybersecurity risks

The NIST CSF for Improving Critical Infrastructure Cybersecurity is not a one-size-fits-all approach to managing cybersecurity risks for critical infrastructure. Organizations will continue to have unique risks—different threat events, different vulnerabilities, and different risk tolerances. They also will vary in how they customize practices described in the NIST CSF for Improving Critical Infrastructure Cybersecurity. Organizations can determine activities that are important to critical service delivery and can prioritize investments to maximize the impact of each dollar spent. Ultimately, the Framework is aimed at reducing and better managing cybersecurity risks [15].

Systems and assets, whether physical or virtual, so vital that the incapacity or destruction of such systems and assets has a debilitating impact on security in general. This requires the use of the NIST CSF for Improving Critical Infrastructure Cybersecurity.

The NIST CSF for Improving Critical Infrastructure Cybersecurity provides a set of activities to achieve specific cybersecurity outcomes, and references examples of guidance to achieve those outcomes. Nevertheless, the NIST CSF for Improving Critical Infrastructure Cybersecurity Core is not a checklist of actions to perform. The NIST Framework comprises four elements: Functions, Categories, Subcategories, and Informative References, depicted as illustrated in Fig. 5.3. Considering the effective and efficient measures taken in the subcategories of the categories of the NIST CSF Functions, the damage resulting from potentially successful cyberattacks, e.g., loss of critical data, loss of reputation of the organization to their customers and other damage caused, can not only be minimized but reduced to a negligible level. However, this can only be achieved through the respective NIST CSF Functions in the associated measures in the subcategories. This is classified and continuously evaluated in the sense of irregular (anomalous) behavior. Ultimately, the comprehensive and continuous monitoring helps to reduce potential cybersecurity risks to such an extent that only a negligibly manageable small residual risk persists.

Based on the NIST CSF for Improving Critical Infrastructure Cybersecurity, their alignment of Functions, Categories, and Subcategories, with their business requirements, risk tolerance, and resources of the organization, the basic requirements achieved to identify their actual cybersecurity profile. Profiles used to identify opportunities for improving cybersecurity posture by comparing a Current-Profile, the as-is-state, with a Target-Profile, the to-be-state. Therefore, the critical infrastructure best practice example uses the NIST CSF profile of the current state, obtained from measures or verification, and the target state, as a result of analytics and improvements.

The approach described in [16], based on the Robust Industrial Control Systems Planning and Evaluation (RIPE) Program, which is a high-level policy document that describes the various activities that organizations perform to manage cybersecurity and robustness. RIPE refers to the current state and any indicators for possible improvement encompassed with the RIPE metrics that can be used to measure and document status and progress in each of the eight RIPE domains. However, the RIPE program itself does not mandate any specific target levels for the cybersecurity capability and performance; it is up to the asset owner to set such targets, thereby defining a target profile. The metrics play a central role within RIPE because they associate framework artifacts with empirical reality, and provide a means for scoring and benchmarking. Scores and benchmarks help an organization to rate if cybersecurity capability within a specific RIPE domain is sufficient or not. In addition, RIPE sub-metrics provide insight on potential reasons why capability scores are above or below expectations, thereby highlighting areas that might need improvement. The objective of RIPE is the implementation of a continuous improvement process, as known from quality management. Current achievements form the basis for further improvement. The result is a sustainable cybersecurity and robustness program that provides for resilience and reliability even with further integration. Compared to the NIST CSF, RIPE is more stringent in the respect that it mandates annual verification and measurement of the actual cybersecurity and definition of a new target security profile for the next RIPE cycle [16].

5.2.1 NIST CSF Critical Infrastructure Best Practice Example

Critical infrastructure includes the vast network of highways, connecting bridges and tunnels, railways, utilities, and buildings necessary to maintain normalcy in daily life. Transportation, commerce, clean water and electricity all rely on these vital critical infrastructure systems. In this regard, critical infrastructure sectors are those, whose assets, systems, and networks, whether physical or virtual, are considered vital that their incapacitation or destruction have a debilitating effect on security, e.g., industrial and economic security, national public security in the digital age, energy supply and public transportation resources security, and others. Thus, critical infrastructure undergoes a major transformation through digitizing information systems, operational systems, infrastructure resources, the automation of infrastructure system processes, and others. Therefore, cybersecurity is required that comprises technologies, procedures, and practices designed to protect data in vulnerable systems, e.g., Railway and Public Transportation Systems (RPTS), Industrial Automation and Control System (IACS), and others from attacks, damage, unauthorized access, and more. The origin of this danger lies in the ubiquitous accessibility and connectivity in industrial, public, and private organizations' computer systems, networks, infrastructure resources, and others. This enables that e.g., railway and public transportation systems and infrastructure resources will become accessible for cyberattacks, making them vulnerable to remote attacks. Remote attacks are malicious actions aimed at unauthorized entry into vulnerable locations in, e.g., railway systems and infrastructure resources to access. Main reasons for remote attacks are illegally stealing data, introducing malicious software to the targeted systems and causing damage. Therefore, cybersecurity teams must identify and analyze weaknesses to gain knowledge to estimate the cybersecurity risk through an efficient cyberattack analysis to attain deeper cybersecurity awareness. However, little reliable information exists about cybersecurity awareness about potential cybersecurity risks, e.g., in the railway and public transport sectors and their infrastructure resources. International studies exist but are rather superficial [17] or broadly designed to gain detailed inside info into these sectors. In cybersecurity, adequate measures are required, which refers to the respective maturity level of cybersecurity in the object under investigation (see Chap. 7). Another essential measure is the Strength, Weakness, Opportunity, and Threats (SWOT) analysis (see Chap. 7) that defines the current operational objectives and goals for cybersecurity in the object under investigation and is a desirable tool to support brainstorming sessions for best practice solutions.

The cybersecurity maturity model (see Sect. 7.3) refers to the state of being complete, comprehensive, ready, or perfect. Therefore, a maturity assessment usage is to measure the current maturity level of a certain aspect of an organization in a meaningful way, enabling organizations to identify strengths and improvement points and prioritize what to do to reach higher maturity levels accordingly. There are over 70 different Maturity Models available, from domains within the spectrum of business information systems, computer science, and others, showing the great variety and width of Maturity Models. Against this background, an organization's maturity defines the effectiveness and efficiency of the organization's business processes and services, to obtain lasting performance. In this regard, the cybersecurity maturity model proposes maturity levels of the organization that gives a global vision of their performance, a gap-analysis of their current capabilities with what it requires to meet its strategic objectives, and others. Some of the more than 70 available different Maturity Models in information security are shown in Table 5.7 [18].

As shown in Table 5.7, most Maturity Models use a five or six scale level maturity that can range, e.g., from one to six or zero (lowest assessment level) to five (highest assessment level), as shown in Table 5.8 [19]. Table 5.2 shows that with each maturity level, cybersecurity maturity increases. In this context, the model of cybersecurity maturity shows which objectives could be further improved and which need further action:

This assessment enables deriving the main constraints for the Cybersecurity Profile to the efficient and effective use of cybersecurity methods for the object under investigation. The starting point of the Cybersecurity Profile determines the potential improvement by the existing Delta (Δ), expressed as difference between the actual achieved maturity level, e.g., 2, and the targeted maturity level, e.g., 4:

$$\Delta_{Maturity\ Level} = Target\ Profile\,(as\text{-}to\text{-}be\text{-}state) - Current\ Profile\,(as\text{-}is\text{-}state)$$
$$= 4\text{--}2 = two\ higher\ maturity\ levels\ must\ be\ achieved.$$

This raises the question on how to achieve the targeted maturity level, answered by the assessment criteria for cybersecurity as shown in Table 5.8. In this regard, the critical infrastructure best practice example uses the NIST CSF cybersecurity profile of the current as-is-state, obtained from measures, surveys, self-assessments, or verification. Based on the current as-is-state, it has to be decided how to achieve

Table 5.7 Maturity Models level context and specific focus

Maturity Model	Level context	Specific focus
Maturity index model	1. Computeriza-tion 2. Connectivity 3. Visibility 4. Transparency 5. Predictive Capacity 6. Adaptability	Specific to production, logistics, sales, marketing, services, and research and development activities
COBIT model	0. Non-existent 1. Initial/ad hoc 2. Repeatable but intuitive 3. Defined pro-cess 4. Managed and measurable 5. Optimize	Framework by information system audit and control association, or for information technology (IT) management mechanism
NIST CSF model	1. Policy 2. Procedure 3. Implementation 4. Testing 5. Integration	Tracking progress implementing information security maturity levels from the current state to the defined target state, using clear, structured documentation
NIST PRISMA model	1. Policy 2. Procedure 3. Implementation 4. Test 5. Integration	Improves information security programs, supports critical infrastructure protection planning, and facilitates the exchange of effective security practices in organizations' communities
SSE-CMM model	1. Conducted informal design 2. Planned and tracked 3. Well-defined 4. Quantitatively controlled 5. Continuous improvement	Provides the industry with a safety standard of the design engineering software with best practices but is no specific guidance on how to achieve security solutions

Table 5.8 Assessment criteria for maturity level

Maturity level	Cybersecurity assessment criteria
0	No activities in cybersecurity
1	Concepts, but no concrete implementation yet
2	Concept partially implemented
3	Full implementation and thorough documentation
4	Continuous state-of-the-art and efficient monitoring
5	Subject to a continuous improvement process

the required improvements to reach the Target Profile as-a-result-state, in the context of the targeted maturity level, as illustrated in Table 5.8. However, this improvement not only is a one-constraint task that has to be satisfied, it is a multi-constraint task, depending on the specific organization's assignments to be achieved and mapped in a spider diagram. This is particularly suitable showing the development status or the characteristics of the defined assignments. In the context of the direction of the narrative of the investigation, the cybersecurity maturity model can cover, but not be limited to, the following criteria, depending on the specific organizational focus:

- Organization and personnel dimension focusing on digitization strategy, qualifications, change management, and others
- IT systems' technological basis dimension focusing on tools in the process, system integration, and others
- IT infrastructure dimension focusing on networks and communication
- New technologies' dimension focusing on Internet of Things, big data and analysis, artificial intelligence, and others
- Applications' dimension focusing on applications in the domain of interest
- Concepts and procedures dimensions focusing on the effective and efficient process description
- Implementation dimension focusing on cybersecurity
- Training focusing on cybersecurity skills

for which the actual cybersecurity maturity level, shown in Table 5.8, is entered in its respective position in the Spider diagram. Gaining the essential knowledge required is achieved by a specific questionnaire. Examples of questions about the implementation status of the organization's cybersecurity-strategy can be as shown in Table 5.9, but not restricted to:

In Fig. 5.1, a Spider diagram is given showing the $\Delta_{\text{Maturity Level}}$ of the Targeted Profile (to-be-state) and the Current Profile (as-is-state) for eight RIPE domains of the RIPE program [16], whereby the RIPE perspective of the NIST CSF functions are as follows:

- *System inventory*: Refers to NIST CSF function **Detect**. A complete and accurate system inventory is mandated by RIPE as prerequisite for identifying rogue hardware and software.
- *Network architecture*: Refers to NIST CSF function **Protect**. In RIPE, this is part of critical infrastructure service delivery.
- *Data flow*: Refers to NSF CSF function **Recover**. In RIPE, this is part of backup of data file, and restoration of data.
- *Policies and standard operating procedures (PSOP)*: Refers to NIST CSF function **Respond**. The RIPE PSOP document lists detailed advice for contingency and reporting procedures.
- *System procurement*: Refers to NIST CSF function **Recover**. Metrics-based sustainability is at the center of RIPE, therefore, necessary system criteria for sustainability and recovery must be included in the procurement process.
- *Plant planning*: Refers to NISR CSF Function **Identify**. In RIPE, foundational step to getting control of the digital ecosystem.

Table 5.9 Possible questionnaire for cybersecurity

Has the organization a cybersecurity strategy implemented with a binding timetable to implement the individual steps?   Please describe the actual situation (as-is-state) and provide a brief explanation
What is missed in the cybersecurity strategy and the associated schedule to implement it efficiently?   Please describe the actual situation (as-is-state) and provide a brief explanation
Is the cybersecurity strategy implemented in the organization as part of a master plan with associated milestones?   Please describe the actual situation (as-is-state) and provide a brief explanation
Has a responsible management body been set up for the cybersecurity strategy?   Please describe the actual situation (as-is-state) and provide a brief explanation
What will be the impact of an unsuccessful implementation of the cybersecurity-strategy?   Please describe the actual situation (as-is-state) and provide a brief explanation
What new quality in the organization's products or services is expected as a result of the implemented cybersecurity strategy?   Please describe the actual situation (as-is-state) and provide a brief explanation
What is the impact of the cybersecurity strategy for the organization's customer loyalty?   Please describe the actual situation (as-is-state) and provide a brief explanation
Which new or additional qualifications are expected for your work implementing the organization's cybersecurity strategy?   Please describe the actual situation (as-is-state) and provide a brief explanation
Which measures can be used to implement the new or additional qualifications as part of the cybersecurity strategy in the organization or the respective work area?   Please describe the actual situation (as-it-state) and provide a brief explanation
What time line is assumed for the measures mentioned to implement the necessary new or additional qualifications as part of the cybersecurity strategy in the organization or at the work area?   Please describe the actual situation (as-is-state) and provide a brief explanation
What is understood by the term malware?   Please select the answer from the answers below that seems most appropriate (multiple entries are possible):   1. Malware is malicious software that penetrates computer systems, networks, infrastructure resources, and others in order to cause disruptions or damage   2. Malware often damages the integrity of sensitive information and data   3. Malware is not only spread out via e-mails and websites, malware can also be found on storage media and appropriately prepared hardware   4. Malware infection is achieved by clicking a crafted link on a web page, which initiates the download of deposited malware—drive-by download   5. Malware infection often occurs by opening malicious attachments in an email sent by a cyberattacker   6. Malware infections are possible in many ways: Backdoor—enables the cyberattacker to access a secured system for further attacks; Rootkit—helps the cyberattacker to hide other malware and attacks on the attacked system; Ransomware—encrypts data against the targeted object in order to extort ransom for decryption; and others
What is meant by the term CIA Triad?   Please select the answer from the answers below that seems most applicable (multiple entries are possible):   1. Cyber security concept to protect information or data from unauthorized access, destruction, disclosure, disruption, alteration, or use   2. Describes the three basic principles of information security, confidentiality (C: Confidentiality), integrity (I: Integrity), and availability (A: Availability)

(continued)

Table 5.9 (continued)

3. Describes a model for guiding security policies in industrial, public, and private organizations

4. The Elements of the CIA Triad are the three most important components for cybersecurity

5. Intrusion detection and intrusion prevention are essential parts of the CIA's triad detecting and preventing cyberattacks

6. Supports the identification of the cybersecurity risk, its probability, and its possible impact

What is understood by the term information technology?

Please select the answer from the answers below that seems most applicable (multiple entries are possible):

1. Information Technology describes the use of computer systems, storage devices, networks, and other physical devices, infrastructure resources, and processes to create, process, store, secure, and exchange all forms of electronic information and data

2. Information Technology encompasses basic computer-based information systems, including computer hardware, operating systems (OS), application software, and the information/data processed to produce useful information

3. Information technology includes the components and services: computer systems, servers, storage devices, operating systems, networks, databases and database management systems, cloud computing, edge computing and IoT, applications, the Internet, websites and web portals, telecommunications technology, and others

4. Information Technology security corresponds to the protection of information technology systems from damage and threats

5. Information Technology Security requires expertise in information security governance, security practices and principles, program development and management, incident management, and risk management

6. Information Technology Security requires advanced technical skills and knowledge to understand information technology governance, risk analysis, and compliance to authorize and operate information technology systems using risk management methods and best practices, policies, and procedures

What do the terms information security and cyber security mean?

Please describe the actual meanings (as-is-state) and provide a brief explanation

And others

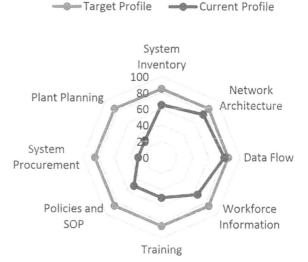

Fig. 5.1 CSF cybersecurity profile comparison referring to the $\Delta_{\text{Maturity Level}}$ after [16]

Therefore, the Maturity Model creates additional value from data records obtained in a self-assessment improving cybersecurity. As part of the self-assessment, in addition to pseudonym sized personal information about the representative of the partaking organization, e.g., role in organization, area of responsibility, and others, and collected information about the organization, e.g., industrial sector, employment numbers, market share, sales, and others. The queried data indicates as an outcome of existing or non-existing roadmap in cybersecurity awareness about malign cybersecurity risk and/or cybersecurity strategy establishment. Such a roadmap ultimately results in strategic activities and responsibilities to increase the current maturity level in the direction of the Target Profile as-to-be-state. Another option to NIST CSF critical infrastructure best practice lies in the SWOT analysis, and described in detail in Chap. 7.

5.3   MITRE Cybersecurity Criteria

The Adversarial Tactics, Techniques & Common Knowledge (ATT&CK)1 project by MITRE is an initiative started in 2015 with the goal of providing a "globally-accessible knowledge base of adversary tactics and techniques based on real-world observations" [11]. Thus, ATT&CK provides a key capability that many organizations have struggled with in the past: a way to develop, organize, communicate, and use a threat-event-informed defensive strategy in cybersecurity risk management. In this regard, MITRE systems engineers supporting government customers in risk management activities have observed the following elements common to the Department of Defense (DoD) and civilian environments. Therefore, cybersecurity risk management represents a fundamental and substantial body of knowledge that includes the development of a risk management approach and plan, identification of components of the risk management process, and guidelines on activities, effective practices, and tools for executing each component [20]. Nevertheless, risk management has to deal with the concept of uncertainty, which includes risk or unfavorable outcomes, and opportunities or favorable outcomes. However, risk management is a formal and disciplined practice for addressing risk. It includes identifying risks, assessing their probabilities and consequences, developing management strategies, and monitoring their state to maintain situational awareness of changes in potential threat events. For this reason the MITRE ATT&CK® approach of risk management contains the steps, as shown in Table 5.10.

MITRE has developed a number of effective and efficient methods defending organizations' computer systems and networks against cyberattacks that require, among other things, advanced technologies and approaches for thwarting cyberattackers' goals. In current industrial organizations' systems and networks, it is unlikely that organizations have the ability or the resources to detect and defend against the manifold of potential cyberattack types cyberattackers might use to gain access to the targeted systems or networks. Even if an organization's patching and software compliance is perfect, a cyberattacker may also use a zero-day exploit, or a

Table 5.10 MITRE ATT&CK® risk management process and objectives

MITRE ATT&CK® steps	Objectives
Risk identification	Its objective is the early and continuous identification of risks, including those within and external to the engineering system project
Risk management approach and plan	Identifies and avoids the potential cost, schedule, and performance/technical risks to a system, takes a proactive and structured approach to manage negative outcomes, responds to them if they occur, and identifies potential opportunities hidden
Risk assessment and prioritization	Involves the overall set of identified risk events, their impact assessments, and their occurrence probabilities, which are processed to derive a most critical to least critical rank-order of identified risks. A major purpose for prioritizing risks is to form a basis for allocating critical resource
Risk mitigation planning, implementation and process monitoring	Involves the development of mitigation plans designed to manage, eliminate, or reduce risk to an acceptable level. Once a plan is implemented, it is continually monitored to assess its efficacy with the intent to revise the course of action, if needed
Selecting risk management tools	

social engineering cyberattack to gain a foothold in a potential targeted system or network. Once intruded, cyberattackers hide in the noise and complexity of their targeted environment, often using legitimate mechanisms and camouflaging their activities in normal network traffic to achieve their objectives. Depending on the security sophistication of the targeted system or network, a cyberattacker is often presented with ample time to execute their work [21]. To combat the extensive growth, evolution, and success of cyberattacks, the MITRE ATT&CK® Framework is ideal to use because it is an open framework and knowledge base of cyberattack tactics, techniques, and procedures (TTPs), which reflects the different phases of a cyberattack lifecycle, the types of actions taken, and the platforms they are known to target. Thus, the different tactics, used by a cyberattacker during an attack, can be interpreted as a sequence of events. Each tactic represents a goal that the cyberattacker is trying to achieve, and leads to the next goal in the sequence. Techniques refer to the specific tools, processes, and steps that the cyberattacker takes to achieve a specific tactic.

There are 14 Tactics in the MITRE ATT&CK® Framework [22]:

1. *Reconnaissance*: Attempt to gather information for a cyberattack
2. *Resource development*: Attempt to create, steal, purchase, or otherwise access resources such as infrastructure, accounts, capabilities, and others that can be used during the cyberattack
3. *Initial access*: Gain a foothold in the network through various means such as spearfishing, exploiting public-facing applications, and others
4. *Execution*: Running adversary-controlled code or modifications to operations

5. *Persistence*: The ability to retain a foothold in the environment through various changes, reboots, and others
6. *Privilege escalation*: Gaining higher levels of permission through vulnerabilities, misconfigurations, and others
7. *Evasion*: Avoid defenses through disabling security software, masquerading malware as approved operations, and others
8. *Credential Access*: Stealing account names and passwords through credential dumping or keylogging, and others
9. *Discovery*: Gaining knowledge of the system that the cyberattacker intends to compromise
10. *Lateral movement*: Move through a remote network once having gained access through legitimate credentials or remote access tools (RATs), and others
11. *Collection*: Gather information from the systems that is either sensitive in itself or provides further information about the defender's environment
12. *Command & control*: The ability to communicate with devices on the network to control their operation
13. *Exfiltration:* Stealing the data that was collected by packaging it and transferring it to adversary-controlled networks or devices
14. *Impact*: Disrupt availability or compromise integrity of the network and systems, such as tampering with or destroying data

In each of these 14 tactics, MITRE ATT&CK® Framework describes the various techniques that cyberattackers can use to achieve the tactical objective. In total, in the MITRE ATT&CK® Framework, there are 188 different techniques, which are not all unique to one tactic. Within each of these techniques, MITRE ATT&CK® Framework also provides a robust set of detailed information. For instance, for the technique External Remote Services found within the Initial Access tactic, MITRE ATT&CK® Framework provides a drill-down option to learn more about [22]:

- Variety of threat events
- Groups known to exhibit specific behaviors
- Detection hints
- References to other information sources
- Related mitigations

The MITRE ATT&CK® Framework has rightfully gained widespread awareness and attention within cybersecurity teams around the world. Its structuring of cyberattackers' behaviors into different steps, based on real-world observations, is a significant step forward for cyberattack defenders worldwide. However, organizations are also familiar with or leveraging other frameworks such as those from NIST CSF, 800, ISA/IEC 62443 in industrial organizations, or the Cyber Kill Chain® introduced by Lockheed Martin, and other. The Cyber Kill Chain® framework is part of the Intelligence Driven Defense® model for identification and prevention of cyberattack intrusions. The Cyber Kill Chain® is based on seven steps [23]:

- *Reconnaissance*: Harvesting email addresses, conference information, and others
- *Weaponization*: Coupling exploit with backdoor into deliverable backdoor

- *Delivery*: Delivering weaponized bundle to the targeted object via email, web, USB, and others
- *Exploitation*: Exploiting a vulnerability to execute code on targeted object
- *Installation*: Installing malware on the targeted asset
- *Command & control*: Command channel for remote manipulation of targeted object
- *Actions on objectives*: With Hands on Keyboard access, intruders accomplish the original goals

As described on the web at attack.mitre.org, the MITRE ATT&CK® Framework enables an organization to search and navigate through the different types of cyberattack techniques, used to enhance, analyze, and test threat event identification and detection efforts. In this context, the MITRE Corporation developed the Common Vulnerabilities and Exposures (CVE) program with the mission to identify, define, and catalog publicly disclosed cybersecurity vulnerabilities in computer systems. In the catalog, there is one CVE record for each vulnerability. The vulnerabilities are discovered then assigned and published by organizations from around the world that have partnered with the CVE Program Partners publish CVE Records to communicate consistent descriptions of vulnerabilities. Information technology and cybersecurity professionals use CVE Records to ensure they are discussing the Identify, define, and catalog publicly disclosed cybersecurity vulnerabilities. Currently, there are 183,334 CVE Records accessible via download or search [24]. All records (cvrf.xml) have the size of at least 280 MB. The web site www.cvedeteil. com provides an easy-to-use web interface to CVE vulnerability data by which one can browse for vendors, products, and versions and view CVE entries, vulnerabilities related to them. In addition, statistics about vendors, products, and versions of products may be viewed. CVE vulnerability data is taken from the National Vulnerability Database (NVD) xml feeds provided by NIST available at www.nvd.nist. gov. Additional data are from several sources like exploits from www.excplot-db. com, vendor statements, and additional vendor supplied data, Metasploit modules are also published in addition to NVD CVE data. Vulnerabilities classified by cvedetails.com using keywords matching and common weakness enumeration (CWE) numbers if possible, but they are mostly based on keywords and on common attack pattern enumeration and classification (CAPEC).

MITRE Corporation operates CWE and the US Department of Homeland Security (DHS), responsible for protecting the critical infrastructure from physical and cyberattacks, with both organizations Cybersecurity and Infrastructure Security Agency (CISA) and US Computer Emergency Readiness Team (US-CERT) supporting it. US-CERT itself is a branch of the Office of Cybersecurity and Communications (CS&C) and the National Cybersecurity and Communication Integration Center (NCCIC). In this context, US-CERT is responsible for analyzing and reducing cyberattacks, vulnerabilities, disseminating cyberattack warning information, fostering and facilitating information sharing and collaboration on cybersecurity issues among, industry, academia, government, and international entities.

CWE categorize software security flaws like implementation defects that can lead to vulnerabilities. It is a community project to understand cybersecurity weaknesses or errors in code and vulnerabilities, and create tools to support in preventing them. CWE has over 600 categories detailing different types of vulnerabilities and bugs [25]. To determine a CWE's frequency, a scoring formula calculates the number of time a CWE is mapped to a CVE within the NVD. Only those CVEs that have an associated weakness used, since using the entire set of CVEs within the NVD would result in lower frequency rate and reduced discrimination amongst the different weakness types [26]

$$FREQ = \{count(CWE_X \in NVD) \text{ for each } CWE\_XinNVD\}$$

$$FR(CWE_X) = (count(CWE_X \in NVD)) - \frac{\min(FREQ)}{(\max(FREQ) - \min(FEREQ))}.$$

The other component in the scoring formula is a weakness severity that is represented by the average CVSS score of all CVEs that map to a particular CWE. The following equation is used to calculate this value:

$$SV(CWE\_X) = (average\_CVSS\_for\_CWE\_X) - \frac{\min(CVSS)}{(\max(CVSS) - \min(CVSS))}$$

with CVSS as Common Vulnerability Scoring System. CVSS is an open framework for communicating the characteristics and severity of software vulnerabilities, which consists of three metric groups [27]:

- Base: *Represent the intrinsic qualities of vulnerability that are constant over* time and across user environments. The base metric group is based on the exploitability metrics and the impact metrics. The exploitability metrics consists of the attack vector, the attack complexity, the privileges required, and the user interaction. The impact metrics are based on the requirements of the CIA Triad: confidentiality, integrity, and availability. The base metrics produce a score ranging from 0 to 10, which can then be modified by scoring the temporal and environmental metrics.
- *Temporal*: Reflects the characteristics of vulnerability that change over time. The temporal metric group contains the exploit code maturity, the remediation level, and the report confidence.
- Environment: Represents the characteristics of vulnerability that are unique to a user's environment. The environmental metric group contains the modified base metrics, and the requirements of the CIA Triad: confidentiality, integrity, and availability (see Sect. 1.6.2).

The level of danger presented by a particular CWE is then determined by multiplying the severity score by the frequency score.

$$SCORE\ (CWE\_X) = FR(CWE\_X) * SV(CWE\_X) * 100$$

There are few properties of the methodology that merit further explanation [27].

- Weaknesses rarely discovered will not receive a high score, regardless of the typical consequence associated with any exploitation. This makes sense, since if developers are not making a particular mistake, then the weakness should not be highlighted in the CWE Top 25.
- Weaknesses with a low impact will not receive a high score. This again makes sense, since the inability to cause significant harm by exploiting a weakness means that weakness should be ranked below those that can.
- Weaknesses that are both common and can cause significant harm should receive a high score.
- Weaknesses that begin with a root cause of a mistake leading to other mistakes, thereby creating a chain relationship. In this year's analysis, the team attempted to capture chains as best as possible without any changes in the scoring. For any chain "X→Y", both X and Y were included in the analysis as if they independently listed. Note that as the CWE Team fleshes out the chain relationships in partnership with the community, it may warrant a change in the scoring for next year and onwards. That said, this year's list is still comparable with previous years' lists, given that they did not treat chains differently than listing of multiple CWEs.

In this context, CWE can be used training developers on building cyber-secure products that are not susceptible to exploitation. Thus, developers use CWE as a resource while writing code to prevent vulnerabilities during the development process. Security orchestration, automation, and response (SOAR) tools use CWE to build policies and workflows to automate remediation. The CWE compatibility program registers products or services as either CWE-Compatible or CWE-Effective. CWE-Compatible products assist organizations in assessing their applications for known weaknesses and flaws, meaning meeting the first four of six requirements, while CWE-Effective products or services must meet all six [28]:

1. *CWE searchable*: Users may search elements using CWE identifiers.
2. *CWE output*: Elements presented to users include, or are obtained associated CWE identifiers.
3. *Mapping accuracy*: Security elements accurately link to the appropriate CWE identifiers.
4. *CWE documentation*: Capability's documentation describes CWE, CWE compatibility, and CWE-related functionality.
5. *CWE coverage*: Capability's documentation explicitly lists the CWE-IDs that the capability claims coverage and effectiveness against.

6. *CWE test results*: For CWE effectiveness, the capability's test results must show an assessment of software for CWEs, and the test results must appear on the CWE website.

The primary difference between CWE and CVE is that CWEs highlight the vulnerabilities, not the specific instance of one within a product. Thus, a CVE might detail a particular vulnerability within an operating system that allows attackers to execute code remotely. This CVE entry only details this vulnerability for a single product. Hence, a CWE outlines the vulnerability independently from any product. CWE has become a common language for discussing, eliminating, or mitigating software security weaknesses. Because developers have access to data regarding weaknesses early in product lifecycles, they can build products without vulnerabilities, eliminating subsequent security issues. This allows developers to keep pace with rapid development lifecycles, build better products, release them faster to customers, minimize attack surfaces, and prevent more cyberattacks [28].

CAPEC™ supports by providing a comprehensive dictionary of known patterns of cyberattacks employed by cyberattackers to exploit known weaknesses in cyber-enabled capabilities, used by analysts, developers, testers, and educators to advance community understanding and enhance defenses. Therefore, CWE and CAPEC are tools to enhance cybersecurity awareness to defend cybersecurity risks.

Furthermore, the MITRE ATT&CK® Framework provides comprehensive details about the TTPs used by ransomware strains like REvil and Conti [28]. REvil, also known as Sodinokibi ransomware, was first identified in April 2019. REvil is a highly evasive and upgraded ransomware that encrypts files and deletes the ransom request message after infection. The message informs the attacked organization that a bitcoin ransom must be paid and that if the ransom is not paid on time, the demand will double. REvil is operated as ransomware-as-a-service (RaaS), used against organizations in different industrial sectors, e.g., manufacturing, transportation, and electrical sectors. Some of the REvil features include [29]:

- Exploits a kernel privilege escalation vulnerability to gain system privileges using CVE-2018-8453.
- Whitelists files, folders, and extensions from encryption.
- Kills specific processes and services prior to encryption.
- Encrypts files on local and network storage.
- Customizes the name and body of the ransom note, and the contents of the background image.
- Infiltrates encrypted information on the infected host to remote controllers.
- REvil uses Hypertext Transfer Protocol Secure (HTTPS) for communication with its controllers.

The MITRE Corporation provides insight into the specific actions taken by REvil, from initial access all the way through encryption and stoppage of impacted services. Figure 5.2 shows an example of those parts of the MITRE ATT&CK Framework that are applicable to REvil, published in [30]:

Initial access	Execution	Privilege escalation	Defense evaluation	Discovery	Commands & control	Exfiltration	Impact
Drive-by compromise	Native API	Process injection	Deobfuscate/decode files or information	File and directory discovery	Ingress tool transfer	Exfiltration over C2 channel	Data destruction
./.	Windows management instrumentation	./.	Modify registry	Query registry	./.	./.	Data encrypted for impact
./.	./.	./.	Obfuscated files or information	System information discovery	./.	./.	Inhibit system recovery
./	./.	./.	Process injection	System service discovery	./.	./.	Service stop

Fig. 5.2 MITRE ATT&CK® Framework part for REvil example

In the same way as with REvil, the MITRE ATT&CK® Framework provides visibility into the equivalent actions taken by Conti during an attack. Figure 5.3 shows how different Conti's techniques look from that of REvil as introduced in [30].

The content shown in Figs. 5.2 and 5.3 are used in the Whitepaper in [30], focusing on two specific techniques that are shared techniques between the two ransomware variants, found under the impact set or tactics.

5.4 MITRE Cybersecurity Taxonomy

Taxonomy is a classification according to a predetermined objective, resulting in a catalog that provides a conceptual framework for discussion, analysis, or information retrieval. A taxonomy is a body of knowledge. Taxonomy is organized to control use of terms in a subject field within a vocabulary to facilitate the storing and retrieving of items from a repository. In this context, the taxonomy development refers to separating elements of a group of objects of a subject area into subgroups of

Execution	Privilege execution	Defense evasion	Discovery	Lateral movement	Impact
Command scripting interpreter	Process injection	Deobfuscate/decode files or information	File and directory discovery	Remote services	Data encrypted for impact
Windows command shell	Dynamic-link library injection	Obfuscated files or information	Network share recovery	SMB/Windows Admin Shares	Inhibit system recovery
Native API	./.	Process injection	Process discovery	Taint shared content	Service stop
./.	./.	Dynamic-link library injection	Remote system discovery	./.	./.
./.	./.	./.	System network configuration discovery	./.	./.
./.	./.	./.	System network connection discovery	./.	./.

Fig. 5.3 MITRE ATT&CK® Framework part for Conti example

objects in the subject area, which are mutually exclusive and unambiguous, and taken together, including all possibilities of the objects in the subject area. In this regard, taxonomy is interpreted as a system of knowledge organization with a controlled vocabulary that focusses on the objects of the subject area. A broader definition of taxonomy was proposed in [31, 32]. According to the definition of these authors, taxonomy consists of seven types of activities:

1. *Recognition*: Description and naming of taxa (taxonomic unit), e.g., species, genera, families, and others, also revision of older descriptions, synonymizations, and others (≈ á-taxonomy)
2. *Comparison of taxa*: Including studies of relationship (phylogeny) (≈ part of á-taxonomy)
3. *Classification of taxa*: (preferably based on phylogenetic analyses) (≈ part of á-taxonomy)
4. *Study of (genetic) variation*: Within species: (≈á -taxonomy)
5. *Construction of tools*: For identification (keys, DNA barcodes)

6. *Identification of specimens*: Referring them to taxa, using the tools
7. *Inventories of taxa*: In specific areas or ecosystems (using the tools for identification)

whereby a taxonomy describes name, revises and synonymizes taxa, following [31].

Besides the foregoing features, the flexibility of a taxonomy model enables organizations to continuously shift based on new data and information within their field of expertise and can leverage this information to scale their processes. This also means that taxonomy management allows finding things easier and faster, but also allows finding linkages within data, which have not been found before. As a result, taxonomy management enables enterprises to confidently tackle data governance, while allowing organizations to further scale their business intelligence and overall ability to innovate in competitive markets [33].

The MITRE Cybersecurity Taxonomy is often described as taxonomy of cyberattacker (adversarial) behavior based on real-world observation of Advanced Persistent Threat (APT) campaigns. APT is a stealthy threat event activity, typically nation state-controlled group of cyberattackers that have unauthorized access to computer systems, networks, infrastructure resources, and others for the purpose of espionage over a longer period of time, possibly moving or spreading within the attacked object(s) and collect information or carry out damages or manipulations. In recent times, APT also refers to non-state-sponsored groups conducting large-scale cyberattack intrusion goals. Therefore, the goal of taxonomy is to standardize knowledge and understanding of cybersecurity from cyberattackers (adversary's) perspective. Specific behaviors or actions, called techniques, classified under categories, called tactics reflect various phases of a cyberattacker's lifecycle, like the Cyber Kill Chain, but with emphasis on finer granularity [34]. Against this background, the MITRE Cybersecurity Criteria ATT&CK contains standardization for threat intelligence, threat event hunting, cyberattacker (adversary) emulation, and analysis and evaluation of computer systems, network, and infrastructure resources defense. Thus, the MITRE ATT&CK knowledge base provides a foundation for developing specific threat event models and methodologies across any sector or industry based on

- 12 tactics
- 244 techniques
- 272 subtechniques
- Comprehensive list of attack behaviors
- Capability management and gap identification
- Vendor performance benchmarks

The tactics and techniques of the, e.g., MITRE ATT&CK Enterprise Matrix contains information for the following platforms: Windows, macOS, Linux, PREm Azure, AD, Office 365, Google Workspace, SaaS, Network, and Containers.

For techniques, the MITRE ATT&CK Enterprise Matrix shows the following elements without sub-techniques, whereby some techniques are classified under multiple tactics as some actions can have multiple functions. For example,

Bypassing User Account Control is both a way to escalate privilege and a way to evade defenses [35]:

- Reconnaissance: 10 techniques
- Resource development: 7 techniques
- Initial access: 9 techniques
- Execution: 12 techniques
- Persistence : 19 techniques
- Privilege escalation: 13 techniques
- Defense evasion: 42 techniques
- Credential access: 16 techniques
- Discovery: 30 techniques
- Lateral movement: 9 techniques
- Collection: 17 techniques
- Command & control: 16 techniques
- Exfiltration: 9 techniques
- Impact: 13 techniques

The same numbers of techniques are available for the sub-techniques elements [35]. Tactics categories include Discovery, Initial Access, Execution, Lateral Movement, and Exfiltration, among others. However, as the framework grows, the matrix for the tactic and technique structure becomes harder to work with. The most obvious structural problem is with inconsistent abstraction, where some techniques are more general than others and some techniques naturally falling under the scope of other techniques. At the top, the MITRE ATT&CK contains three matrices, MITRE ATT&CK for Enterprise, Pre-MITRE ATT&CK, and MITRE ATT&CK Mobile, a list of cyberattacker (adversarial) groups mapped to techniques, and a list of software use by those groups. Nevertheless, the scheme for which the work put into MITRE ATT&CK already represents an excellent foundation, because MITRE ATT&CK is a burgeoning ontology of cyberattacker (adversarial) behavior and taken up incrementally and in varying matters of degree [34].

5.5   Exercises

5.5.1   NIST CSF

What is meant by the term *NIST?*
Describe the duties of the NIST Organization.
What is meant by the term *NIST Cybersecurity Framework?*
Describe the main characteristics of the NIST Cybersecurity Framework.
What is meant by the term *CIS Critical Security Controls?*
Describe the main characteristics of the CIS Critical Security Control.
What is meant by the term *ISA/IEC 62443 Cybersecurity Standard?*
Describe the main characteristics of the ISA/IEC 62443 Cybersecurity Standard.

What is meant by the term *Identify?*
Describe the main characteristics of the function Identify.
What is meant by the term *Protect?*
Describe the main characteristics of the function Protect.
What is meant by the term *Detect?*
Describe the main characteristics of the function Detect.
What is meant by the term *Respond?*
Describe the main characteristics of the function Respond.
What is meant by the term *Recover?*
Describe the main characteristics of the function Recover.
What is meant by the term *Level Content of the NIST CSF?*
Describe the main characteristics of the Level Content of NIST CSF.
What is meant by the term *Target Profile?*
Describe the main characteristics of the Target Profile.
What is meant by the term *Current Profile?*
Describe the main characteristics of the Current Profile.
What is meant by the term *Radar Chart?*
Describe the main characteristics of the Radar Chart.

5.5.2 MITRE ATT&CK

What is meant by the term *MITRE?*
Describe the duties of the MITRE Organization.
What is meant by the term *Risk Identification?*
Describe the main characteristics of Risk Identification.
What is meant by the term *Risk Management?*
Describe the main characteristics of the Risk Management Approach and Plan.
What is meant by the term *Risk Assessment?*
Describe the main characteristics of Risk Assessment and Prioritization.
What is meant by the term *Risk Mitigation?*
Describe the main characteristics of the Risk Mitigation Planning, Implementation, and Process Monitoring.
What is meant by the term *MITRE Cybersecurity Taxonomy?*
Describe the seven main characteristics of the MITRE Cybersecurity Taxonomy.

References

1. https://en.wikipedia.org/wiki/The_CIS_Critical_Security_Controls_for_Effective_Cyber_Defense. 2022 (Accessed 12.2022)
2. https://learn.cisecurity.org/cis-controls-download (Accessed 12.2022)

3. Brooks, R.: Top 20 Critical Security Controls for Effective Cyber Controls. 2018 https://blog. netwrix.com/2018/02/01/top-20-critical-security-controls-for-effective-cyber-defense/ (Accessed 12.2022)
4. CIS Benchmarks™ https://www.cisecurity.org/cis-benchmarks/ (Accessed 12.2022)
5. ISA/IEC 62443 https://ldra.com/iec-62443/ (Accessed 12.21022)
6. Purdue Model for ICS Security https://www.checkpoint.com/cyber-hub/network-security/what-is-industrial-control-systems-ics-security/purdue-model-for-ics-security/ (Accessed 12.2022)
7. Pavleska, T., Aranha, H., Masi, M., Grandry, E., Selitto, G.P.: Cyber Security Evaluation of Enterprise Architectures: The e-SENS Case. In: Proceedngs 12th IFIP Working Conference on The Practice of Enterprise Modeling (PoEM), pp. 226–241, 2019
8. Röhrig, S.: Using Process Models to Analyze IT Security Requirements. PhD Thesis, University of Zürich, 2003
9. Anton, A.I., Earp, J.B., Reese, A.: Analyzing Website Provacy Requirements using a Privacy Goal Taxonomy. In: Proceedings IEEE Joint international Conference on Requirements Engineering, pp. 23–31, 2003
10. Pfleeger, C.P., Pfleeger, S.L.: Security in Computing, 4th Edition. Prentice Hall Publ., 2006
11. Hubbard, J.: Measuring and Improving Cyber Defense using the MITRE ATT&CK Framework. SANS Whitepaper, 2020
12. NIST Special Publication, Revision 5, Joint Task Force, 2020 https://doi.org/10.6028/NIST.SP. 800-53r5 (Accessed 12.2022)
13. NIST Framework for Improving Critical Infrastructure Cybersecurity, Version 1.1, 2018 https:// nvlpubs.nist.gov/nistpubs/CSWP/NIST.CSWP.04162018.pdf (Accessed 12.2022)
14. Achieving NIST CSF Maturity with Verve Security Center. Verve Use Case Report https:// verveindustrial.com/resources/case-study/achieving-nist-csf-maturity-with-verve-security-center/ (Accessed 12.2022)
15. NIST Framework for Improving Critical Infrastructure Cybersecurity, 2018 https://nvlpubs.nist. gov/nistpubs/cswp/nist.cswp.04162018.pdf (Accessed 12.2022)
16. Pederson, P.: A RIPE Implementation of the NIST Cybersecurity Framework. Whitepaper from the Langner Group, 2014 https://www.langner.com/wp-content/uploads/2017/04/A-RIPE-Implementation-of-the-NIST-CSF.pdf (Accessed 12.2022)
17. Liveri, D., Theocharidou, M., Naydenov, R.: Railway Cybersecurity: Security Measures in the Railway Transportation Sector. ENISA, 2020
18. Möller, D.P.F., Iffländer, L., Nord, M., Leppla, B., Krause, P., Czerkewsky, P., Lenski, N., Mühl, K.: Cybersecurity in the German Railway Sector. In: CRITIS 2022
19. Becker, J., Knackstedt, D., Pöppelbuß, J.: Developing Maturity Models for IT Management: A Procedure Model and its Application. In: Business Information Systems Engineering, Vol. 1, pp. 213–222, 2009
20. Systems Engineering Guide: An Introduction to Risk Management. MITRE Corporation, 2020 https://www.mitre.org/news-insights/publication/systems-engineering-guide-introduction-risk-management (Accessed 12.2022)
21. Strom, B.E., Battaglia, J.A., Kemmerer, M.S., Kupersanin, W., Miller, D.P., Wampler, C., Whitley, S.M., Wolf, R.D.: Finding Cyber Threat with ATT&CK™-Based Analytic. The MITRE Corporation, 2017# https://www.mitre.org/sites/default/files/2021-11/16-3713-finding-cyber-threats-with-attack-based-analytics.pdf (Accessed 12.2022)
22. Livingston, J.: What is MITRE ATT&CK? The Definite Guide. Verve Whitepaper, 2022 https://verveindustrial.com/resources/what-is-mitre-attck-the-definitive-guide-thank-you/? submissionGuid=06f2b320-058f-401d-9517-17663c6c65b9 (Accessed 12.2022)
23. Cyber Kill Chain https://www.lockheedmartin.com/en-us/capabilities/cyber/cyber-kill-chain. html (Accessed 12.2022)
24. Downloads | CVE https://www.cve.org/Downloads (Accessed 12.2022)
25. 2022 CWE Top 25 Most Dangerous Software Weaknesses https://cwe.mitre.org/top25/ archive/2022/2022_cwe_top25.html#cwe_top_25 (Accessed 12.2022)

26. Supplemental Details – 2022 CWE Top 25 https://cwe.mitre.org/top25/archive/2022/2022_cwe_top25_supplemental.html#comparison_mssw (Accessed 12.2022)
27. Common Vulnerability Scoring System version 3.1: Specification Document https://www.first.org/cvss/specification-document (Accessed 12.2022)
28. CWE – Why it is Important https://www.hackerone.com/vulnerability-management/cwe-common-weakness-enumeration-why-it-important (Accessed 12.2022)
29. REvil Ransomware-as-a-Service – An Analysis of a Ransomware Affiliate Operation. INTEL471 Whitepaper, 2020 https://intel471.com/blog/revil-ransomware-as-a-service-an-analysis-of-a-ransomware-affiliate-operation/ (Accessed 12.2022)
30. Smith, R.F., Coulson, B., Kaiser, D., Vincent, S.: Using the MITRE ATT&CK® Framework to Boost Ransomware Defenses. LogRhythm Whitepaper, https://gallery.logrhythm.com/white-papers-and-e-books/logrhythm-na-ransomware-as-a-service-white-paper.pdf (Accessed 12.2022)
31. Enghoff, H., Seberg, O.: A Taxonomy of Taxonomy and Taxonomists – The Systematist. In: Newsletter of the Systematics Association, Vol. 27, pp. 13–15, 2006 http://www.systass.org/newsletter/TheSystematist27.pdf (Accessed 12.2022)
32. Enghoff, H.: What is Taxonomy – An Overview with Myriapodological Examples. In: Soil Organisms, Vol. 81, No. 3, pp. 551–451, 2009
33. Taxonomy & Thesaurus Management. Pool Party Whitepaper, 2021 https://www.poolparty.biz/taxonomy-thesaurus-management (Accessed 12.2022)
34. ATT&CK Structure Part 1: A Taxonomy of Adversarial Behavior https://www.tripwire.com/state-of-security/mitre-framework/attck-structure-taxonomy-adversarial-behavior/ (Accessed 12.2022)
35. MIRE ATT&CK Enterprise Matrix, 2022 https://attack.mitre.org/matrices/enterprise/ (Accessed 12.2022)

Chapter 6
Ransomware Attacks and Scenarios: Cost Factors and Loss of Reputation

Abstract Cyberattacks and thus cybersecurity risks have accelerated over the past years. Cyberattacks are based on threat event attack types, as described in Chap. 2. Besides other threat event attack types, ransomware is probably the No. 1 challenge of threat event attacks that industrial, public, and private organizations are facing. Ransomware is a type of malware that typically locks the data on a targeted computer system or user's files by encryption. This cyberattack demands a payment (ransom) before the ransomed data is decrypted and access returned to the targeted user, but ransomware comes in many forms. In this regard, ransomware is a type of malware used by cybercriminals for financial gain. Typically, a ransom note is installed on a targeted computer system at the same time the data/files are encrypted. They not include information on the ransom demands, meaning the amount of ransom a deadline for payment, and instructions how to reach and pay the ransom providing details on the cryptocurrency wallet or other wiring information to complete the transaction. In this context, ransomware is a two-step-extortion: Step 1 is to encrypt and extract the data/information; Step 2 is to negotiate the ransom. However, over the past years, ransomware has emerged to Ransomware-as-a-Service (RaaS), because ransomware has proven to be an effective approach for cybercriminals to hit it big, in terms of both payouts and notoriety. One of the cases was the 2020 Solar Winds supply chain attack. Cybercriminals targeted Solar Winds by deploying malicious code into its Orion IT monitoring and management software platform used by thousands of industrial organizations and government agencies worldwide, which creates a backdoor through which cybercriminals access and impersonate users and accounts of the targeted organizations' systems. The SolarWinds supply chain attack was a major cybercriminal event because not a single company was attacked by a breach-in, but it triggered a much larger supply chain incident that affected thousands of organizations, including the US government. In this cyberattack, the cybercriminals used tools used for many years, developed, and adjusted them with new attack pattern, and cybercriminals hit it big in terms of payout and notoriety. Such ransomware attacks led to an evolution capitalizing on a growing number of cybercriminals who want to get in. These successful cybercriminals started as cybercriminal entrepreneurs offering RaaS, which makes carrying out ransomware much easier by other cybercriminals, lowering the barrier to entry, and expanding the reach of ransomware. In this,

D. P. F. Möller, *Guide to Cybersecurity in Digital Transformation*, Advances in Information Security 103, https://doi.org/10.1007/978-3-031-26845-8_6

cybercriminal business model gains the RaaS entrepreneur a percentage of the ransom paid to the new cybercriminal or a group of cybercriminals using RaaS in a license model, who attack organizations for a ransom. Against this background, Chap. 6 introduces Sect. 6.1 in ransomware attacks and the ransomware landscape, whereas Sect. 6.2 focuses on ransomware attacks and scenarios in Sect. 6.2.1 and ransomware attacks on OT systems in Sect. 6.2.2. Section 6.3 refers to Cost Factors of Ransomware Attacks (CFoRA) and introduces a useful design of the approaches in Recovery Point Objective (RPO) and Recovery Time Objective (RTO) in the Sects. 6.3.1, 6.3.2, and 6.3.3. The focus in Sect. 6.4 is on Loss of Reputation (LoR) and preventing it. Section 6.5 contains comprehensive questions of the topics ransomware, Cost Factors of Ransomware Attacks and Loss of Reputation through ransomware attacks. Finally, "References" refers to the used references for further reading.

6.1   Introduction

Threat event attack is a sensitive issue in the era of digital transformation. The impact and prevalence of threat event attacks has grown, and need action toward computer system and network security for better protection of computer systems, networks, and infrastructure resources against them. Strategies, defending threat event attacks, such as Intrusion Detection and Prevention (see Chap. 3), strict firewall policies, penetration tests, and access controls, are some common approaches defending threat event attacks. Therefore, cybersecurity dependence of computer systems, networks, and infrastructure resources has driven up the demand for the analysis of pre-emptive threat event attacks to help in the early discovery of potential threat event attacks or vulnerabilities. One reason for this is that white-hat cyber-hackers actually attack computer systems and networks to discover vulnerabilities. Unfortunately, currently there are no standard methods for measuring the effectiveness of the risk(s) of threat event attack(s) and no standardized effective defense strategies available. However, a number of options are available, based on which a strategy can be formulated to defend against threat event attacks (see Chaps. 2 and 3). Thus, industrial, public, and private organizations expend enormous efforts to secure their valuable data and business assets against threat event attacks. They use various types of Threat Intelligence Management Platform (TIMP) framework tools and techniques to keep their organizations daily work possibly undisturbed, while threat event actors are trying to breach cybersecurity and infiltrate malicious software to access valuable data and business assets. However, the situation with threat event attacks is getting worse because of new types of emerging malware to attack computer systems, network, infrastructure resources, and others. Thus, it is important to gain knowledge and skills to understand threat event attacks before and after they happen, in order to provide better insight into possible scenarios of threat event attacks. In this context, threat event attacks refer to cyberattackers who attempt unauthorized access into computer systems, networks, infrastructure resources, and others, using data communications pathways by making use of various techniques, tactics, or exploits (TTE), to attract the targeted objects in their direction, or to steal or manipulate valuable as well as sensitive data and business assets. These exploits can be directed through various conducts, for instance from remote locations by unknown persons using the Internet. From a general perspective, threat event attacks can originate from different types of cyberattackers (also termed adversaries), including:

- *Cybercriminals*: Commit or support criminal acts classified as either digital technology-based crimes, or digital-assisted crimes that use technology in a supporting capacity to commit the criminal act. Cybercriminal profiling techniques including inductive and deductive profiling, and the need employing a hybrid technique that incorporates both inductive and deductive profiling [1]
- *Cyber terrorists*: Breach-in electronically into computer systems, networks, infrastructure resources, and others to create violence against persons or property, or at least cause enough harm to generate fear
- *Cyber activists*: Using Internet-based socializing and communication techniques to create, operate, and manage activities of any type, e.g., to reach out to and gather followers, broadcast messages, and progress a cause or movement

as described in Chap. 2. Of all the threat event attacks described in Chap. 2, which affect industrial, public, and private organizations, ransomware was the most common one occurring in the last 2 years. Ransomware also had the highest impact on targeted objects: production, reputation, and finances. Therefore, ransomware attack deserves special attention today, more than other types of threat event attacks, described in Chap. 2. In this regard, ransomware is probably the no. 1 challenge of threat event attacks that industrial, public, and private organizations are facing. Thus, ransomware has become the primary threat event attack in the field of cybersecurity risk(s) due to the high profitability (ransom) of the operation and the perpetrators' virtual impunity. Hence, it is important to understand how organizations are managing ransomware risks, because ransomware attackers try to cyberattack vulnerable systems opportunistically. They are also known to study targeted organizations to try identifying exploitable weaknesses. Furthermore, mass phishing attacks are sent out to steal and encrypt the data from whomever is gullible enough to respond. Moreover, spear phishing is used by cyberattackers targeting key personnel in an organization they know they have access to business assets and business-critical systems.

6.2 Ransomware Attacks

A rapid growth in cybercrimes happen that cause tremendous financial loss to organizations ransomware. Ransomware is a complex malicious software; however, it is more just cybercriminal groups who develop the malware and conduct attacks and extortion. Recent studies reveal that cybercriminals tend to collaborate or even transact cyberattack tools like ransomware via the dark net markets. Hence, ransomware is also a business model of cybercriminal entrepreneurs operating Ransomware-as-a-Service (RaaS), which allows unprecedented opportunities gaining ransom payment. Furthermore, cybercriminal ransomware attackers choose their targets based on the assailed target's solvency, its operational status, or the fragility of its information systems [2] as well as the criticality of infrastructure resources. Thus, the only way organizations can defend themselves against these cyberattacks is to make their systems resilient enough to avoid becoming successfully attacked in the first place. Fortunately, organizations that actively defend their organizations can and do avoid such attacks altogether or can respond quickly and mitigate the damage [3].

6.2.1   Introduction in Ransomware Attack Scenarios

Ransomware is a type of malware attack in which the cyberattacker typically locks and encrypts the targeted object's data, important files, and then demands a payment (ransom) to unlock and decrypt the attacked data. The ransomware-attack process is as follows: The cyberattacker demands a payment (ransom) before the ransomed data or files are decrypted and their access returned to the targeted object/user, but ransomware comes in many varieties. Thus, ransomware is a type of malware used by cybercriminals for financial gain. Typically, a ransom note is installed on a targeted computer system at the same time the data/files are encrypted. They not include information on the ransom demands, meaning the amount of ransom demand a deadline for payment, and instructions on how to pay the ransom providing details on the cryptocurrency wallet or other information the targeted object/user need to complete the transaction. In this context, ransomware is a two-step-extortion: Step 1 is to encrypt and extract the data/files; Step 2 is to negotiate the ransom. However, over the past years, ransomware has emerged to Ransomware-as-a-Service (RaaS), because ransomware has proven to be an effective approach for cybercriminals to hit it big, in terms of both payouts and notoriety. One of the cases was the 2020 Solar Winds supply chain attack. Cybercriminals targeted Solar Winds by deploying malicious code into its Orion IT monitoring and management software platform, used by thousands of industrial organizations and government agencies worldwide, creates a backdoor through which cybercriminals access and impersonate users and accounts of the targeted organization. The SolarWinds cyberattack was a major cybercriminal event because it attacked not a single company but it triggered a much larger supply chain incident that affected thousands of organizations, including governments. In this cyberattack, the cybercriminals used tools used for many years, developed, and adjusted them with new cyberattack pattern, and cybercriminals hit big in terms of payout and notoriety. Such ransomware attacks led to an evolution capitalizing on a growing number of cybercriminals who want to get in. These successful cybercriminals started as cybercriminal entrepreneurs offering Ransomware-as-a-Service (RaaS), which makes carrying out ransomware cyberattacks much easier by other cybercriminals, lowering the barrier to entry, and expanding the reach of ransomware. In this, cybercriminal business model gains the RaaS entrepreneur a percentage of the ransom paid to the new cybercriminal or a group of cybercriminals using RaaS in a license model, who attack organizations for a ransom. Besides the Solar Wind ransomware cyberattack there are many other strains or ransomware malware that made global impact and caused widespread damage [4, 5], as shown in Table 6.1.

The ransomware attacks referred to in Table 6.1 use different methods to spread out ransomware to infect a targeted device. Therefore, understanding how ransomware spreads is important to avoid being a targeted device to a ransomware attack. The most common ways ransomware infects targeted devices [4] are shown in Table 6.2.

Table 6.2 shows that it is important for an organization to recognize how their computer systems, networks, infrastructure resources, and others can be targeted and intruded upon with malicious code; which require proactively taken steps through a layered security approach to keep themselves protected and to safeguard their business continuity. Therefore, protection against ransomware is an important issue that can happen in a variety of ways. Some of which [5] are shown in Table 6.3.

In case a ransomware infection is detected, there are immediate steps to be taken to mitigate the ransomware threat, as described in [5] and shown in Table 6.4.

However, due to manifold ransomware possibilities, organizations may struggle with ransomware attacks, but cyberattackers have still a clear view today. Thus, organizations must take additional steps defending against ransomware from becoming more successful and avoiding them also in the years ahead [3].

6.2.2   Ransomware Attacks on Operational Technology Systems

Ransomware attacks on Operational Technology (OT) systems represent a dangerous type of cyberattack because OT assets include critical industrial equipment or infrastructure that assure the running of operations in an industrial environment. When cybercriminals target unsecured OT systems with ransomware or other malware, they can directly disrupt real, physical processes, can affect the process flow on the production job shop floor and bring operations to a standstill [12], and others. But why are OT systems more vulnerable to attack than their IT counterparts? One major reason is that patching and updating is substantially harder, if not impossible, in the OT systems domain. As reported in [12], organizations may overlook updating legacy OT systems because it takes much time and resources. Shutting down a factory for a day or even just several hours in order to install updates or patches may not be an economically viable option. Unfortunately, this lack of regular updates to OT systems can leave them exposed to bad actors such as

Table 6.1 Characteristics and effect of ransomware attacks

Ransomware attack type	Characteristics and effects
Cerber	RaaS available for use by cybercriminals, who carry out cyberattacks and spread their loot with the malware developer. The malware runs silently while it is encrypting files. It tries to prevent antivirus and Windows security features from running, to impede users from restoring the targeted system. After successfully encrypting files on the targeted system, it displays a ransom note on the desktop.
Cryptolocker	Released in 2017, it affected over 500,000 computer systems. Typically, infects computer systems through email, file sharing sites, and unprotected downloads. Not only encrypts files on the local targeted system but can also scan mapped network drives, and encrypt files and has permission to write. New variants of Cryptolocker are able to elude legacy antivirus software and firewalls.
CryptoWall	Cryptowall ransomware attacks are used in an attempt to bypass antivirus detection. Success debated since this practice is not uncommon among malware developers and security products account for it. Therefore, computer users should themselves protect against malvertising and drive-by-download attacks by keeping the software installed on their computers up-to-date, especially the Web browsers and their plug-ins.
CryptXXX	When executed, CryptXXX encrypts user's files and adds the .crypt extension to the filename, and does the same on all mounted drives. Furthermore, it steals Bitcoins from the infected computer systems, as well as user data.
CrySis	Family of ransomware that has been evolving since 2016, targets Windows systems, and this family. Primarily targets businesses, and distributed as malicious attachment in spam email. It can intrude as disguised installation files for legitimate software. If CrySis has infected a computer system, it creates registry entries to maintain persistence and encrypts practically every file type, while skipping systems and malware files. It performs the encryption routine using a strong encryption algorithm (AES-256 combined with RSA-1024 asymmetric encryption), which is applied to fixed, removable, and network drives [6]. When the encryption routine is completed, it drops a ransom note on the desktop for the infected user, providing two email addresses to contact the ransomware attacker(s) for paying the ransom. Some variants include one of the contact email address in encrypted file names. Ransom demand is usually around 1 Bitcoin, but there have been cases where pricing seems to be adapted to match the revenue of the affected company. Financially sound companies often have to pay a larger ransomware sum [6].
Grand Crab	Released in 2017; it encrypts files on targeted system and demands a ransom. Used to launch ransomware-based extortion attacks, where cyberattackers threatened to reveal personal knowledge of the targeted victims to the public. There are several versions, all of which target Windows-based computer systems. Free decryptions are available today for most versions of GrandCrab.
Locky	Released in 2016, it is able to encrypt 160 file types, primarily files used by designers, engineers, and testers. Typically distributed by exploit kits or phishing, meaning threat event attackers send emails that encourage the targeted user to open a Microsoft Office Word or Excel file with malicious

(continued)

Table 6.1 (continued)

Ransomware attack type	Characteristics and effects
	macros, or a ZIP file that installs the malware upon extraction. Locky has become one of the largest ransomware threats today, mainly hitting users, but already infecting computer systems in over 100 countries. Recently, Locky began code modification to prevent cybersecurity teams from efficiently monitoring its activity.
LowLevel04	Old malware threat, initially identified in October 2015. Attack stopped a few months after its initial release, but new waves have been identified recently against various other targets, and surprisingly, it has not undergone any improvements or code modifications. Upon infection, the virus starts to encrypt target user file extensions. An advanced AES cipher is used which has a very strong security, making decryption impossible without decryption key. This strain has the ability to affect also the connected cloud storage services such as Dropbox and OneDrive. However, it does not delete the Shadow Volume Copies which makes file recovery possible [7].
Nemucod	First discovered in December 2015, it is associated with downloading malware including Teslacrypt, a variant of ransomware. Commonly spread through spam or phishing emails that contain malicious attachments. These emails are normally disguised as department of an organization claiming that the attached file is vital information that requires opening to view. When an unsuspecting user opens the attachment, malicious code runs and downloads further malware on the affected computer system [8].
Petya and NotPetya	Released in 2016, it infects a targeted system and encrypts an entire hard drive by accessing the Master File Table (MFT), which makes the entire disk inaccessible, although the actual files are not encrypted. It requires the targeted user to agree to give it permission to make admin-level changes. After the targeted user agrees, it reboots the computer system and shows a fake system crash screen, while it starts encrypting the disk and thereafter shows the ransom notice. It only affected Windows-based computer systems. The Petya virus was not highly successful, but a new variant, named NotPetya by Kaspersky Labs, proved to be more dangerous. NotPetya is equipped with a propagation mechanism, and is able to spread without human intervention. It not only encrypts the MFT but also other files on the hard drive. While encrypting the data, it damages it in such a way that recovery is not possible. Users who pay the ransom cannot actually get their data back.
PrincessLocker	Cybercrime campaign that uses the RIG exploit kit to infect computer systems, and demands a ransom from the targeted user. Ransomware encrypted files increases the ransom from 3 bitcoins to 6 bitcoins if payment not made in the specified time frame. RIG spread out through suspicious advertisements that can be inserted into legitimate websites.
Ryuk	Infects computer systems by phishing emails or drive-by downloads. Uses dropper, which extracts a trojan on the targeted computer system and establishes a persistent network connection as a basis for Advanced Persistent Threat (APT) attack, installing additional tools like keyloggers, performing privilege escalation and lateral movement. Once the attackers have installed the Trojan on as many targeted computer systems as possible, they activate the locker ransomware and encrypt the files. The ransomware

(continued)

Table 6.1 (continued)

Ransomware attack type	Characteristics and effects
	issue is only the last stage of the attack, after the cyberattackers have already done damage and stolen the files they need.
SamSam	Was responsible for significant damage in 2018 in the City of Atlanta, Colorado Department of Transportation, Hospitals, and other organizations. A recent report estimated that SamSam authors made $5.9 million of revenue.
WannaCry	Released in 2017, an encrypting ransomware that exploits a vulnerability in the Windows SMB protocol, and has a self-propagation mechanism that lets it infect other computer systems. Packaged as dropper, a self-contained program that extracts the encryption/decryption application, files containing encryption keys, and Tor communication program. Not obfuscated and relatively easy to detect and remove. In 2017, it spread rapidly across 150 countries, affecting 230,000 targeted computer systems and causing an estimated $4 billion in damages.
And others	... [9–11]

cybercriminal attackers. In addition to knowing that OT systems are relatively easy to cyberattack, cybercriminals are aware that compromising an OT device can cause severe, tangible consequences that operators are desperate to avoid. With OT systems, there is more than financial losses at stake. The physical cost of shutting down a production process or a water treatment plant or stopping oil production or logistic processes and others is that big that compromised businesses may be more likely to pay a hefty ransom to prevent serious disruptions to their operations [12] but accept losses in revenue. Many organizations were affected by the NotPetya ransomware attack, which originated in Ukraine before rapidly spreading internationally. Two examples are described in [12]. One attacked organization was FedEx, which paid $300 million in a NoTPetya ransomware attack on its TNT Express division in 2017. The company was one of several to have its computer systems severely disrupted by the NotPetya ransomware outbreak in 2017. TNT had to restore all its IT operations and deliveries, and sales suffered as a result. Another organization hit by a NotPetya ransomware attack was shipping company Maersk that paid up to $300 million ransom. After learning of the cyberattack, the Maersk company said it immediately worked to shut down infected networks to help contain the malware. Maersk also stated the company has put in place different and further protective measures to defend against such ransomware attacks.

While cybercrime is quickly becoming one of the biggest threat event attacks to industry, these kinds of threat event attacks keep increasing, and organizations or companies need to make sure they have the right defenses. Nevertheless, industrial cybersecurity may soon be a more pressing concern than physical security in some industrial areas because information security professionals feel their organizations have not made the necessary security improvements to thwart future attacks. In this regard, the National Institute of Standards and Technology (NIST) has developed

Table 6.2 Distribution techniques and their characteristics and capabilities

Distribution techniques	Description
Phishing email	Spread ransomware through phishing emails to trick a targeted user into opening an attachment or clicking on a link that contains a malicious file. Cyberattack can come in a number of different formats, including PDF, ZIP file, Word document, or JavaScript. In case of a Word document, the cyberattacker most commonly tricks the user into enabling macros upon opening the document, which enables attacker to run a script that downloads and Execute a Malicious eExecutable File (EXE) from an external web server. EXE includes functions necessary to encrypt data on a targeted user's computer system. Once data is encrypted and ransomware is installed on one targeted computer system, ransomware can spread to other computer systems or servers. All it takes is for one user to naively open an email attachment in the phishing email, and an entire organization can be infected. Popular ransomware exploiting attacked users using phishing emails include: Cerber, Locky, and Nemucod.
Email attachment	Opening email attachments enabling malicious macros or downloading a document with a Remote Access Trojan (RAT) or downloading a ZIP file containing a malicious JavaScript or Windows Script Host (WSH) file.
Remote Desktop Protocol	Popular mechanism in which cyberattackers are infecting targeted computer systems through Remote Desktop Protocol (RDP). Originally, RDP was created to enable IT administrators to securely access a user's computer system remotely to configure it, or to simply use the computer system. Typically, RDP runs over port 3389. Cyberattackers can search for those computer systems using search engines such as Shodan.io to find devices that are vulnerable to infection. Once the target computer systems are identified, attackers commonly gain access by brute-forcing the password to log on as an administrator. Open source password-cracking tools help achieve this objective. Popular tools, including Cain and Able, John the Ripper, and Medusa, allow cybercriminals to quickly and automatically try multiple passwords to gain access. Once cyberattackers are in as an administrator, they have full control of the computer system and can initiate the ransomware encryption operation. To create additional damage, some attackers disable the endpoint security software running on the computer system or delete Windows file backups prior to running the ransomware. This creates even more reason for infected user to pay the ransom, as the Windows backup options may no longer exist. Popular RDP ransomware attacks includes CrySis, LowLevel04, and SamSam.
Drive-by downloads	Another entry path intruding ransomware are drive-by downloads, which are malicious downloads that happen without the user's knowledge when they visit a compromised website. Ransomware-attacks initiate often drive-by download, taking advantage of known vulnerabilities in the software of legitimate websites, then use these vulnerabilities to either embed malicious code on a website or to redirect the user to another site that they control that hosts software known as exploit kits. Exploit kits enable ransomware-attackers to silently scan the visiting device for its specific weaknesses, and, if found, execute code in the background without the user clicking anything. However, the unsuspecting user may then suddenly be faced with a ransom note, alerting them of infection and demanding payment for returned files. Drive-by-downloads cyberattack

(continued)

Table 6.2 (continued)

Distribution techniques	Description
	happened to some most popular sites in the world, including the New York Times, the British Broadcasting Corporation (BBC), and the NFL that were targeted in a ransomware campaign through hijacked advertisements [5]. Popular ransomware exploiting victims through drive-by downloads attacks includes CryproWall, CryptXXX, and Princess Locker.
USB and removable media	Another way of introducing ransomware uses USB devices containing malicious software to penetrate an environment. USB drive masquerades as a promotional application, once opened, and deploys ransomware on unsuspecting user's computer system. The mighty Spora Ransomware even added the capability to replicate itself onto USB and Removable Media drives (in a hidden file format), jeopardizing subsequent machines in which the USB device is plugged in [5]. Ransomware becomes the go-to attack of choice for cybercriminals to generate revenues, which they simply buy on the dark web through RaaS., and attacks are relatively easy to launch.
Self-propagation	Spreading malicious code to other devices through network and USB drives.
Social media	Clicking a malicious link on Facebook, Twitter, social media posts, instant messenger chats, etc.

guidelines industrial cybersecurity companies can follow (see Chap. 5). The NIST CSF focusses on Industrial Automation and Control Systems (IACS) and corporate networks. In the context of IACS, and thus OT system components, ransomware attacks can manipulate the process flow through a theft of process data or damage of production machines or completely shut down, and others. Thus, the impact of a potential ransomware attack onto OT systems have two big issues: Besides the ransom and the Loss of Reputation, a higher level of damage, through downtime, which costs the manufacturing company an additional loss in revenue through downtime in productivity. Looking at the technology used in OT systems, the challenge with OT systems is that they often use outdated technologies, which not regularly updated and patched. Thus, ransomware attacks on OT systems follow often a multiple-motivated goal: (1) gain high ransom; (2) steal secret data or business assets; (3) initiate a loss of reputation; (4) production downtime; (5) loss in revenue.

The Cybersecurity and Infrastructure Security Agency (CISA) leads a ransomware awareness campaign [13]. This fact sheet is a valuable guidance for protecting OT assets by containing actions that organizations should implement to help prevent and mitigate ransomware attacks. Therefore, organizations need to identify their critical equipment and essential processes and continue operations without interruptions [12], which require maintaining an inventory of valuable data and business assets, and evaluate them against cybersecurity risk(s). Another issue is to develop backup procedures and test them regularly. This plan should be ready soon so that operations do not come down in case of a ransomware attack. Further-more, organizations should have manual controls ready so that IACS networks can

Table 6.3 Ransomware protection characteristics and capabilities

Ransomware protection	Characteristics and capabilities
Application whitelisting & control	Establish device controls that allow to limit applications installed on devices to a centrally controlled whitelist. Increase browser security settings, disable Adobe Flash and other vulnerable browser plugins, and use web filtering to prevent users from visiting malicious sites. Disable macros on word processing and other vulnerable applications.
Data backup	Regularly backup data to an external hard-drive, using versioning control and the 3-2-1 rule (create three backup copies on two different media with one backup stored in a separate location). If possible, disconnect the hard-drive from the device to prevent encryption of the backup data.
Email protection	Train employees to recognize social engineering emails, and conduct tests that employees are able to identify and avoid phishing. Use spam protection and endpoint protection technology to automatically block suspicious emails, and block malicious links if user ends up clicking on them.
Endpoint protection	Antivirus is an obvious step in ransomware protection, but legacy antivirus tools only protect against some ransomware variants. Modern endpoint protection platforms provide Next-Generation Antivirus (NGAV) that protects against evasive or obfuscated ransomware, fileless attacks like WannaCry, or zero-day malware, whose signatures are not yet found in malware databases. They also offer device firewalls and Endpoint Detection and Response (EDR) capabilities, which help security teams detect and block ransomware attacks occurring on endpoints [4].
Network defenses	Use a firewall or web application firewall, or Intrusion Prevention/ Intrusion Detection Systems (see Chap. 3), and other controls to prevent ransomware from communicating with Command & Control centers.
Patch management	Keep the device's operating system and installed applications up-to-date, and install security patches. Run vulnerability scans to identify known vulnerabilities and remediate them quickly.
Ransomware detection	Use real-time alerting and blocking to automate identifying ransomware-specific read/write behavior and then blocking users and endpoints from further data access. Use deception-based detection, which strategically plants hidden files on file storage systems to identify ransomware encryption behaviors at the earliest attack stage. Any write/rename actions on the hidden files automatically triggers a block of the infected user or endpoint, while continuing to allow access by uninfected users and devices. Use granular reporting and analysis to provide detailed audit trail support for forensic investigations into who, what, when, where, and how users access files [5].

be isolated, if needed. Moreover, cybersecurity check of IoT devices deployed in OT environments should be done before being embedded. Further, third-party penetration tests should be done from time to time to check whether OT systems are secure. Finally, a zero-trust security model [14] to manage permissions and authorization should be implemented to ensure only authorized users can access critical OT

Table 6.4 Mitigate ransomware infection

Mitigate active ransomware infection by	Activities
Isolation	Identify infected computer systems, disconnect from networks and lock shared drives to prevent encryption.
Investigation	Check what backups are available for encrypted data. Check impact on ransomware hit, and if decryptions available. Understand if paying the ransom is a viable option.
Recovering	If no descriptor tools are available, restore data from backup. In most countries, authorities do not recommend paying the ransom, but this may be a viable option in some extreme cases. Use standard practices to remove ransomware or wipe and reimage affected systems.
Reinforcement	Run lessons learned session to understand how internal systems were infected and how to prevent a recurrence. Identify key vulnerabilities or lacking security practices that allowed attackers in, and remediate them.
Evaluation	After cyberattack crisis has passed, evaluation is important. Evaluate what happened and lessons learned, e.g., how ransomware was successfully executed? Which vulnerabilities made penetration tests possible? Why did antivirus or email filtering fail? How far did the infection spread? Was wipe and reinstall infected systems possible? Was infected organization able to successfully restore from backup? Address weak points in the security posture to be better prepared for the next attack.

applications and systems. The zero-trust model is a security framework that requires continuous authorization and validation of all users before they are granted access to business assets or applications. The zero-trust model deals with modern cyberattack types on OT networks, allowing OEMs and vendors to access the network, while simultaneously reducing the risk of breaches and cyberattacks [14], and thus reduce costs and downtime. Also, controlling and monitoring users with Multi-Factor Authentication (MFA) enables to ensure good governance and feel more confident with a technologically advanced cybersecurity measure. These few methods help organizations become more resilient to OT attacks and continue operations even if there is a breach in cybersecurity [14].

However, ransomware attackers can become cybercriminal entrepreneurs selling their expertise as RaaS to broaden their cybercriminal business opportunities to gain more ransom money. With their extensive knowledge and expertise, very damaging malwares may be used for the ransomware attacks. One possible impact is false control commands, which can have disastrous consequences for production processes and cause immense cost and time-consuming damage. Therefore, two measures become important to recover from ransomware attacks, the Recovery Time Objective (RTO), and the Recovery Point Objective (RPO). The RTO is the maximum tolerable length of time that an attacked system or network can be down after a failure or disaster occurs. The RPO is the maximum acceptable amount of data loss after an unplanned data loss incident, expressed as an amount of time. However, it is also possible, that incoming threat event attacks on IT and OT systems can happen simultaneously.

6.3   Costs of a Ransomware Attack

Ransomware has moved out of the realm of a medium threat event attack to top threat event attacks. Two out of three organizations have reported ransomware attacks in 2021, in which attacks have almost doubled from the year before. On average, the ransom paid by organizations infected by successful ransomware attacks, in which their data were encrypted, has increased enormously. Two recently published studies by Sophus [10] and PaloAlto [11] describe the average ransom payment cost of a successful ransomware attack between $570,000 and $812,360. Besides the amount of ransom payment itself, the aim of ransomware leads to far-reaching issues with the increasing expense that keeps growing and can cost an organization millions of dollars. However, there are more cost factors than only ransom payment cost, e.g., costs of disruption and downtime, which are enormous losses in terms of productivity, disruptions in services and downstream impact that causes revenue as well as loss of reputation. Nevertheless, paying the ransom doesn't mean getting all the data back because some decryption tools end up being faulty, while in other cases, ransomware cyberattackers walk away without providing decryption keys. Furthermore, sometimes, data stolen through ransomware cyberattacks is sold on the dark web and these eventually open new entry points for deeper levels of cybercriminal attacks, which can be stated as another gray area to be considered when calculating ransomware cost. This is perhaps the most durable and the hardest damage to quantify, which costs a loss in reputation. It could take years for businesses to recover their customers because of a damaged reputation. However, not only is this essential, a new upcoming law may lead to fines to the attacked organization, e.g., the NIS2 directive of the EU (see Sect. 1.6.3), which also is costly and has to be taken into account.

The recently published Gartner Report [15] states, "CIOs and their boards must start treating cybersecurity as a business decision, because total security cannot be guaranteed at any cost," which requires outcome-driven metrics and protection-level target forecast to estimate cybersecurity and cybersecurity costs. This can be achieved by (1) estimating the true cost factors of a ransomware attack; (2) decisions about ransomware attacks' impact and how to effectively mitigate ransomware risk (s); (3) combating the ransomware attack with prevention and recovery mechanisms; and finally (4) damage or loss in reputation impacts must be identified and quantified in case of a successfully intruded ransomware attack. The three most expensive industries to experience a data breach in 2022 are top–down, healthcare, finance, and technology.

Calculating costs of ransomware attacks, e.g., in the manufacturing industry, mostly considers only the cost of the ransom payment and the cost of getting the infected system back to regular operating conditions [16]. Besides these costs, there are other costs caused by a ransomware attack as described above that can be classified in a general sense as current, future direct and indirect costs. Estimating the current and future cost of a cybersecurity protection level includes costs for internal and external labor, consultants costs, application costs, infrastructure costs, and others [15]. However, cybersecurity cost varies with different levels of protection, meaning that higher or lower protection levels align with greater or lower cost. As reported in [12] "the granularity for costing security services is on a spectrum. One extreme would be to simply take the cost of all direct security personnel and carry out a high-level apportionment to each security service. The other end of the spectrum is to attempt to extract every single dollar spent on security, including embedded security-related costs in every application or piece of infrastructure. CIOs should aim somewhere in the middle. The intent here is to provide a level of costing that is good enough to make an informed decision, without spending an inordinate amount of time on unnecessary rigor in the costing process." The following objectives enable a realistic cost estimate [17]:

- *Persistent costs*: Present even if a ransomware attack occurred or not:

  - Ransomware mitigation measures to prevent an attack from occurring in the first place
  - Increasing costs for data backup
  - Cost of cybersecurity insurance, if existing

- *Costs in the event of a ransomware attack*:

  - Ransom payment
  - Recovery costs
  - Losses due to business interruption
  - Law enforcement and investigative costs
  - Personnel changes, if any
  - Lost productivity due to new procedures and safeguards
  - Reputational damage
  - Additional defensive preparations to mitigate the next attack
  - Cyber insurance deductible, if any

Considering the overall cost factors after a potential ransomware attack, it is obvious that they go much further than only paying a ransom and recovering the system for further use. As mentioned earlier, paying the ransom doesn't mean getting all data back because some decryption tools end up being faulty, while in other cases, ransomware cyberattackers walk away without providing decryption keys. Therefore, a much better option is to prevent the ransomware attacks from the very beginning, as far as possible, and to be well prepared defending against a ransomware incident, but this includes the costs to ensure a ransomware attack doesn't happen again. These typically include [18]:

- Infrastructure costs that help mitigate the risk of a ransomware attack
- Backup costs and labor costs relating to ransomware preparedness and incident response
- Cybersecurity insurance premiums, if any

It's also imperative to beef up defenses. Identify and remedy the root causes of the initial attack. Studies show that 80% of victims who paid ransoms are hit by ransomware again.

Using a backup strategy in case of a ransomware attack, the data of the targeted organization are restored quickly. Another benefit of a backup strategy is that it avoids paying ransom. Especially effective for preventing a ransomware attack is the consequent maintenance of the 3-2-1 backup strategy [19]. The 3-2-1 backup strategy simply states that three copies of the valuable assets, e.g., production data, and two backup copies on two different media with one copy off-site for disaster recovery should be available. The three copies mean that in addition to the last backup, it is always possible to fall back on the second from the last backup at any time. This is useful because even if the last backup was defective or already contains files encrypted by malware, there is still the second to last one, which normally does not have this defect. Two different volumes are useful because one can be lost just due to a defect, theft, or external damage.

Another defense strategy against a ransomware attack is the preparation of a disaster recovery plan or strategy. Disaster recovery involves restoring important IT services, servers, networks, data storage, and others after a disruption [20]. However, disaster recovery is not achievable in any case, but disaster recovery is essential to survive in business, and sometimes organizations have experienced ransomware attacks twice. An example is the CryptoLocker ransomware attack against Tencate, a multinational textile company in the Netherlands. The first time, one of its manufacturing facilities was hit with CryptoLocker, which infected all its file servers. At that time, Tencate was only able to restore from disk and, as a result, experienced 12 h of data loss, and was not able to recover for 2 weeks. The second time, a more advanced form of CryptoLocker hit a manufacturing facility, but this time, Tencate had a disaster recovery solution based on continuous data protection (CDP). However, CDP is not simply integrated, it is integral, providing a data protection that is perfectly suited for minimizing the disruption caused by any ransomware attack. In this regard, it offers a fast Recovery Point Objective (RPO) and a fast Recovery Time Objective (RTO) with always-on replication, application, and recovery, offered by Zerto [21]. The result was amazing: 10 s of data loss and full recovery under 10 min. In this context, Zerto tools enable a rapid maturity of data protection and recovery and a continuous data protection solution that integrates cyber-resilience. These features are required due to the actual situation that ransomware attacks are increasing in costs and frequency. Industry, public, and private organizations are yet not prepared for such attacks and dominate the news for not just paying high amounts of ransom but for the disruption of their business and services. There is no organization that is safe from ransomware attacks, and these attacks will only continue to escalate [22]. To determine whether an organization is

Table 6.5 Ransomware readiness checklist after [22]

Ransomware Prevention	
☐	**Active antivirus solution:** Have an up-to-date antivirus scanning solution to detect infected incoming files.
☐	**Active firewall protection:** Have active firewalls in place restricting port access on the network.
☐	**Multi-factor Authentication (MFA):** Use MFA where possible to restrict unauthorized access attempts.
☐	**Latest operation systems (OS) security updates:** Eliminate known vulnerabilities in the OS by applying available patches.
☐	**Latest application security updates:** Keep applications updated to eliminate known vulnerabilities.
☐	**Regular security training for users:** Make sure users are trained on how to avoid phishing and other malware attacks and how to keep their personal devices secure.
☐	**Regular security training for IT staff:** Make sure IT staff are trained and up-to-date on security best practices and technologies.
Disaster Recovery Solution	
☐	**RPO of seconds:** Have the ability to recover data in seconds before the attack happened.
☐	**RTO of minutes:** Have the ability to bring systems back online within minutes of the attack.
☐	**Individual file recovery:** Have the ability to recover individual files that were encrypted.
☐	**Application consistent failover and recovery:** Have the ability to recover entire applications that may span multiple servers with consistent data.
☐	**Full site failover and recovery:** Have the orchestration and automation to recover an entire site of data and servers.
☐	**Immutable data copies:** Have the option to make some data copies immutable to encryption.
☐	**Non-disruptive disaster recovery testing:** Have the ability to test frequently without affecting production servers.
Ransomware Response and Recovery Plan	
☐	**Ransomware incident response team:** Identify and document the IT staff responsibility for responding to an attack.
☐	**Network isolation plan:** Have a plan to isolate infested servers, data, and users from the network.
☐	**Malware detection:** Have a plan for finding and identifying the malware causing the attack.
☐	**Recovery detection:** Have the ability to test the recovery in isolation before recovering to production.
☐	**Fully documented disaster recovery runbook(s):** Make sure all procedures and plans are documented for disaster recovery (DR) testing, training, and for a DR response.

ready to conquer ransomware, Zerto developed a checklist based on ransomware prevention, disaster recovery solution, and ransomware recovery and response plan [22], as shown in Table 6.5.

Against this background, a successful ransomware attack means, besides ransom payment and loss of reputation, a significant downtime that causes Operational Expenditure (OpEx) costs, the money that is spent on a day-to-day basis to run the business.

As reported in [9], a significant downtime results in a loss in productivity because sensitive data has been stolen or corrupted and cannot be used. Thus, manufacturing products becomes more expensive and a potential delay in market entry can cause the loss of potential customers and business opportunities, as well as attracting new customers and business partners. In this context, the National Cyber Security Alliance (NCSA) [10] reported that 60% of Small and Medium Businesses (SMB) go out of business within 6 months of falling prey to a breach-in or ransomware attack. Also 90% of companies said the ransomware attack had affected their ability to function, and 86% of those infected in the private sector claimed they had lost business and/or revenue due to the ransomware attack. Therefore, better ransomware attack prevention is necessary in industrial organizations, which costs a higher amount of money spent on systems that are more sensitive. Such systems are manufacturing systems or secure data, information, and business assets that require higher levels of protection. However, cybersecurity decision makers often lack the knowledge and experience to deliver cybersecurity based on varying business needs, so they struggle to determine whether funding is enough to meet Protection-level Agreements or business expectation, which is one key finding in [15]. As indicated in [20], in case of a ransomware attack, the fast Recovery Point Objective (RPO) and the fast Recovery Time Objective (RTO) are the two important metrics to evaluate and enhance the maturity level in disaster recovery in ransomware prevention.

As conclusion, cost of data breaches diminished dramatically in organizations that follow RPO and RTO as well as implementing solid automated Machine and Deep Learning cybersecurity tools (see Chap. 8), zero trust systems, and regularly tested response plans against threat event attack to increase the cybersecurity maturity level.

6.3.1 Recovery Point Objective in Disaster Recovery

The Recovery Point Objective (RPO) is the age of files that must be recovered from a backup storage for normal (regular) operations to resume after a successfully executed ransomware attack, expressed as an amount of time. This time is thought of as the point in time before the event at which data can be successfully recovered, which means the time elapsed since the most recent reliable backup and can be specified in seconds, minutes, hours, or days. Thus, the purpose of RPO is to determine

- What is the minimum backup schedule frequency
- How much data can be lost after a disaster
- How far back the IT administration team should go to employ sufficient restoration without delaying data loss against expected Recovery Time Objective (RTO)

Thus, the RPO refers to the age of the backup data. The backup is required to restore operations after a hardware, program, or communication failure or a ransomware attack. This can vary from organization to organization. A smaller business might only need a backup since the most recent close of business, while many enterprises have a very short RPO, requiring a backup from the point of failure [23]. Therefore, the RPO has to meet the necessary requirements. However, the shorter the time in RPO, the higher the expected costs for data recovery.

With a specified RPO of 4 h, a new backup at least must be made within 4 h of the last data backup. If backups are made later, the loss of data is too high, and if backups are made too frequently, the backup costs increase. Thus, the RPO determines the loss tolerance, meaning the costs to compensate its potential losses. This specifies how much data is allowed for loss, and for this loss the time intervals for data backup must be correct [23].

RPO can determine

- How much data will be lost after a disaster
- How frequently data backup for disaster recovery is required

Let's assume a ransomware attack occurs; if the most recent data backup copy is from 10 h ago and the standard RPO for the business is 15 h, then the backup is still within the bounds specified by the RPO.

Essentially, RPO answers the question: "Up to what point in time can the recovery process move tolerably given the volume of data lost during that interval?" [24]. Factors that can affect RPOs include [25]:

- Maximum tolerable data loss for the specific organization
- Industry-specific factors—businesses dealing with sensitive information such as financial transactions or health records must update more often
- Data storage options, such as physical files versus cloud storage, can affect speed of recovery
- The cost of data loss and lost operations
- Compliance schemes include provisions for disaster recovery, data loss, and data availability that may affect businesses
- The cost of implementing disaster recovery solutions

Another factor that is additionally included in the evaluation of the RPO is the Time to Data (TtD). The TtD describes the required time to retrieve the backup data and deliver it to the restore location [26]. For backup, an interval is specified called snapshots. Defining the Snapshot Interval (SSI), RPO must not exceed SSI. In best case be equal [16]. Calculating RPO for a business organization considers the following five steps [27], as shown in Table 6.6.

6.3.2 Recovery Time Objective in Disaster Recovery

The Recovery Time Objective (RTO) is the maximum acceptable amount (length) of time that a computer system, network, or application can be down after a successful ransomware attack, a failure, or a disaster. Thus, the RTO defines the time span between the occurrence of the ransomware attack and the recovery to normal

Table 6.6 Calculating RPO after [27]

Calculation step	Calculation activity	Objectives
1	Look at how often files update	Set up RPO to match the update frequency that ensures the most up-to-date information is retrievable. For example, digital files and transactions update every 30 min, setting an RPO for every 30–40 helps to ensure have continuous access to recent information with minimal data loss.
2	Review the goals of the business continuity plan (BCP)	BCP: System of processes and technology helps businesses continue to deliver products and services. RPO often supports goals of BCP, reviews each element carefully to determine separate RPOs of various time allotments for each business unit. For example, financial transactions and critical data processes of a national bank are vital, requiring much shorter RPO times than the human resource files of personnel records, which get updated less frequently and can sustain a longer RPO time.
3	Consider industry standards	Institutions have unique RPO needs, can consider industry standards to guide calculating RPOs for business units. Common intervals for RPO: **0–1 h:** Use the shortest time frame for critical operational elements that can't afford to lose an hour of data, typically because they're high volume, dynamic, or difficult to recreate. For example, online banking transactions, patient records, or stock market trading activities. **1–4 h:** Use this time frame for business units deemed semi-critical, where only some data loss is acceptable. For example, customer online chat logs, file servers, or social media records. **4–12 h:** A time frame of this length might get used for business units that update daily or less frequently, like advertising and marketing, sales or operational statistics data. Typically, these units rarely have a grave impact on a business if affected. **13–24 h:** Setting longer RPO time frame for important, but not critical data and business units rarely exceed 24 h. This settings can be used for things like purchase orders, inventory control, or personal files.
4	Establish and approve each RPO	After factoring in all concerns for each element of data management, establish the RPOs and have them approved by leadership for IT teams or business partners to implement. Properly document the process and keep the records to refer to and use as a baseline when reassessing or

(continued)

Table 6.6 (continued)

Calculation step	Calculation activity	Objectives
		adjusting them. Depending on the role in an organization, follow existing RPO processes and procedures or help establish and create them.
5	Analyze RPO settings consistently	As a company grows or business continuity plans change, so might the RPO objectives. Consider establishing a routine and frequent review and analysis of how well existing RPO settings perform and if any need adjusting. This is an important step even though it comes last. If a data loss or system failure happens, an ad hoc review and in-depth analysis of how RPO and RTO performed can help to learn meaningful insights into the overall data management system.

(regular) operation and the amount of revenue lost per unit time because of the ransomware incident. These factors, in turn, depend on the affected objects and application(s). RTO measures are in seconds, minutes, hours, or days [28]. It is an important consideration in a Disaster Recovery Plan (DRP) [29]. The DRP provides step-by-step instructions to be followed during an incident in order to restore service and/or operation. DPR generally outlines [30]

- The nature of the disaster
- The information about the systems such as server build documentation, architectural designs, and details about templates used for cloud server provisioning
- Any tools or services that can aid in the recovery process, such as backup tools
- The team responsible for recovering the failed systems
- The protocols for communicating with team members and documenting recovery steps

Once an organization has defined the RTO for an application, the team responsible for cybersecurity can decide which DRP methods are best suited to the situation. The DRP covers an organization's ability to respond to and recover from an unexpected cyberattack like ransomware that negatively affects business operations. Thus, DRP methods enable organizations to regain use of critical systems and IT infrastructure resources as soon as possible after a disaster recovery occurs. The DRP process involves

- Identifying potential threat event attack to IT operations
- Establishing procedures for responding to and resolving outages
- Mitigating the cybersecurity risk and impact of outages to operations

In this context, DRP is a subset of Business Continuity Planning (BCP). BCP focusses on restoring the essential function of an organization as determined by a Business Impact Analysis (BIA). Best practice is if RTO for a given application is

1 h, redundant data backup on external drives may be the best solution. If the RTO is 5 days, then tape or off-site cloud-storage may be more practical [26].

Calculating RTO requires determining how quickly the recovery process for a given application, service, system, or data needs to happen after a ransomware incident, based on the loss tolerance the organization has as part of its BIA. The BIA is a systematic process to determine and evaluate the potential effects of an interruption to critical business operations as a result of a disaster, e.g., through a ransomware attack and others. Furthermore, a BIA is an essential component of an organization's BCP. It includes an exploratory component to reveal any threat event attacks and vulnerabilities and a planning component to develop strategies for minimizing cyberattack risk exposure in business, which refer to the quantified potential loss from current business activity. The result is a BIA report, which describes the potential risks specific to the respective organization [31]. Therefore, for a ransomware attack, defining the loss tolerance is required, which involves how much operational time an organization can afford to lose after a ransomware attack before normal (regular) business operations must resume. In this regard, complete inventory of the organization's (digital) environment is the first step in the RTO process, including all systems, business-critical applications, data and business assets. Without an accurate inventory, there is no way to accurately determine RTO [29].

After completing the inventory part, the next step is to evaluate the value of each service and business-critical application in terms of how much it contributes to how an organization operates and conducts business. That value should be determined based on duration of time and as granular as possible. The application value linked to any existing Service Level Agreement (SLA), which defines how available a service needs to be, and may include penalties if those service levels are not met [29].

Understanding what is running and what operational value of the running systems and applications is important is necessary to be able to calculate RTO. However, there can be different RTO requirements based on application priorities as determined by the value the application brings to the organization [29].

Calculating RTO requires determining how quick the recovery process for a given application, service, system, or data needs to happen after, e.g., a ransomware incident, based on the organization's loss tolerance as part of its BIA. If a regular business operation is infected by a ransomware attack, RTO decision is required, based on the time the event occurs, the elapsed time, and the recovery to regular business operation. However, defining RTO is a critical component of a DRP, as the goal of disaster recovery is to have a strategy in place that helps the business recover and restore regular business operations. Hence, with RTO in place as top-level goal, an organization can align its data backup and failover policies and have the required level of additional services available for deployment to ensure the desired speed of recovery. Without RTO, an organization does not know recovery speed from downtime and data loss after a ransomware attack. Therefore, DRP is about being prepared for unexpected outages, but being prepared requires having some guidelines or a plan to know how long it will take to recover [26]. In case a ransomware attack causes an application or service outage, the objective set for RTO may be variable. Against this background, a mission-critical application or production process may have lower RTO, while the associated loss tolerance is lower. Due to the existing correlation of costs, downtimes, and losses, RTO design focuses on where the costs compensate the downtimes and losses during the recovery. This must be considered separately for each subsystem or subcomponent, because each of which has its own costs and loss curves when defining RTO [16]. In this regard, a daily backup enables to have a copy of process data and business assets as it was at the end of the previous business day. This requires determining that a data loss at any point in the day be replaced making use of the previous day's backup, meaning protecting data with a daily backup. Nevertheless, if an organization has determined that it cannot afford to lose more than a few hours' data, then a daily backup is not enough, and the organization needs an ongoing data backup through the day. For example, a data hard drive needs a mirrored drive, while an SQL database needs a transaction-level backup which runs every 10 or 15 min [32].

6.3.3   Design of Recovery Point Objective and Recovery Time Objective in Disaster Recovery

For the design of RPO and RTO, the goal is to implement a Disaster Recovery Plan (DRP), which, in practice, means a backup strategy that is as efficient as possible. This backup strategy should bring the disturbed system back to a normal (regular) operational or functional state.

In order to design the backup system practically, the possible downtime costs are considered together with the running costs for the implementation of the required Disaster Recovery Plan (DRP). It is particularly important to notice that time-to-data has a negative impact on both RPO and RTO. The reason is that RPO is not completed until backup data is available for disaster recovery. RTO in disaster recovery can only happen after backup data is available. Therefore, the Time to Detect (TtD) should be as small as possible when designing RPO and RTO in order to keep the negative impact as small as possible.

When designing RTO and RPO for organizations' computer systems, networks, and others, they may not necessarily have the same scope. In IT systems, the aim is to implement the smallest possible RPO, as no data should be lost. For OT systems, the most important aim is to get the systems after a disruption back to a functional state as quickly as possible. In this context, the term functional does not describe achieving 100% of the original operational state before disruption. An acceptable partial state defined as adequate for some applications. With this approach, the necessary system functionality is restored and the RTO is shortened. The remaining missing features are not required for the necessary functionality restored afterwards, after the required partial state has been achieved [16]. Nevertheless, the key recovery objectives that define how long the business can afford to be offline and how much data loss it can tolerate is determined by

- RTO, which sets the maximum length of time it should take to restore normal (regular) operations following an outage or data loss
- RPO, which sets the maximum amount of data the organization can tolerate losing, measured in time from the moment a failure/ransomware occurs to the last valid data backup. For example, if a failure occurs now and the last full data backup was 24 h ago, the RPO is 24 h

Thus, the DRP defines how—and how quickly—to recover from a ransomware incident that unexpectedly renders critical apps or operations and make data inaccessible. As such, it prepares for getting back online fast and minimizes damage to the business.

6.4 Loss of Reputation and Its Prevention

Damage of reputation after a successful intruded ransomware attack, ranks number two in the list of cybersecurity risk(s), as stated in Aon's 2019 Global Risk Management Survey [32]. The survey interviewed thousands of risk managers across 60 countries and 33 industries and revealed reputational damage through cyberattacks rated as businesses' top three risks. Top risk #1 is a globally ranked slowdown in economy, followed by reputation/brand as the #2 concern. A Forbes Insight report found that 46% suffered reputational damage(s) as the result of a data breach and 19% of organizations suffered reputation and brand damage as the result of a third-party security breach [33]. If a ransomware attack happens, the response to the data breach can break down the reputation of the targeted organization(s). According to Forbes, those that stay in the headlines are the breaches questioning an organization's response and criticizing its communication. In this regard, the growing amount of threat events risk(s) surrounding critical access is not a matter of if, but rather a matter of when an organization is targeted by cybercriminals. Therefore, understanding how to cyber-secure an organization's most valuable assets by identifying critical assets and access points is required, and thereafter implementing optimized cybersecurity tools, to support minimizing losing reputation after a ransomware attack.

A strong counterexample is Norsk Hydro. When the Norwegian energy company experienced a ransomware attack in 2019, it refused to pay. Instead, the company decided to consult supply chain cybersecurity experts to inspect 30,000 employee credentials and get to the root of the attack. By taking responsibility and better protect their systems in the future, the company saved reputational damage and put themselves in a better position if another threat event attack occurs. Now the question is: How other organizations can follow best practices to prevent both organizational and reputational damage? The answer, given in [33] is: "The best defense is a good offense, meaning to protect before a breach-in even becomes a possibility." This requires, on the one hand, Critical Process Management (CPM) or, on the other, management of all critical or sensitive access points and critical assets within the organization. Critical access points are entryways to critical assets, like computer systems, networks, databases, infrastructure resources, information and operational technology, and others. Access is considered as critical when two of the following three factors are at high risk: identities, assets, or privileges [34]. Therefore, a Critical Access Management (CAM) is required minimizing reputational damage from ransomware attacks. Implementing the CAM requires identifying the organization's critical access points [34]:

- *Identify user*: User needs access. To determine if user(s) has a high-risk level, their identity traits are important to know. Is the user an employee? Has the user followed access policy rules in the past or has he broken access rules and exhibits poor behavior? Is the user a third party representative? Is the watching organization able to track and control the third party access, or does this access fall outside of the organization's watching systems?
- *Assets*: Protected items owned by an organization. An asset is considered at high-risk level based on what would happen if misused in any way. For example, there is minimal damage if an email account is breached, which is not at high risk. If the server of a software provider is breached, there are consequences that affect not only the organization but many of the customers that rely on the server for daily operations.
- *Privileges*: Rights or permissions needed to access an asset. Access of assets categorized as critical if the privileges needed for accessing are high. Privileges not only need a password, but also a high level of clearance and authentication to access the asset.

After identifying the critical access points within an organization, implementing the pillars of the access management occur, which work together to fully secure access and create a comprehensive cybersecurity strategy [34]:

- *Access governance*: Consists of the systems and processes put in place to ensure an access policy followed as closely as possible. Access policies are rules laid down by an organization that state who have access, and what privileges are needed to access an asset. Access governance best works when applying role-based access control, whereby access is distributed based on the responsibility and the principle of least privilege, whereby the minimum amount access needed to run the business and nothing more. The access governance best practices are described in [35] and are followed here:

  - *Apply the principle of least privilege when defining access policies*:
    Implementation of access governance work needs clear and defined access policies. Access policies are rules that state who should have access to what and what privileges a user(s) should have when accessing an asset.
    As an access governance best practice, policies should align with least privilege access, meaning user(s) only have the minimum access needed to do their job. Access governance ensures that only those with permitted access be granted access to a critical asset. These access policies are a framework, a

standard, and a best-case scenario for what assets a user should have access to as part of their job.

Under the umbrella of critical access, these policies must enforce heavily, meaning no loopholes or secret passageways into those access points are allowed. Whether access is limited by a job role, time, or a variety of other factors, following the least privilege principle will create strong access governance and robust access policies.

– *Create close linkage between human resources data and access rights*:

Human Resources systems are inherently equipped to help an organization define access rights and implement access governance, because they already track job responsibility, function, and changes to employment status. Human Resources systems can set up access by a given job role, and often have the ability to provision and de-provision access when a given role changes.

In addition, these systems have all the data needed for fine-grained access controls. This allows an organization to define access rights by key points in a job cycle like hiring, termination, promotion, and others that build an automatic fail-safe within an access policy.

However, Human Resource systems cannot help when it comes to third-parties' issues. Investing in a bespoke system for managing the identities and access of third parties is a best practice for implementing full access governance, as third parties are often high risk.

– *Regular user access review.*

Make use of user access reviews as a double-checking task. This enables to know who has access on the access controls previously set, which helps to know whether those access controls are still in place? Have users been properly de-provisioned or is access creep becoming an inside threat?

Access reviews are the process of reviewing all identities' access rights and ensuring adhering to the concept of least privilege. Another way to think of an access review is as a process to identify gaps between an access policy and access rights, and ensure those gaps are legitimate and still required.

Conducting regular access reviews for those critical assets and access points is an access governance best practice that helps ensure that any access policies and controls are working the way an organization intended. Access reviews are especially important when dealing with third parties, whose data and user access will not be a part of an organization's internal Human Resources systems.

• *Access control*: Mechanism(s) used to reduce risk, increase visibility, and increase friction in granting access. When access controls zero trust network access and fine-grained access controls are established, it adds friction to a user's movement through a network or system, and helps minimize their exposure and lateral movement, and thus the amount of damage they could possibly cause. Basics of access control in network security are described in [36] and are shown here:

– *Fine grained access controls*:

Allow an organization, even a user, or a department like Information Technology (IT) or Human Resources (HR), to control and limit a user's access rights. These kinds of controls affect how a user accesses assets, whether it is adding time-based controls, a monitoring measure or an allowed limit on how often to access.

– *Zero Trust Network Access (ZTNA)*:

Implementing a full zero trust network removes any implicit trust, regardless of the access or the assets. With this model, both insider and outsider access needs to be verified and authenticated every time they request access. ZTNA is just one part of a Zero Trust framework that every organization should employ.

– *Multi-Factor Authentication (MFA)*:

Common access controls applies to the specific user requesting access. Think of the two-factor authentication you need to log into your bank account. It employs multiple methods (password, a phone notification, an email, or a fingerprint), to double-check the user's identity before granting access.

– *Privileged credential management*:

Credentials can become major threat events if they are not properly stored and managed. Privileged credential management is exactly that—a system that allows one to vault and obfuscate privileged credentials.

– *Access control best practices*:

Understanding access control is important, but implementing it on top of access governance is better. In case an organization has identified critical access points and assets that need some extra security, a few access control best practices are available that can employ to ward off cyberattacks:

Focused use of access controls:

Implementing access controls can be daunting, especially for an organization with limited resources or capacity. One access control best practice is to focus on what's most critical, and make sure that is the area with the metaphorical security cameras and keypads and laser beams. Implement as much access control as you need, where you need it.

Combination of access controls:

A longer password is harder to hack, and more access controls are harder for a bad actor to work through it. For critical assets, employing more than one control to add layers of security is another access control best practice. Maybe it is multi-factor authentication and a time limit, or a limited number of accesses over a quarter, plus a time limit on that access.

Implement zero trust for critical cases:

It's easier to say you don't trust users—especially internal ones—than it is to actually remove that trust when it comes to access. For critical access, an organization should make sure that every user, no matter how much they can theoretically be trusted, has to go through the same procedures to access critical assets. No special privileges, no one-off cases, and no slacking on

access controls. Everyone is treated like a threat event incident to make sure
every asset is safe.

- *Access monitoring*: Observation and analytics of a user's behavior while they are
  accessing an asset. The components of asset monitoring are described in [34]
  which are taken as a basis here:

  - *Proactive monitoring*:
    Observation or analysis of a session with no pre-defined reason. This kind
    of monitoring is often conducted in real time, or as close as possible to a broad
    set of sessions. This kind of monitoring is a real-time, multitude of angles kind
    of watching that offers up a broad, thorough view of what is happening in a
    system.
  - *Reactive monitoring*:
    Observation or analysis after a session with specific reason happens.
    Reactive monitoring requires systems and tools to be in place to record
    sessions. Generally applied to a single session or a small subset of sessions,
    and most commonly used as part of an incident investigation. It is after the
    fact, and very targeted in what the monitoring is watching for.
  - *Observation*:
    Collection of informationor reviews of sessions. Observation is required for
    analysis but not vice versa. Strong access monitoring does not exist without
    observation, which can take forms such as a video recording of a session, a
    text-based audit, or a collection of session data.
  - *Analysis*:
    Interrogation of the information or data collected. Used in both proactive
    and reactive use cases. Once an observation is complete, an analysis of a given
    session or data can occur.

    Best practices in access monitoring are described in [37] and are
    followed here:
  - *Complement analysis with observation*:
    As stated, one can have observation without analysis, but cannot have
    analysis without observation. As a best practice, a strong access monitoring
    strategy uses both, and both can work together to create a full picture of what's
    happening within a system. A proactive analysis of EHR records may flag a
    suspicious event, but a reactive observation of a session can provide additional
    context, and highlight the details of what happened.
  - *Use proactive observation sparingly*:
    Because it often occurs in real-time, proactive observation is the most time-
    consuming, and often ineffective, form of access monitoring. Without param-
    eters in place, a user could be real-time observing too much for too long
    without understanding what they are observing. However, it does have bene-
    fits if used sparingly and strategically. For high-risk, low-frequency access
    points and assets, employing another set of eyes can protect what is most
    critical for an organization.

- *Proactive monitoring for high frequency, high risk accesses*:
    High-frequency and high-risk accesses, like those to patient files, should have proactive access monitoring utilized as a best practice. By using proactive analysis of the session data, cases of anomalies, threat events, or misuse have quickly been identified. In addition, subsequent reactive observation can confirm or deny the suspicion and provide more critical context as part of an investigation.

6.5 Exercises

What is meant by the term *Ransomware?*
Describe the main characteristics and capabilities of Ransomware.
What is meant by the term *Ransomware Attack?*
Describe the main characteristics and capabilities of at least five Ransomware Attacks.
What is meant by the term *Cyber Activists?*
Describe the main characteristics and capabilities of a Cyber Activist.
What is meant by the term *Cybercriminal?*
Describe the main characteristics and capabilities of Cybercriminals.
What is meant by the term *Petya Attack?*
Describe the main characteristics of a Petya and NotPetya Attack.
What is meant by the term *Spreading Ransomware?*
Describe the characteristics and capabilities of Spreading Ransomware.
What is meant be the term *Drive-by Download?*
Describe the characteristics and capabilities of Drive-by Download.
What is meant by the term *Distribution Technique?*
Describe the main characteristics and capabilities of at least four Distribution Techniques.
What is meant by the term *Ransomware Protection?*
Describe the main characteristics and capabilities of at least five Ransomware Protection techniques.
What is meant by the term *Ransomware IT/OT Attack?*
Describe the main characteristics and differences between IT and OT Ransomware Attacks.
What is meant by the term *Cost of Ransomware Attack?*
Describe the main characteristics of Costs in Ransomware Attacks.
What is meant by the term *Recovery Point Objective?*
Describe the main characteristics and capabilities of the Recovery Point Objective.
What is meant by the term *Recovery Time Objective?*
Describe the main characteristics of the Recovery Time Objective.
What is meant by the term *Loss of Reputation through a Ransomware Attack?*
Describe the main characteristics and capabilities of Loss of Reputation through a Ransomware Attack.
What is meant by the term *Preventing Reputation Loss in case of a Ransomware Attack?*

Describe an example on how to avoid Loss of Reputation in case of a Ransomware Attack.

What is meant by the term *Access Governance?*

Describe the main characteristics of Access Governess.

What is meant by the term *Privilege Access?*

Describe the main characteristics of Privilege Access.

What is meant by the term *Access Control?*

Describe the main characteristics and capabilities of Access Control.

What is meant by the term *Multi-Factor Authentication?*

Describe the main characteristics and capabilities of Multi-Factor Authentication.

What is meant by the term *Privileged Credential Management?*

Describe the main characteristics and capabilities of Privileged Credential Management.

References

1. Warikoo, A.: Proposed Methodology for Cyber Criminal Profiling. In: Information Security Journal: A Global Perspective, Vol. 23, No. 4–6, pp. 172–178, 2014
2. Billois, G., Lahoud, M.: Cybercrime – Ransomware: Number One Cyber Threat. In: Whitepaper Institute Montaigne, 2021. https://www.institutmontaigne.org/en/analysis/cybercrime-ransomware-number-one-cyber-threat (Accessed 12.2022)
3. State of Ransomware: Invest now or pay later. CRA Business Intelligence Study, 2022. https://resources.menlosecurity.com/reports/state-of-ransomware-invest-now-or-pay-later (Accessed 12.2022)
4. Ransomware. Imperva Whitepaper, 2022. https://www.imperva.com/learn/application-security/ransomware/A (Accessed 12.2022)
5. Challita, A.: The Four Most Popular Methods Hackers use to Spread Ransoware. Whitepaper ITProPortal, 2022. https://www.itproportal.com/features/the-four-most-popular-methods-hackers-use-to-spread-ransomware/ (Accessed 12.2022)
6. Arntz, P.: Threat Spotlight: CrySis, aka Dharma ransomware, causes a Crisis for businesses. Malwarebytes Lab Whitepaper, 2019. https://www.malwarebytes.com/blog/news/2019/05/threat-spotlight-crysis-aka-dharma-ransomware-causing-a-crisis-for-businesses (Accessed 12.2022)
7. Beltov, M.: LowLevel04 Ransomware Virus – Removal Steps and Protection Updates. 2016. https://bestsecuritysearch.com/lowlevel04-ransomware-virus-removal-steps-protection-updates/ (Accessed 12.2022)
8. Malware Analysis Report: Nemucod Ransomware. Center for Internet Security Whitepaper. 2022. https://www.cisecurity.org/insights/blog/malware-analysis-report-nemucod-ransomware (Accessed 12.2022)
9. Threat Landscape Dashboard RIG Exploit Kit. https://www.mcafee.com/enterprise/en-us/threat-center/threat-landscape-dashboard/exploit-kits-details.rig-exploit-kit.html (Accessed 12.2022)
10. The State of Ransomware 2022 – Sophos News. https://news.sophos.com/en-us/2022/04/27/the-state-of-the-ransomware-2022/ (Accessed 12.2022)
11. Unit 42 Ransomware Threat Report 2022. https://www.paloaltonetworks.com/content/dam/pan/en_US/assets/pdf/reports/2022-unit42-ransomware-threat-report-final.pdf (Accessed 12.2022)

12. Why Ransomware Attacks on OT Systems are Growing. Cylo Team Blog, 2022. https://cyolo. io/blog/ot/why-ransomware-attacks-on-ot-systems-are-growing/ (Accessed 12.2022)
13. https://www.cisa.gov/publication/ransomware-awareness-campaign-fact-sheet   (Accessed 12.2022)
14. Shmuely, H.: How to increase the Security of OT Systems with Zero Trust. Cyolo Blog, 2022. https://www.cyolo.io/blog/ot/how-to-icrease-the-security-of-ot-systems-with-zero-trust/ (Accessed 12.2022)
15. Buchanan, S., Proctor, P., Hayes, B.: Measure the Cost of Cybersecurity Protection. Gartner Report ID G00764671, 2022
16. Dukin, J., Stellwag, D.: Cost Factors of a Ransomware Attack –A Description of the possible Threat of a Ransomware Attack on IT and OT systems with reference to the possible Costs with Consideration of reasonable RTO and RPO. Student Project t the Course IoT and IIoT at TU Clausthal, Germany, 2022
17. Grimes, R.A.: Ransomware Protection Playbook. John Wiley & Sons Inc., 2021
18. Sjouwerman, S.: Seven Factors Analyzing Ransomware's Cost to Business. In: Forbes Technology Council Post, July 29, 2021. https://www.forbes.com/sites/forbestechcouncil/2021/07/2 9/seven-factors-analyzing-ransomwares-cost-to-business/?sh=65dc91a92e98   (Accessed 12-2022)
19. Möller, D.P.F.: Cybersecurity in Digital Transformation: Scope and Application. Springer Nature 2020
20. Luber, S., Schmitz, P.: Security Insider – Definition Disaster Recovery. In: Vogel Communications Group, 2020. https://www.security-insider.de/was-ist-disaster-recovery-a-732206/ (Accessed 12.2022)
21. Ransomware Recovery. Zerto Whitepaper, 2022. https://www.zerto.com/resources/essential-guides/ransomware-recovery-guide/#5 (Accessed 12.2022)
22. Ransomware Readiness Checklist. In: Zero Whitepaper, 2022. https://www.zerto.com/wp-content/uploads/2022/02/Ransomware_Iceberg-Infographic_Checklist-1.pdf   (Accessed 12.2022)
23. Litone, M.: Mission-critical Network Planning. Artech House Publ. 2003
24. Marget, A.: RPO and RTO: What are they and How to Calculate Them. In: Unitrends Whitepaper 2022. https://www.unitrends.com/blog/rpo-rto (Accessed 12.2022)
25. https://www.druva.com/glossary/what-is-a-recovery-point-objective-definition-and-related-faqs/ (Accessed 12.2022)
26. A Salamanca, F., Jimenez, J.: Implementing Automated Replication for Cost Effective Disaster Recovery. 2011. https://dsimg.ubmus.net/envelope/157842/313522/1332863421_3_21_ Implementing_automated_replication_for_cost_effective_disaster_recovery   (Accessed 12.2022)
27. What is a Recovery Point Object and How to Calculate one. In: Indee Editorial Team, 2022. https://www.indeed.com/career-advice/career-development/recovery-point-objective (Accessed 12.2022)
28. https://www.f5.com/services/resources/glossary/recovery-time-objective-rto   (Accessed 12.2022)
29. Kerner, S.M.: Recovery Time Objective: In: techtarget Notes, 2022. https://www.techtarget. com/whatis/definition/recovery-time-objective-RTO (Accessed 12.2022)
30. https://www.gremlin.com/community/tutorials/testing-disaster-recovery-with-chaos-engineer ing/ (Accessed 12.2022)
31. Kirvan, P., Sliwa, C.: What is Business Impact Analysis?. Techtarget Whitepaper, 2022. https:// www.techtargete.com/searchstorage/definition/business-impact-analysis (Accessed 12.2022)
32. Global Risk Management Survey. Aon PLC, 2019
33. Taylor, T.: How Reputational Damage from a Data Breach affects Consumer Perception. Securelink, 2022. https://www.securelink.com/blog/reputation-risks-how-cyberattacks-affect-consumer-perception/ (Accessed 12.2022)

34. Secure your Mission-Critical Systems and fill Security Gaps in Access Management. https://www.securelink.com/why-choose-securelink/ (Accessed 12.2022)
35. Taylor, T.: What is Access Governance. SecureLink, 2021. https://www.securelink.com/blog/what-is-access-governance/ (Accessed 12.2022)
36. Taylor, T.: What is Access Control. SecureLink, 2021. https://www.securelink.com/blog/what-is-access-control/ (Accessed 12.2022)
37. Taylor, T.: What is Access Monitoring. https://securelink.com/what-is-access-monitoring/ (Accessed 12.2022)

Chapter 7
Cybersecurity Maturity Models and SWOT Analysis

Abstract Digital transformation has become an integral part of everyday life. With innovative digital technologies such as Artificial Intelligence, Big Data and Analytics, Cloud Computing and Services, Industrial Internet of Things, Machine Learning, and others, industrial, public, and private organizations face challenges and pressure in adapting their business models, processes, procedures, services, and others to the digital reality. In this regard, digitization is becoming the engine of far-reaching digital transformation that encompasses and changes all organizations, from business to society. For digital transformation, it is particularly characteristic that it not only influences production in industry, but also the entire corporate organization, their corporate culture, and the conditions of employees' work in the organization. Therefore, digital transformation needs, besides pure digital technologies, maturity of skills for their successful usage, which requires skilled executives and employees. Thus, maturity in digital transformation enables organizations to reveal its transformative power due to resultant new and innovative business models, products, services, and others, which exert enormous pressure on traditional business models. However, complexity and connectivity in digital transformations have a negative side effect through cybercriminals' attack possibilities, which make maturity in cybersecurity awareness one of the most important topics. Cybersecurity requires a cybersecurity plan or strategy of organizations because cybersecurity spans many areas, including, but not limited to, data security, information security, operational security, and others. Nevertheless, the enormous changes of digitization require methods enabling mapping organizations current state of their cybersecurity awareness and/or cybersecurity strategy, to defend cyber-criminal attacks. Therefore, measures required are quantifying the actual state in cybersecurity awareness and/or cybersecurity defense strategy to identify the essential actions to achieve a target state in cybersecurity. This is where cybersecurity Maturity Models come into play. Cybersecurity maturity models describe an anticipated desired or necessary development path of criteria in consecutive discrete ranks, starting with an initial state of zero up to a complete cybersecurity maturity level, e.g., five. Therefore, the Maturity Model is a suitable methodology for the systematic development and gradual improvement of skills, processes, structures, and further essential conditions to organizations in the context of cybersecurity awareness and—defense strategy. The prerequisite for this is that the characteristics of the individual maturity development stages be clearly defined beforehand, so user(s) get an overview of what actually is necessary to achieve the next maturity level. This is precisely why cybersecurity maturity models are a suitable instrument that enables management to recognize the necessary changes in organizations and to approach the transformation process in a structured manner. Besides the cybersecurity maturity model, another method is available, recording the economic and technical initial situation in an organization, the Strength-Weakness-Opportunities-Threats (SWOT) analysis. SWOT is a pragmatic approach capturing the current state of specific and relevant organizational characteristics, to initiate further improvement. This is the case for existing omission in the IT area, insufficient implementation of digital technologies, and others. In this context, considering the internal and external impact on organizations is an important issue, archived through the SWOT analysis. In this context, Chap. 7 introduces in Sect. 7.1 Cybersecurity Maturity Models and SWOT Analysis from a general perspective. Section 7.2 refers to Maturity Index and Maturity Models, Sect. 7.2.1 Maturity Index, and Sect. 7.2.2 Maturity Models, and all their different approaches are considered. Section 7.2 focusses on Maturity Models after ISO 9004:2008. In Sect. 7.2.3, the focus lies on Cybersecurity Models, followed by Sect. 7.4 with the topic Cybersecurity Maturity Best Practice Model. In Sect. 7.5, the SWOT Analysis method is introduced from a general perspective, whereby Sect. 7.5.1 focusses on SWOT Best Practice Analysis, and Sect. 7.5.2 refers to a SWOT Company Analysis, while Sect. 7.5.2.2 shows a SWOT Cybersecurity Analysis. Section 7.6 contains comprehensive questions from the topics Cybersecurity Maturity Models and SWOT Analysis, followed by section "References", which covers references for further reading.

7.1 Introduction

The advances in digital transformation show an enormous potential for change in all areas of industrial, public, and private organizations. This fact has its origin in the fourth technological wave, termed Industry 4.0, with its availability and use of digital and new technologies. As a measure of an organization's ability to create value through digital transformation, digital maturity is a key predictor of success launching digital transformation. Business with high digital maturity level has a competitive advantage along with multiple performance indicators. These indicators are: revenue growth, time to market, cost efficiency, product quality, customer satisfaction, and others. In this context, businesses with low levels of digital maturity struggle to achieve these indicators. However, with the availability of digital transformation in all areas of today's life, cyberattacks also have grown over the last years to a level that constantly negatively affects business performance by leveraging vulnerabilities of safety-critical industrial infrastructure networks and business models. This development led to increased needs in cybersecurity awareness, which requires that organizations must identify and evaluate their attack surfaces and the impact of potentially successful cyberattacks to defend their critical data and business assets. A measure of an organization's ability to defend their critical data and business assets from cybersecurity risks is the Maturity Index and the Cybersecurity Maturity Model (CSMM). The CSMM provides a directive to organizations in evaluating their preparedness against continuously evolving threat event attacks. Assessing the maturity level of organizations' cybersecurity status helps to establish an effective approach addressing and managing cybersecurity risks in the industrial sectors. Assessing the outcome of cybersecurity maturity models results in different criteria. Cybersecurity Maturity Models (CSMMs)

- Enable identifying the areas of improvement in organizations' cybersecurity controls and processes
- Benchmark to measure performance to continuously improve cybersecurity activities in alignment with organizations' compliance objective
- Enable a continuous improvement in cybersecurity activities allowing organizations to react to the ever-growing different types of cyberattacks
- Assessment help to design an actionable guideline that makes the organizations' compliance process achievable
- Assess to enable organizations to estimate their cybersecurity posture to prioritize activities that need immediate action(s).

With this Cybersecurity Maturity Model Assessment (CSMMA) organizations gain better control over their cybersecurity strategies and/or programs to defend successfully against cyberattacks. In order to achieve an appropriate Cybersecurity Maturity Level (CSML), at first, the organization must meet all criteria described in the initial maturity levels from 0 to 2, before the organization can try to operate its production and business activities at maturity level 3 and higher, as explained in Sect. 7.2. For this purpose, key process areas (KPAs) are derived that characterize each level of the Cybersecurity Maturity Model. KPAs are a set of related practices that, when implemented together; satisfy goals to improve a given area of activities to enhance cybersecurity. Against this background, the Maturity Model is a suitable method for the systematic development of maturity levels from level 0 to level 5 and the required gradual improvement of skills, processes, structures, and further essential conditions to organizations' cybersecurity plan, based on maturity levels. The prerequisite for this is that the characteristics of the individual development stages from maturity level 0 to maturity level 5 clearly defined beforehand, so user(s) get an overview of what actually is necessary to achieve the next maturity level. However, there is another important method available recording the economic and technical initial situation in an organization, the Strength-Weakness-Opportunities-Threats (SWOT) analysis. SWOT is a pragmatic approach capturing the current state of specific and relevant organizational characteristics like cybersecurity risk(s), to initiate further measures for improvement.

7.2 Maturity Index and Maturity Models

The term maturity refers to the state of being complete, comprehensive, ready, or perfect. Therefore, a maturity assessment is to measure the current maturity level of a certain aspect of an organization in a meaningful way, enabling stakeholders to identify strengths and improvement points and prioritize what to do to reach higher maturity levels accordingly.

7.2.1 Maturity Index

The term maturity index is a measurement used to determine an organization's operational business cybersecurity preparedness and capabilities is mature. Thus, the maturity index must consistently relate to all operational activities and businesses of an organization and must be consistent and objective. In this regard, an index is made up of a series of questions, broken into several categories such as technical responses, organizational threats, capacity building, organization wide cybersecurity plan or strategy, cybersecurity readiness, and others. In a report published by the International Telecommunication Union [1], fifteen indexes described and categorized into three groups. These indexes include Cyber Readiness Index (CRI), Accenture Security Index (ASI), National Cybersecurity Index (NCSI), Information Risk Maturity Index (IRMI), Kaspersky Cybersecurity Index (KCSI), and others. All maturity indexes classified their capabilities in different domains, such as business alignment, cyberattack response readiness, resilience readiness, strategic content of threat event attack, and others. Nevertheless, each index relies on a different process to gauge cybersecurity maturity.

The approach presented in [2] builds a cybersecurity index, based on an audit for measuring formative variables for the object of interest to generate the dimensions. For the use case in [2], these variables are planning, performing, and reporting. Thereafter, assessing the content validity of the object is conducted. To ensure content validity, the maturity index was evaluated by literature, professional standards and interviews. The next step is to specify a formal measurement model of the considered dimensions. Afterwards, a formal measurement model of the considered dimensions is considered.

The dimension planning comprises the indicator's reactiveness and strategic orientation in planning [3], risk-based planning [4] and cybersecurity frameworks, which considers using any cybersecurity framework that exhibits a greater effectiveness than no framework because of their systematic map of processes. In Table 7.1, indicators are given of the planning dimension, taking the use case in [2] as a basis here.

With regard to Table 7.1, the planning dimension expressed by [2] is taken as a basis here:

$$Planning = \left[\left(\frac{\sum_i Proact_i - Proact_{min}}{Proact_{max} - Proact_{min}} \right) * 0.3 + \left(\frac{\sum_i Risk_i - Risk_{min}}{Risk_{max} - Risk_{min}} \right) * 0.3 + Frame * 0.4 \right]$$

As described in [2], "$\sum_i Proact_i$ is the total score of pro-activeness questions; $Proact_{max}$ is the maximum score (4,5), if all responses to nine items equal to 5; $Proact_{min}$ is the minimum score (9), if all responses to nine items equal to 1. $\sum_i Risk_i$

Table 7.1 Indicators of the planning dimension

Abbreviation	Planning	Weight	Description
Proact	Proactiveness in planning	0.3	Nine items in proactiveness in planning assumed, assesses on a Likert scale from 1 not at all, 2 lightly, 3 moderately, 4 considerably to 5 completely
Risk	Risk assessment	0.3	Four items in actual activities in risk assessment assumed, assesses on a Likert scale from 1-not at all, 2-lightly, 3-moderately, 4-considerably to 5-completely
Frame	Frameworks used	0.4	1-if participants use a framework, 0-otherwise

is the total score of the actual activities regarding risk assessment in planning of a cybersecurity plan or strategy; $Risk_{max}$ is the maximum score (20); $Risk_{min}$ is the minimum score (4); *Frame* is a dummy variable that takes the value of one respondents use any framework and equals zero; *i* is the frequency." The weights shown are the results of the investigation in [2].

With regard to Table 7.1, the performance dimension, or execution of cybersecurity audit, comprises two sets of indicators. The indicators measure the use of specific audit procedures such as inquiry, observation, inspection, analytical procedures, and re-performance, covering 12 cybersecurity areas such as prevention, detection, response and recovery, program management, data protection, identity and access management, infrastructure, cloud and software security, third-party and workforce management, threat and vulnerability management, monitoring, crisis management, and organization's resiliency [4]. They largely align with the dominant cybersecurity frameworks. As a combination of procedures ensures more effective testing of areas, assigned a higher score for the use of more procedures and allocated a maximum score for the use of three or more procedures. Some types of procedures, by their nature, are more effective than others. Since reperformance involves the auditor's independent execution of the entity's internal controls, this audit procedure is considered to provide the most reliable type of evidence. Accordingly, the use of this procedure alone was assigned maximum score points. Another set of indicators measures the usage of any of the specified cybersecurity tools by the first and the second line of defense. Admittedly, not all of the tools are appropriate for all organizations, but assign a higher score to those that demonstrate a higher usage of tools [5]. Table 7.2 shows the indicators of the performing dimension with the corresponding weights and description. The performing dimension is presented in the Cybersecurity Audit Index (CSAI).

The performing dimension shown in Table 7.2, expressed in [2], taken as a basis here:

$$Performing = \left[\left(\frac{\sum_i Proced_{i,j}}{Proced_{max}} \right) * 0.8 + \left(\sum_i Tools_i \right) * 0.2 \right]$$

Table 7.2 Indicators of the performing dimension

Abbreviation	Performing	Weight	Description
Proced	Audit procedures that have been performed on each audit cycle to check 12 cybersecurity areas	0.8	Audit procedures used for each of the twelve cybersecurity risk areas (measured as ordinal variable: Not reviewed, inquiry, observation, inspection, analytical procedures, and re-performance) in each audit cycle
Tools	Checks the usage of cybersecurity tools	0.2	Fourteen cybersecurity tools

As described in [2], "$\sum_i Proced_i$ is the total score of audit procedures performed on each audit cycle for each of the 12 areas; $Proced_{max}$ is the maximum score (12) and 0 is the minimum score; i is the number of cybersecurity areas; j is the number of audit procedures. The total score of 1 reached either by using 3 procedures or by using re-performance. For less than three procedures, the score is computed proportionally (0.33 and 0.67). $\sum Tools_i$ is the total score for the number of tools checked in an audit cycle. For each tool checked, a score of 0.1 is ascribed. The minimum score is 0, and the maximum is 1."

The third dimension of the CSAI is reporting that comprises two indicators. The first captures whether independent overall opinions provided to the management board, regarding the organizations cybersecurity governance, risk management and internal controls. The second one measures the frequency of communication. The overall opinion, defined by Standard 2450 Overall Opinions, presents the final phase of the assurance process to the management board, assuring that risk management and internal controls are managed within the acceptable risk level. As an investigation outcome in [2], which was agreed, have allocated the weights accordingly in Table 7.3.

The reporting dimension in Table 7.3, expressed in [2], taken as a basis here:

$$Reporting = \left[\left(\frac{Freq_i - Freq_{min}}{Freq_{max} - Freq_{min}} \right) * 0.3 + Opinion * 0.7 \right]$$

where "$FREQ_i$ is score that measures communication frequency with the management board on findings about cybersecurity risk management effectiveness with $FREQ_{max}$ is maximum score of frequency (5) and $FREQ_{min}$ is minimum score (1). $OPINION$ is a dummy variable that takes value of 1 if an independent overall opinion issue is communicated to the management board, and 0 otherwise, i is the frequency."

Finally, the three dimensions combined in the CSAI in the following way whereby dimensions, *planning* and *performing*, have a weight of 0.4 each, and *reporting* has a weight of 0.2 because it depends on the two preceding dimensions, which constitute the substance of the Cybersecurity Audit (CSA). The weights are

Table 7.3 Indicators of the reporting dimension

Abbreviation	Performing	Weight	Description
Freq	Frequency of reporting	0.3	Frequency of communication with management board about findings: 1-never; 2-less frequently than 2 years; 3-every 2 years; 4-annually; 5-quarterly or at every audit committee's meeting
Opinion	Provision of an overall comprehensive opinion	0.7	1-an independent overall opinion to management board as part of providing ongoing assurance that the internal controls are adequate and operate effectively; 0-otherwise.

based on the professional judgment in [2] and supported by prior research. Performing engagement has been identified as the most crucial dimension not only by the criterion of time allocation but also in terms of its impact on CSA outcomes [6, 7]. Planning is the most important dimension of CSA [8], especially in cybersecurity strategic planning [9]. By most accounts, reporting is seen as a less independent dimension, and therefore it was decided to allocate the weights expressing the relative importance of the three dimensions. In a robustness analysis, also the hypothesis tested with a measurement of effectiveness in which no weights are considered, which allow to compute *CSAI* as follows [2]:

$$CSA\ I = (0.4 * Planning + 0.4 * Performing + 0.2 * Reporting) * 100$$

7.2.2 Maturity Models

The Software Engineering Institute (SEI) popularized the term Maturity Model when they developed the Capability Maturity Model® in 1986. It helps software development teams measure how well their project is doing and how capable they are for continuous improvement. In this context, a Maturity Model is a framework for an area of interest that describes a number of levels of sophistication at which activities in this area may be carried out. Therefore, Maturity Models define a structured collection of elements that describe characteristics' actual state of processes of interest, whereby the model refers to what is done, but it does not specify how it is done, and is used mainly to achieve two objectives [10]:

* Helps set process improvements and priorities, improves processes, and provides guidance for ensuring stable, capable, and mature processes
* Appraise organizations for the sake of improvement

Maturity Models aim to see if organizations are maturing, which means they are constantly testing, growing, and improving. Furthermore, Maturity Models define

Table 7.4 Assessment criteria for maturity indicator level

Maturity level	Characteristics
0	No activities
1	Concepts, but no concrete implementation yet
2	Concepts partially implemented
3	Full implementation and thorough documentation
4	A continuous state-of-the-art and efficient monitoring
5	Subject to a continuous improvement process

different levels of effectiveness (see Table 7.4) and pinpoint a person, team, project, organization and other current positions within the Maturity Model.

As shown in Table 7.4, most Maturity Models use a five- or six-scale level maturity that can range, e.g., from one to five or zero (lowest assessment level) to five (highest assessment level), as shown in Table 7.4 [12]. In the Maturity Model, shown in Table 7.4, the lowest (zero) level needs the most improvement and level five being a fully realized and effective mature operation. In this regard, Table 7.4 shows that with each level the maturity increases, in other words, which business objectives could be further developed and need action to achieve the next maturity level.

There are over 70 different Maturity Models available, from different domains within the spectrum of business information systems, computer science, and others, showing the great variety and width of Maturity Models. Against this background, an organization's maturity defines the effectiveness and efficiency of the organization's business processes and services, to obtain lasting performance. In this regard, the Maturity Model proposes maturity levels of the organization that gives a global vision of their performance, a gap-analysis of their current capabilities with what it requires to meet its strategic objectives, and others. Some of the manifold Maturity Models are listed in Table 7.5.

As Table 7.5 shows, there is no uniform model for measuring the maturity level. Table 7.6 [15] shows seven currently available Maturity Models focusing on digital maturity that differ from Table 7.5. Parallel to these Maturity Models, continuously new Maturity Models are developed for measuring the digital maturity level.

7.2.3   Maturity Models After ISO 9004:2008

Digital transformation significantly improves organizations' process performance and their innovation. However, often, organizations are not aware or not capable of assessing their digital maturity levels and their more specific cybersecurity maturity levels. Many considerations on maturity successfulness include elements of the quality of a quality management system (QMS). However, the sustainable success, according to which an organization should consistently and over the long term meet the needs and the expectations of all involved parties in a balanced way, is an

Table 7.5 Maturity Models specific focus

Maturity Model	Specific focus
CERT Resilience Management Model (CERT RMM)	Promotes the convergence of security, business continuity, and IT operations activities supporting organizations to actively direct, control, and manage operational resilience and risk
Circular Product Design Maturity Matrix (CPDMM)	Provides a diagnosis and broader understanding of how circular the new product development process in an industrial organization is
Cybersecurity Capability Maturity Model (CCMM, C2M2)	Cybersecurity maturity and knowledge integration that ranges from limited awareness and application of security controls to pervasive optimization of the protection and defense of critical assets based on 6 levels
Cybersecurity Maturity Model Integration (CMMI, CM2I)	Describes best practices used in industry for development, maintenance, and procurement of products and services. Framework that makes it possible to assess organization's level of maturity to put improvements into practice. There are three Maturity Models: *CMMI for development*: Aimed at organizations that develop and maintain products and services for systems development *CMMI for acquisition*: Aimed at organizations that outsource development services and maintain products and services for systems development *CMMI for services*: Aimed at organizations that provide services to other companies
Organizational Project Management Maturity Model (OPM3)	Provide to understand organizational project management and to measure their maturity against a comprehensive and broad-based set of organizational project management best practices. OPM3 supports organizations wishing to increase their organizational project management maturity to plan for improvement
Project Management Maturity Matrix (ProMM)	Framework built on top of the regular project management system. Use project management Maturity Model an organization can systematically plan its project management capabilities and benchmark its performance against competitors and industry standards. Bridges the gap between strategy and individual projects
And others	...

essential issue too. For sustainable success, the ISO 9004:2008 provides a self-assessment tool to identify systematically organizations' own strengths and weaknesses, e.g., a reference-based maturity determination and defining prioritized measures for improvement. However, each organization must define its own data values and business assets, mission and vision. Nevertheless, the initiative for the maturity assessment should come from top management. The results and progress illustrated graphically and transparently, e.g., using a radar chart (Fig. 7.1), called a

Table 7.6 Overview of some available digital Maturity Models

Designation	Originator	Link
Digital Excellence Maturity Benchmark	DT Associates	http://dt-associates.com/digital-excellence-maturity/
Digital Maturity Assessment	Arrk Group	http://www.arrkgroup.com/digital-maturity-assessment/
Digital Maturity Matrix	MIT Center for Digital Business Capgemini Consulting	http://sloanreview.mit.edu/article/the-advantages-of-digital-maturity/
Digital Maturity Model	University St. Gallen Crosswalk AG	https://aback.iwi.unisg.ch/kompetenz/digital-maturity-transformation/
Digital Transformation Maturity Index	University of applied Sciences Neuland Consulting	http://www.researchlab-db.com/
KPMG Digital Readiness Assessment	KPMG AG accounting Firm	https://home.kpmg.com/de/de/home/themen/2015/08/digital-readiness-assessment.html
Organizational Transformation Assessment	Adapt2Digital	http://www.adapt2digital.com/digital-maturity-assessment-1/

spider diagram. Success provided by self-assessment, carried out regularly. However, sustainable success can only be achieved if all employees of an organization are involved in the implementation of a Maturity Model after ISO 9004:2008. The steps, elements, and criteria of an ISO 9004:2008 approach are shown in Table 7.7.

The elements of ISO 9004.2008 in Table 7.7 assigned graded requirements that enable identifying the current assessment status and possible improvements. ISO 9004:2008 maturity levels show achieved success: from Level 1 to Level 5, indicated by the criteria achieved from Level 1 to Level 5 (best practice). Compared to generic Maturity Models, application-specific Maturity Models have the advantage of precise measurement of the improvement status due to the specific indicators.

ISO 9004:2008 includes in general the following features [23, 24], as shown in Table 7.8.

The above structure allows improving a sustainable development of a maturity level model based on quality management standards, because many considerations on maturity successfulness include elements of a quality management system (QMS). Therefore, ISO 9004:2008 is a guidance for achieving sustained success for assessment of a maturity level. The thesis in [18] describes a Maturity Model assigned to five maturity levels, using defined criteria of ISO 9004:2008. For the ISO 9004:2009 Management towards a Sustainable Organization's Success, a digital management approach update, a model for self-assessment of organization's maturity with respect to key elements of organization's functioning, is published [25]. This publication grouped the key elements in the following areas, focusing on suitable organization's success strategy and policy, resource and process

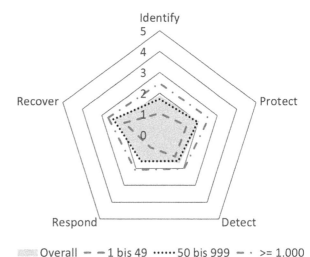

Fig. 7.1 Radar chart with five axes (dimensions) of the NIST CSF and six maturity levels, from zero to five, as an example comparing maturity levels of organizations from the same industrial sector but with different numbers of employees

Table 7.7 ISO 9004:2018 steps, elements, and criteria

Step	Element	Criteria
0	Below Basic Level 0	No activities
1	Basic Level 1	Basic process description created
2	Enhanced Level 2	It only partially takes into account the needs, goals, and values of interest
3	Midterm Level 3	Refinement of process description; taking risks and opportunities into account
4	Advanced Level 4	Objectives of interest are reviewed and monitored and improvement achieved
5	Best Practice Level 5	All objectives of interest are analyzed, evaluated to improve sustainable performance and understanding of goals and values

management, monitoring, measurements, analysis and review improvement innovation, and learning.

Other ISO 9004:2008 Maturity Models published, like building optimal management for the maturity level of company's business processes, whereby the level of maturity describes a linear dynamic control system [26]. Developing the model, the method of Analytical Construction of Optimal Regulation (ACOR) is used, based on Kalman-Letvos's approach. The model constructed shows plausible behavior in predicting the process of managing an organization's maturity and reproduces the effect of accelerating growth of controlled indication, identified in the model as a-priority. As published in [27], organizations are not aware or not capable of assessing their Industry 4.0 maturity level. Thus, a Maturity Model, based on ISO 9004:2018 for assessment of maturity levels of quality in Industry 4.0, framed as Quality 4.0, developed based on a questionnaire with a sample size of 335 companies to explore the topic. Confirmatory Factor Analysis (CFA) and Structural Equation

Table 7.8 ISO 9004:2018 features and capabilities

Criteria no.	Features	Capability
I	Foreword	./.
II	Introduction	./.
1	Scope	Document gives guidelines for enhancing organization's ability to achieve sustained success. Guidance is consistent with the quality management principles given in ISO 9000:2015. Document provides a self-assessment tool to review the extent to which the organization has adopted the concepts in this document. Document is applicable to any organization, regardless of its size, type, and activity.
2	Normative references	The following documents are referred to in the text in such a way that some or all of their content constitutes requirements of this document. For dated references, only the edition cited applies. For undated references, the latest edition of the referenced document (including any amendments) applies.
3	Definition of sustainable development	3.1 Introduction 3.2 Organization is able to maintain and develop its performance over a longer period of time.
4	Management of sustainable development	4.1 Introduction 4.2 Key features of sustainable development (organization) 4.3 Management of sustainable development 4.4 Evaluation of sustainable development
5	Environment organization	5.1 Introduction 5.2 Monitoring 5.3 Analysis
6	Strategies, policies and communication	6.1 Strategic orientation 6.2 Mission and vision 6.3 Aspects of the strategy 6.4 Policies and objectives 6.5 Strategic planning 6.6 Risk management 6.7 Review strategies for sustainable development 6.8 Communications 6.8.1 General 6.8.2 Effectiveness and efficiency of the process of communication
7	Resources	7.1 Resource management 7.2 Planning 7.3 Resource allocation 7.4 Resources staff 7.4.1 General 7.4.2 Motivation and involvement of employees 7.5 Infrastructure 7.5.1 General 7.5.2 Working conditions 7.6 Knowledge

(continued)

Table 7.8   (continued)

Criteria no.	Features	Capability
		7.7 Financial resources
		7.8 Natural resources management and life expectancy
8	Processes	8.1 Process approach
		8.2 Types of processes
		8.3 Management processes of the organization
		8.4 Responsibility and authority for the process
9	Measurement and analysis	9.1 Measurement approach
		9.2 Performance matrix
		9.3 Measurement of achieving the goals
		9.4 Key indicators
		9.5 Tools for measuring
		9.6 Internal audit
		9.7 Evaluation
		9.8 Review and evaluation process
10	Learning	10.1 Introduction
		10.2 Learning
		10.2.1 Types of learning
		10.2.2 Sources of learning
		10.2.3 Factors that influence the effectiveness of learning
		10.2.4 Planning of learning
		10.3 Improvement
		10.4 Innovation
		10.4.1 Generally
		10.4.2 Types of innovation
		10.4.3 Factors that influence the effectiveness of innovation
		10.4.4 Planning process innovation
Appendix A	Tools for assessing the maturity	A.1 Introduction
		A.2 Description of the level of maturity
		A.3 Assessment strategy
		A.4 Mark work
		A.5 Tools for evaluation
		A.6 Results evaluation and improvement planning
Appendix B	Forms estimation of the maturity	

Modelling (SEM) used to verify the ISO 9004:2018 model potential, used for assessing Quality 4.0 maturity level. Resource management proved to have a positive impact on performance analysis and evaluation as well on improvement, learning, and innovation constructs. The findings indicate the model is usable in the context of Quality 4.0. Such model may be utilized as a basis for developing a sustainable Quality 4.0 system roadmap.

7.3   Cybersecurity Maturity Models

Cybercrime is increasingly sophisticated, which requires cybersecurity awareness and practice to become more efficient at defending or mitigating cybersecurity risk (s). The elements of cybersecurity and their capabilities is shown in Table 7.9.

Organizations' cybersecurity managers identified cybersecurity risk(s) as an essential business issue that needs to be taken into account. To identify an organization's actual cybersecurity risk(s), cybersecurity maturity models help organizations to identify their actual cybersecurity awareness as well as their cybersecurity strategy planning. A cybersecurity maturity model is a framework of security practices, guidelines, and controls that provide an organization with a roadmap for creating effective and compliant cybersecurity capabilities and characteristics. In this regard, cybersecurity supports improving security posture to defend or mitigate cybersecurity risk(s). The several maturity frameworks considered as standards when it comes to cybersecurity maturity models, as shown in Table 7.10, three of which are with their specific focus.

These models require management experiences, advanced employee skills, change management strategies, cybersecurity defense plan or cybersecurity defense strategy, and others. Therefore, the assessment through the Maturity Model enables deriving the main obstacles to the efficient and effective use of cybersecurity strategies or methods. The Maturity Model creates additional value from data records obtained in a self-assessment improving cybersecurity. As part of the self-assessment, in addition to pseudonym personal information, e.g., role in the organization, area of responsibility, and others, information collected about the organization, e.g., industrial sector, employment numbers, market share, sales, and others. The queried data indicates the existing or non-existing roadmap in the cybersecurity defense strategy or the cybersecurity awareness of organizations' employees. Such a roadmap ultimately results in strategic activities and responsibilities to increase the current maturity levels. Furthermore, the aim of improvement of the current (actual) state of activities, preventing or defending cybersecurity risk(s), needs to consider the organization's focus on a standardized cybersecurity framework like NIST CSF (see Chap. 5) or integrating cybersecurity guidelines to prevent or defend cybercriminal attacks, as shown in Table 7.6. This issue is associated with critical development paths for the organization and its staff, who have the necessary expertise and essential approved budget [13, 14]. Besides these dimensions, other dimensions are taken into account, such as IT infrastructure, OT infrastructure continuous improvement processes, and others. To operationalize the dimensions with regard to the evaluation criteria to use the Maturity Model, the current (actual) state will be determined to decide about the next maturity level achieved. For the maturity levels in Table 7.4, some maturity level characteristics are

- *Maturity levels below 3*: Represent the perseverance of one or more significant obstacles to attaining the objective, e.g., cybersecurity plan or cybersecurity defense strategy

Table 7.9 Elements of cybersecurity and capabilities

Cybersecurity element	Capabilities and characteristics

Application security	*Types are*:
	  Authentication means data can be traced back to its origin at any time.
	  Authorization means the process of giving someone permission to do or have something.
	  Encryption means data/information is converted into secret code that hides the information's true meaning.
	  Logging means recording of all or defined events during a processing operation in a log file.
	  Application security testing means making applications more resistant to security threats, by identifying security weaknesses and vulnerabilities in source code.

Information security	*Standards and principles are* Confidentiality, integrity, and availability (CIA triad (see Sect. 1.6.2).

Network security	*Types are*:
	  Network security strategies such as firewalls, antivirus programs, email security, web security, wireless security.
	  Network security software such as network firewalls, cloud application firewall, web application firewall, and others.

Disaster recovery Planning	*Primary objectives of disaster recovery planning include* [28]:
	  Protect the organization during a disaster.
	  Giving a conviction security.
	  Limiting the risk of postponements.
	  Ensuring the dependability of backup systems.
	  Giving a standard to testing the plan.
	  Limiting decision-production during a disaster.
	Disaster recovery planning categories are:
	  Data center disaster recovery.
	  Cloud applications disaster recovery.
	  Service-based disaster recovery.
	  Virtual disaster recovery.
	Steps of disaster recovery planning are:
	  Acquire top management commitment.
	  Planning panel establishment.
	  Performing risk management.
	  Establish priorities for handling and tasks.
	  Decide recovery strategies.
	  Data collection.
	  Record a composed plan.
	  Build testing rules and methods.
	  Plan testing.
	  Support the plan.

Operational security	*Practices are*:
	  Carry out double control.
	  Implement exact change management processes.
	  Limit access to network devices.
	  Minimum access to employees.
	  Reaction and disaster recovery planning.
	  Task automation.

(continued)

Table 7.9 (continued)

Cybersecurity element	Capabilities and characteristics
	Steps are: Characterize the organization's valuable assets and data. Distinguish the types of dangers. Evaluation of risks. Execution of accurate countermeasures. Investigate security openings and weaknesses.
End-user security	*End-user dangers can be achieved in the following ways*: Applications download. Creation and irregular use of passwords. Text messaging. Utilization of email. Utilization of social media.

Table 7.10 Cybersecurity Maturity Models' specific focus

Maturity Model	Specific focus
NIST CSF	National Institute of Standards and Technology Cybersecurity Framework (NIST CSF) accommodates a rapidly evolving landscape of threat event attacks and advises security teams that adopt this model to adjust monitoring techniques and remediation strategies to match the ongoing threat event attack environment. It tracks progress implementing information security maturity from the current state to the defined target state, using clear structured documentation. The model contains five maturity levels.
ISO/IEC 27 k	International Standard Organization/International Electrotechnical Commission standard 27 k specifies the requirements for Information Security Management Systems (ISMS) to improve information security programs, support critical infrastructure protection planning, and facilitate the exchange of effective security practices in organizations' communities. ISO/IEC 27 k includes: ISO/IEC 27000: Information Security Management Systems: Overview and Vocabulary. ISO/IEC 27000: Information Security Management Systems; Requirements ISO/IEC 27002: Code of Practice for Information Security Management ISO/IEC 27003: Information Security Management Systems: Implementation Guidelines ISO/IEC 27004: Information Security Management Measurements ISO/IEC 27005: Information Security Risk Management
CIS 20	Center for Internet Security developed a series of 20 critical controls for protecting organizations' networks and data from cyberattack vectors. CIS 20 model is designed to be all-encompassing and requires attention to organizations' cybersecurity management processes that also include cloud and mobile technologies.

- *Maturity levels of 4*: Represent certain sovereignty in attaining the objective, e.g., cybersecurity plan or cybersecurity defense strategy
- *Maturity levels of 5*: Show a deep embedment of the objective, e.g., cybersecurity plan or cybersecurity defense strategy

Using the scale given in Table 7.4 and focusing on a standardized cybersecurity framework like NIST CSF (see Chap. 5), the assignments of the evaluation criteria are illustrated in a radar chart. The radar chart is the graphical representation of multivariate data in a two-dimensional plot with multiple variables, shown on the axes from the same point. This is particularly useful for the explicit representation of the current state of objectives under investigation. In this context, Fig. 7.1 shows an example of a radar chart referring to the National Institute of Standards and Technology Cybersecurity Framework (NIST CSF).

The radar chart in Fig. 7.1 is a 2D chart presenting multivariate data of quantitative variables mapped onto an axis and plotting the data as a polygonal shape over all axes. All axes have the same origin, and the relative positions and the angles of the axes usually are not informative. The equi-angular spokes, from the origin to the point on each axis represented by the variable, are called radii. Radar charts are used to plot a series of observations or cases with multivariate data for comparisons, plotting a series of observations. Each observation or case is represented by a polygon; and if they are shaded opaquely, it is easy to see how they overlap and in which direction. However, graphing multiple observations or cases can become messy, as it can be hard to visually distinguish more than three or four stacked polygons. Using opaque colored layers can help, but there is a limit as to how many can be included in one chart.

The NIST CSF describes cybersecurity outcomes organized in a hierarchy of core functions [16]. These core functions are included in the radar chart in Fig. 7.1 and are

- *Identify* (*ID*): Supports understanding and managing cybersecurity risks
- *Protect* (*PR*): Supports the ability to limit the impact of a potential cybersecurity event
- *Detect* (*DE*): Enables timely discovery of cybersecurity events
- *Respond* (*RS*): Includes appropriate activities taking action regarding cybersecurity incidents
- *Recover* (*RC*): Supports timely recovery to normal operation to minimize the impact of a cybersecurity incident

Interpreting the representation of maturity levels of the NIST CSF core functions in the radar chart in Fig. 7.1 shows that none of them achieve the maturity level of 3 and higher. Maturity levels below 3 represent the perseverance of one or more significant obstacles to attaining the objective. This means for the case shown in Fig. 7.1 that, e.g., the employees of the organizations under consideration do not have sufficient cybersecurity knowledge to support adequately the NIST CSF core functions in their implementation. Hence, the cybersecurity maturity model shows the objectives that must be further developed and need action to lift them up to mean maturity level 3. As a secondary condition, when determining the level of maturity, it is necessary to estimate the effort that arises from the NIST CSF criteria achieving a higher level of maturity by eliminating the identified weak point(s) in the current state. This has also an economic impact in terms of assessing its usefulness. Furthermore, besides the financial considerations for introducing an organization-wide cybersecurity defense strategy, the responsible persons should not forgot that there is usually a change at the employees' workplace concerned, which may result in additional, targeted qualification measures. Nevertheless, considering this secondary condition, the Maturity Model offers a readily applicable and practical approach to determine the current maturity level status, e.g., the cybersecurity objective. In this regard, the current maturity level is merely a part of intermediate state in order to determine the skills and/or activities to acquire the next maturity level.

With progress towards the desired target state, it is possible to check the achievability of the target state. At the same time, determining the degree of the current implementation by early detection of possible deviations from the desired target state is achievable, which corresponds to an actual-target state comparison. The deviation expressed by the difference Δ between the actual state, representing the already reached level, and the target state, representing the next level to be achieved. Let

$$actual\ state = X$$

and

$$target\ state = Y$$

and

$$X \neq Y$$

then

$$\Delta > 0.$$

If $\Delta = 0$, the target state is reached.

From the radar chart in Fig. 7.1, it can be concluded there is a need to move towards highly effective cybersecurity analytics to identify, protect, detect, and respond to prevent cybersecurity breaches or finally to recover from, by evaluating the cybersecurity maturity level. This finally reduces cybersecurity risks by cyberattacks and vulnerabilities, and improves the cybersecurity capabilities by evaluating their cybersecurity maturity levels and making recommendations for improvements.

Cyberattacks like data breaches are one of the biggest information security risks organizations face. Sensitive data are used across all areas of businesses these days, increasing its value for legitimate and illegitimate use. Daily countless incidents occur, whether cyber-criminals hacking or misappropriating information. Wherever the data goes, the financial loss and reputational damage caused by a breach can be devastating. Therefore, organizations are investing heavily in their defenses, using ISO/IEC 27 k as a guideline for effective security. ISO/IEC 27 k may be applied to organizations of any size and in any sector, and the framework's breadth means its

Table 7.11 ISO/IEC 27 k security standards

ISO/IEC	Number of documents	Security element
27000	1	Information security management systems—Overview and vocabulary
27002	1	Guidelines for cybersecurity
27004	1	Information security management—Monitoring, measurement, analysis, and evaluation
27005	1	Information security risk management
27033-X	7	Network security
27034-X	7	Application security
27035-X	4	Information security incident management
27036-X	4	Information security for supplier relationships
27039	1	Intrusion prevention
27043	1	Incident investigation
TS27110	1	Incident information technology, cybersecurity and privacy protection—Cybersecurity framework development guidelines
27701	1	Information technology—Security techniques—Information security management systems—Privacy information management system (PIMS)

implementation will always be appropriate to the size of the business. Some of the published ISO/IEC 27 K standards are related to information technology security techniques with numbers of documents given in Table 7.11.

The CIS 20 are a set of actions for cyber defense that provide specific and actionable ways to stop today's most pervasive and dangerous cyberattacks, so organizations can reduce chances of compromise by moving from a compliance-driven approach to a risk management approach. The security controls of CIS 20 include [17]:

1. Inventory and Control of Hardware Assets
2. Inventory and Control of Software Assets
3. Continuous Vulnerability Management
4. Controlled Use of Administrative Privileges
5. Secure Configurations for Hardware and Software on Mobile Devices, Laptops, Workstations, and Servers
6. Maintenance, Monitoring, and Analysis of Audit Logs
7. Email and Web Browser Protections
8. Malware Defenses
9. Limitation and Controls of Network Ports, Protocols, and Services
10. Data Recovery Capabilities
11. Secure Configuration for Network Devices, such as Firewalls, Routers, and Switches
12. Boundary Defense
13. Data Protection
14. Controlled Access Based on the Need to Know

15. Wireless Access Control
16. Account Monitoring and Control
17. Implement a Security Awareness and Training Program
18. Application Software Security
19. Incident Response and Management
20. Penetration Test and Red Team Exercises

Another essential issue is the maturity evaluation strategy. The evaluation strategy depends on the constellation of the evaluators and the approach used for evaluating the maturity level. The group of evaluators can be composed of corporate (internal) or external people. Moreover, an external evaluator can be a suitably designated employee of the customer. Furthermore, an internal evaluation can be performed by independent or involved evaluators (in relation to the object of the evaluation). In case of involved persons, evaluation carried out by the consensus of several persons or by a designated individual. The last constellation corresponds to the self-assessment approach. Although it requires less effort, it is associated with a higher degree of subjectivity and thus uncertainty in the assessment result. Therefore, comprehensive assessments by external people have a higher level of objectivity due to their unbiased approach. However, this constellation is also associated with a higher effort.

Moreover, as a matter of principle, the determination of the maturity level is associated with uncertainties, since an assessment of the fulfillment of the requirements of the maturity level by the object of interest always uses subjective opinions or non-quantifiable variables, and a quantified degree of an alleged accuracy suggested to be convicted [18]. The consideration and specification of the information uncertainty in the maturity level analysis ensures valid and differentiable data on the development status, which means that improvement measures selected items in a targeted manner. For this purpose, a feature-based uncertainty model is set up parallel to the development of an application-specific Maturity Model, which represents a stage model with factors relevant to insecurity and thus enables the multi-criteria recording of information insecurity [18]. The uncertainty factors deduced on the basis of a system model that captures the basic elements in the Maturity Model development and application as well as the uncertainties involved. In system modeling, structure diagrams or interrelation diagrams are used as methodological support. The subsequent derivation of the uncertainty factors are carried out efficiently using the Metaplan-Technique [19] and the Categorizing Affinity Diagrams [20].

The Metaplan-Technique is a group communication and decision-making methods in which opinions are developed, a common understanding is built, and objectives, recommendations, and action plans are formulated to focus on a problem and its solution(s).

The Categorizing Affinity Diagram organizes and categorizes ideas into categories based on their similarities. After weighting the factors, the uncertainty-characteristic value matrix can be set up, which defines the step-by-step requirements for the respective factors, based on a maturity degree characteristic value matrix. The uncertainty characteristics depend of the complexity of the object of

Table 7.12 Fuzzy elements and characteristics

Fuzzy elements	Characteristics
Fuzzyfication	Process mapping crisp input $x \in U$ into fuzzy set $A \in U$ with different types of fuzzifier, including singleton fuzzifiers, Gaussian fuzzifiers, and trapezoidal or triangular fuzzifiers. These fuzzifiers map crisp input x into fuzzy set A with different membership functions $\mu_A(x)$.
Fuzzy inference engine	Key unit of a fuzzy system having decision making as its primary issue it uses the *IF...THEN* rules along with connectors of *fuzzy OR, fuzzy AND* for drawing essential decision rules.
Knowledge base	Represent the facts of the rules and linguistic variables based on the fuzzy sets so that the knowledge base will allow approximate reasoning
Defuzzification	Inverse process of fuzzification where mapping is done to convert the fuzzy results into crisp results. It maps from a space of fuzzy actions defined over an output universe of discourse into a space of crisp (non-fuzzy) actions.

interest, the level of knowledge and the recognizable features of the object of interest. Finally, combining the uncertainty model and the Maturity Model in a fuzzy system is possible [21]. Their main characteristic involves symbolic knowledge representation in a form of fuzzy conditional if-then rules. A fuzzy system architecture consists of four functional elements, as illustrated in Table 7.12.

Linguistic values, defined by fuzzy sets, and crisp or numerical data can be used as inputs for a fuzzy system. The values of input variables are mapped into linguistic values of the output variable by means of the appropriate method of approximate reasoning by the inference engine, using expert knowledge, which is represented as a collection of fuzzy conditional rules of the knowledge base. In addition to linguistic values, numerical data may be required as fuzzy system output. For that, defuzzification methods are used, which assign the representative crisp (numerical) data to the resultant output fuzzy set. Therefore, the linguistic variable is one of the fundamental concepts of fuzzy set theory. More in general, its values are statements of natural language, in which the terms that are labels are the descriptions of fuzzy sets, defined on a given universe of discourse. From the mathematical perspective, a linguistic variable is defined as a quintuple as follows [21, 22]:

$$X = (N, L(G), X, G, S)$$

where N is name of the linguistic variable, while $L(G)$ denote the family of values of the linguistic variable, which is a collection of labels of fuzzy sets defined on the universe. Moreover, G is the set of syntactic rules defined by a grammar determining all terms in $L(G)$, and S represents the semantics of the variable X which defines the meaning of all labels. In this regard, the maturity levels are mapped as linguistic input terms, whose membership function is designed dependent on uncertainty [21].

As an example using a linguistic variable describing the Cybersecurity Risk (CSR), where the linguistic variable is defined as

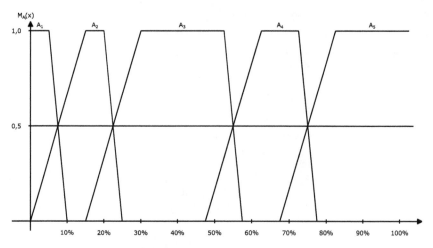

Fig. 7.2 Examples of membership functions of fuzzy sets defining the values of the linguistic variable $X = mean\ CSR$, with A_1: rare, A_2: unlikely, A_3: possible, A_4: likely, A_5: highly likely

$$N = \text{mean CSR.}$$

The set of possible linguistic values of the likelihood of cybersecurity is a collection of five states:

$$L = \{\text{highly likely, likely, possible, unlikely, rare}\}.$$

To each of the labels a fuzzy set A is assigned as follows:

$$A_i : i = 1.2.3. \ldots, 5,$$

defined on

$$X = [0,100]\%,$$

which represents the range of a possible cybersecurity risk (CSR) through a range of possible cyberattacks. The examples of membership functions $\mu_A (X)$ of the fuzzy sets A are shown in Fig. 7.2.

An expression for the linguistic variable X is

$$X \text{ is } L_A$$

where L_A is a label from the collection $L(G)$, defined by a fuzzy set A on the universe X. The logical value of the expression is determined based on the membership function $\mu_A (X)$ of the fuzzy set A. In the preceding example, a realistic statement is:

$$X = \text{mean CSR is possible,}$$

which value for the measurement of 50% that is equal to

$$\mu_A(X) = 0,5,$$

which can be seen in Fig. 7.2.

A more complex fuzzy expression for a cybersecurity maturity level can be obtained combining two or more expressions, which can be described in the conjunctive (and), termed MIN-Operator

$$(X_1 \text{ is } L_{A1}) \text{ and } (X_2 \text{ is } L_{A2}),$$

or disjunctive (or) form, termed MAX-Operator,

$$(X_1 \text{ is } L_{A1}) \text{ or } (X_2 \text{ is } L_{A2})$$

where X_1, X_2 are linguistics variables of the cybersecurity risk(s) with labels L_{A1}, L_{A2}, defined by the fuzzy sets A_1 and A_2, respectively, on the universe of X_1 and X_2.

The values of a complex fuzzy expression for

$$x_1 \in X_1$$

and

$$x_2 \in X_{2\#}$$

is determined by the membership function of fuzzy sets A_1 and A_2 [21].

Elementary fuzzy statements are expressed in the form of an implication output aggregation, the MAX-Operator n that forms a fuzzy *if **then** rule* (fuzzy conditioning statement) like

$$\text{if } (X \text{ is } L_A) \text{ then } (X \text{ is } L_B)$$

The input aggregation is carried out by the Gamma-Operator, which is suitable for modeling the human decision process [24]. The Gamma-Operator is parameterized to $\gamma = 0.3$, what leads to a compensatory character of the transfer characteristic [24]. The Gamma-Operator values selected are between 0.025 and 0.975. For the output aggregation, the MAX-Operator is selected. The Max-Operator is the union of the two fuzzy sets A and B, the logical or, defined by the following membership function

$$\mu_{A \cup B}(x) = max \ \{\mu_A(x), \mu_B(x), x \in X\}.$$

Based on the maturity level matrix and the cause-and-effect matrix, a six-sigma tool, to prioritize the input variables of fuzzy systems linguistic variables, linguistic terms, membership functions, and inference matrices, the progress and status indicators of the maturity level can be determined by using software tools, such as fuzzyTECH 5.71d [23]. For each progress indicator and its characteristic as well as for the status indicators and their operative parameters, linguistic variables are derived. Their linguistic terms are implemented primarily with trapezoidal and/or triangle membership functions.

For defuzzification, the Center-of-Area (CoA) method is used to gain the best compromising value of the fuzzy output variable [24]. The CoA method determines the center of area of fuzzy set and returns the corresponding crisp value. Thus, the CoA defuzzification method calculates the area under the scaled membership functions and within the range of the output variables. The membership function can be defined as triangular, trapezoidal, singleton (vertical line shape), sigmoid (wave shape), and Gaussian (bell shape).

The feature-based uncertainty model calculates the geometric center of this area using the following equation:

$$CoA = \frac{\displaystyle\int_{x_{min}}^{x_{max}} f(x)^* x \ dx}{\displaystyle\int_{x_{min}}^{x_{max}} f(x) dx}$$

where CoA is the center of area, x is the value of the linguistic variable, and x_{min} and x_{max} form the range of the linguistic variable. Thus, the CoA defuzzification method computes the best possible compromise between multiple linguistic terms for output variables.

7.4   Cybersecurity Maturity Best Practice Model Example

Cybersecurity incidents have increased in all industrial sectors, causing serious breach-in attacks. In the context of the NIST CSF (see Chap. 5), cybersecurity is a process that involves identification, protection, detection, response, and recovery from cyberattacks, which includes cybersecurity awareness as well as cybersecurity strategies. In this regard, a primary challenge is in developing an adaptive balance between security, privacy, costs, and efficiency. Therefore, cybersecurity maturity models are used to gain insight into organizations' actual status in cybersecurity awareness and cybersecurity strategy in order to decide about the steps on how to

Table 7.13 Maturity Models and methodology

Maturity Models	Methodology and specific models
Capability Maturity Model (CMM, CM2)	Used to denote the levels of corporate performance model (CPM). CMM describes a five-level evolutionary path of increasingly organized and systematically more mature processes. Specific CMM are: Software Capability Maturity Model (SW-CMM, SW-CM2). Systems Security Engineering Capability Model (SSE-CMM, S2E-CM2). Service Integration Maturity Model (SIMM, SIM2) an outgrowth of CMM/CM2.
Cybersecurity Capability Maturity Model (CCMM/C2M2)	Intended to provide guidance on the necessary considerations and requirements to achieve and maintain cybersecurity advantage. It illustrates the stages of readiness or preparedness to respond to threats, vulnerabilities, and technological advancement that exists within the continuously evolving cybersecurity environment, which draw on maturity processes and cybersecurity best practices, based on multiple standards. It includes five factors of the cybersecurity environment: Dynamisms and rivalries of the cyber environment, capabilities, technological development, threats, and vulnerability measures. Specific CCMM/C2M2 are: Community Cybersecurity Capability Maturity Model (CCSCMM, C2SCM2) National Initiative for Cybersecurity Education-Capability Maturity Model (NICE) Railway Cybersecurity Capability Maturity Model (R-CCMM, R-C2M2)

achieve an adaptive balance level. Thus, cybersecurity maturity models are useful in guiding organizations in the development of processes leading a state of maturity indicator level in their area for which the Maturity Model was developed.

In the industrial sectors, specific cybersecurity maturity model approaches are used, assessing compliance to cybersecurity standards and guidance or assessing specific information systems' components. Two examples of maturity level models are shown in Table 7.13, which make use of a questionnaire to test the achieved maturity level at the maturity indicator level (see Table 7.4).

The Maturity Models in Table 7.13 make use of a questionnaire to evaluate the achieved actual maturity level at the maturity indicator level (see Table 7.4). For this reason, in the first part of the questionnaire, the actual status of cybersecurity awareness and cybersecurity strategical planning is identified, based on the NIST CSF model described in Chap. 5, which fit with the Cybersecurity Capability Maturity Model (CCMM, C2M2). The wording of the questions ensures that the answers are clearly assigned. In order to add another dimension to the evaluation, the questions are divided into topics like organization, Information Technology (IT) systems, IT infrastructure, Operational Technology (OT) systems, and others

Table 7.14 Number of questions at the intersections of topic and NIST CSF core functions

Topic	NIST CSF core functions				
	Identify	Detect	Protect	Respond	Recover
Organization	xxx	xxx	xxx	xxx	xxx
IT-systems	xxx	xxx	xxx	xxx	xxx
IT-infrastructure	xxx	xxx	xxx	xxx	xxx
OT-systems	xxx	xxx	xxx	xxx	xxx

in the context to the five NIST CSF core functions. Collecting information about the organization for a NIST CSF core function, e.g., *Protect*, a possible question is: "Does the organization ensure that all employees are trained in cybersecurity?" A breakdown of the questions to the different NIST CSF core functions in regard to the various topics included in the cybersecurity maturity model is required, and used in an industrial survey, as shown Table 7.14. The answer pattern is the same for each question. The participants of the questionnaire are able to choose one of the answer options. The answer options refer to the status quo of cybersecurity based on the maturity indicator levels, as shown in Table 7.4. For a cybersecurity status quo statement, questions are about cybersecurity incidents that have already occurred, such as number of cyberattacks and damages. Also type and number of cyberattacks recorded and the hurdles that stand in the way of an adequate cybersecurity strategy are important. The Maturity Model enables a linear evaluation.

Some question examples of the topics in Table 7.14 are shown in Table 7.15.

The questions are asked in closed form, which means that the participating organizations have no opportunity for additional comments. Therefore, at the end of the questionnaire, the participating organization could voluntarily provide contact information for a subsequent interview, to discuss uncertainties in the answers in the different topics in detail. The results of the best practice example are recently published in [25, 26].

Cyberattacks are increasing in railways with an impact on railway stakeholders, e.g., threat event attacks against the safety of employees, passengers, or the public in general; loss of sensitive railway information; reputational damage; monetary loss; erroneous decisions; loss of dependability, and others. This requires advanced security analysis, to improve the cybersecurity capabilities of railways by evaluating their cybersecurity maturity levels and making recommendations for improvements. After assessing various cybersecurity maturity models, C2M2 was selected to assess the cybersecurity capabilities of railway organizations, which resulted in the Railway-Cybersecurity Capability Maturity Model (R-C2M2), described in detail in [27], and the Cybersecurity Framework in Railway for Improvement of Digital Asset Security [28]. Extending this approach to other (critical) infrastructures is possible with necessary adaptations.

Table 7.15 Question examples to identify maturity level

Question topic	Question	Answer options
Organization #1	Has the organization a cybersecurity strategy implemented?	☐ No activities ☐ Concepts but no concrete implementation yet ☐ Concept partially implemented ☐ Fully implemented and thorough documentation ☐ A continuous state-of-the-art and efficient monitoring ☐ Subject to a continuous improvement process
Organization # n-1	. . .	. . .
Organization #n	Does the company have a strategy to counteract attacks in the long term?	☐ No activities ☐ Concepts but no concrete implementation yet ☐ Concept partially implemented ☐ Fully implemented and thorough documentation ☐ A continuous state-of-the-art and efficient monitoring ☐ Subject to a continuous improvement process
IT-systems #1	Are the assets and their potential need for protecting organization's IT systems identified?	☐ No activities ☐ Concepts but no concrete implementation yet ☐ Concept partially implemented ☐ Fully implemented and thorough documentation ☐ A continuous state-of-the-art and efficient monitoring ☐ Subject to a continuous improvement process
IT-systems #n-1	. . .	. . .
IT-systems #n	Are there suitable indicators in the organization for the early and/or proactive detection of security incidents in IT systems implemented?	☐ No activities ☐ Concepts but no concrete implementation yet ☐ Concept partially implemented ☐ Fully implemented and thorough documentation

(continued)

Table 7.15 (continued)

Question topic	Question	Answer options
		☐ A continuous state-of-the-art and efficient monitoring ☐ Subject to a continuous improvement process
IT-infrastructure #1	Is an end-to-end network segmentation of the organization's IT infrastructure implemented?	☐ No activities ☐ Concepts but no concrete implementation yet ☐ Concept partially implemented ☐ Fully implemented and thorough documentation ☐ A continuous state-of-the-art and efficient monitoring ☐ Subject to a continuous improvement process
IT-infrastructure #n-1	. . .	. . .
IT-infrastructure #n	Has a threat analysis of the organization's IT infrastructure been carried out?	☐ No activities ☐ Concepts but no concrete implementation yet ☐ Concept partially implemented Fully implemented and thorough documentation ☐ A continuous state-of-the-art and efficient monitoring ☐ Subject to a continuous improvement process
OT-systems #1	Is a regular back-up of the organization's OT-systems implemented?	☐ No activities ☐ Concepts but no concrete implementation yet ☐ Concept partially implemented ☐ Fully implemented and thorough documentation ☐ A continuous state-of-the-art and efficient monitoring ☐ Subject to a continuous improvement process
OT-systems #n-1	. . .	. . .
OT-systems #n	Are measures defined in the organization to be able to restore the functionality of an OT system after a cybersecurity incident?	☐ No activities ☐ Concepts but no concrete implementation yet

(continued)

Table 7.15 (continued)

Question topic	Question	Answer options
		☐ Concept partially implemented ☐ Fully implemented and thorough documentation ☐ A continuous state-of-the-art and efficient monitoring ☐ Subject to a continuous improvement process

7.5 SWOT Analysis

The SWOT analysis is a framework to identify and analyze an organization's strengths, weaknesses, opportunities, and threats. These words make up the SWOT acronym.

7.5.1 Introduction to SWOT Analysis

The SWOT analysis is a powerful tool to identify opportunities for improvement. To do this, SWOT analyzes the internal and external factors that can affect the viability of a decision. A factor is a relevant data or information. Internal factors are strengths and weaknesses, external factors are opportunities and threats. Furthermore, SWOT analysis supports to improve processes, drive prioritization, and plan for continuous improvement of the most important operations in organizations. While similar to competitive analysis, which is the process to identify competitors in the organization's sector and researching their different strategic plans, the SWOT analysis differs in that it assesses both internal and external factors. Analyzing the key areas surrounding these opportunities and threats gives the necessary insights to thrive successfully and make strategic decisions with confidence. Hence, the SWOT analysis is used either at the start of, or as part of, a strategic planning process that enables organizations to uncover opportunities for success preciously disregarded. Applying SWOT analysis in information security can be useful for developing a better understanding of the security environment. Furthermore, SWOT can support an organization's overarching strategy by giving insight into security assets, security risk(s), security issues, and security challenges to the organization as a whole. Against this background, the SWOT analysis has many advantages and disadvantages. Some advantages are:

- Applicable at different depths, meaning lightweight treatment for simpler circumstances, through to highly detailed treatment for complex or larger issues

Strength	Weaknesses
• What the organization is doing well? • What unique innovations the organization can draw on? • What other organization see as strengths of the organization? • And other;	• What should the organization improve? • Why other organizations come up with innovations the organization investigated fails? • What other organizations see as weakness of your organization? • And other;
Opportunities	**Threats**
• What opportunities are open to the organization? • What is needed to turn strength into opportunities? • What trends the organization can take advantage of? • And other;	• What threats could harm the organization? • What threats the weaknesses expose to the organization? • What are the organizations competitors doing better? • And other;

Fig. 7.3 SWOT analysis matrix with possible questions

• Applicable to many levels in organizations
• Easy to understand by a simple diagram
• Highly visual and consequently easy to communicate to other stakeholders

Some disadvantages are:

• Easy to ignore the underlying principles that lead to factors assigned to the wrong area of analysis and consequently resulting in an invalid strategy and/or result
• Not separating the analysis elements of data collection, its evaluation, and the consequent decision-making
• Using data which is biased by perceptions, beliefs, personality types, and preferences
• Using poor data and factors expressed as generalizations

The SWOT analysis is typically represented by a 2 × 2 grid matrix, with one square for each of the four SWOT dimensions, strengths, weaknesses, opportunities, and threats. In Fig. 7.3, the SWOT analysis matrix is shown with the associated dimensions headed by some possible strategic questions.

As an outcome of the questionnaire examples in Table 7.17 for the four dimensions, applied to information security analysis, answers that can occur are illustrated in Table 7.16.

In general, SWOT analysis used to identify necessary measures in order to abstract the future target state from the current organization's state. For this purpose, the SWOT analysis requires data or information, usually extracted from questions, asked with a view to the dimensions and the influencing factors, as illustrated in Table 7.17. From the answers strategic decisions occur, which, in addition to the

Table 7.16 Outcome in information security SWOT analysis

SWOT dimension	Answers
Strength	Strong data encryption practices Regular technological innovations Robust access policy updated regularly
Weaknesses	Lack of a cybersecurity plan Lack of well-trained, skilled employees Lack of progress in product innovations
Opportunities	Embed options for cloud storage for data Keeping data secure and backed up off-site regularly Embed more innovative technologies in products
Threats	Computer viruses Non-compliance with regulations Organization-wide cybersecurity strategy

Table 7.17 Questions in SWOT analysis dimensions

SWOT dimension	Questions
Strength	What are the reasons behind the successes of the organization so far? What unique selling points does the organization have? Which core competencies and resources in the organization are decisive for the Unique Selling Point (USP)? Which special expertise and which unique informationpromote it? And others
Weaknesses	Which skills are missing in the company to secure the current position or to achieve more in the future? Which products of the company are in a late-life phase (life cycle)? Where is there a lack of budget for the upcoming tasks to improve the cyber-security of the organization? What are the reasons behind the success of competitor organizations implementing the digital transformation?
Opportunities	What cost savings are gained reducing the organization's downtime in produc-tion through cyberattacks by defending cyberattack risks? Which innovative solutions will increase cybersecurity in the organization? Which innovative technologies in the organization's products will extend the actual market position of the organization? Will cloud services offer better customer relationship to the organization?
Threats	How to act after new competitors have infiltrated the organization's market segment? How to avoid a lack of content filtering which leads to liability and a risk of hacking? How to act after being more often targeted by cyber-criminal attacks compared to other organizations? How to act to avoid losing money and reputation by ransomware attacks?

organizations focus and responsibilities, contain specific activities required to achieve the organization's targeted goals. From Table 7.17 is concluded that the basic principle of the SWOT analysis follows the style of expanding strengths and reducing weaknesses, as well as seizing opportunities and avoiding risks. Since this is a complex interplay of the four dimensions, their sorting and prioritizing depend on the expected degree of difficulty to achieve the relevant organization's goal(s), in order to derive the associated achievable measures based on this. The measures are taken geared towards achieving the future organizations goal(s). Primarily and foremost, this provides a simple way of examining the organization's current state. Furthermore, the SWOT analysis enables a reliable methodical starting position, which is necessary for the development of measures together with the strategy required to achieve the target state. For this purpose, goal-oriented questions examples in the four SWOT dimensions are given in Table 7.17.

A conclusion of the previous discussion is that the primary goal of the SWOT analysis is to increase awareness of factors that go into making business decisions, establishing a new business strategy, building up a cybersecurity plan, and others. For this purpose, the SWOT analyzes the internal and external factors that influence the viability of a decision.

- *Internal Factors*: Are those an organization has control over, which are strengths and weaknesses. The manifold of internal factors are [29]:

  - Core skills, authenticity, expertise, and experiences
  - Product range and R&D activities
  - Similarities to competitors. Are there things that are commonly used in your industry, but you missed out?
  - Differences. What is it you do differently that will give you a competitive advantage? What makes you different from the competition?
  - Trade secrets, patents, and trademarks
  - Value, benefit, price, quality, geographic advantages
  - Technology, finance
  - Sales and marketing
  - Customer service, PR, and management
  - Vision, mission, ethos
  - Organizational structure and management

- *External Factors*: Are the ones which an organization has little or no control over, which are opportunities and threats. The manifold of external factors are [29]:

  - Social, economic (macro and micro), and political environment
  - Price setting and market-driven pricing
  - Government or industry rules and regulations
  - Supply and distribution
  - Market and product trends and innovations
  - Customer bias and consumer behaviors
  - Technological changes and advances
  - Competition and branding practices in the market

Furthermore, the SWOT analysis matrix distinguishes between factors that are helpful, and those that are harmful, in respect of the SWOT analysis objective. Helpful factors are those that support an organization's success; strength and

Table 7.18 Objectives in SWOT analysis

Strengths	Weaknesses
Strength is internal and helpful factors of the SWOT objective, and a factor that supports an opportunity or overcomes a threat. Strengths include: Customer relation: Marketing, sales, service, reputation Employees: Talented, dedicated, skilled, well trained Financial: Robust balance sheet, cash flow; credit rating Technology: Expertise, innovation, market leader/pioneer, high potential	Weaknesses are internal but harmful factors of the SWOT objective that rely on being unable to take advantage of an opportunity or are vulnerable to a threat. Weaknesses include: Customer relation: Long delivery time, poor customer communication Employees: Skill shortages, poor enthusiasm Financial: High debit-liquidity ratio Technology: Inflexible technology or processes. Slow innovative cycle
Opportunities	Threats
Opportunities are external and helpful factors of the SWOT analysis that rise from different sources such as: Competitors: Withdraw from or not enter a market Restrictive legislation: An opportunity, if it is a threat to competitors Technology: Innovative potential, first in market	Threats are external and harmful factors of the SWOT analysis over which the organization has no control. Threats are tangible or nontangible. Tangible threats: Hostile takeover bid, new competitors, thefts Nontangible threats: Potential loss of reputation, brand damaging factors

opportunities are helpful factors. Harmful factors are those that impede or block the organization's success; weaknesses and threats are harmful factors.

As shown in Fig. 7.3, the SWOT analysis demonstrates its usefulness in recording factors of each of the four dimensions, which contain a list of factors that are essential data or information for a SWOT analysis. This mostly is a bullet point list with supporting rationale in an enclosed document, as shown in Table 7.18 after [30]. The SWOT list factors represent a classification of input factors, and using the analysis matching and converting suggestions, responses to the situation in the context or objective of the SWOT investigation.

The SWOT analysis is often used either at the start of, or as part of, a strategic planning process. SWOT is a powerful support for decision-making because it enables an organization to uncover opportunities for success that were previously unarticulated. It also highlights threats before they become overly burdensome. The SWOT analysis is most effective when used to pragmatically recognize and include business, technical issues, and concerns. The SWOT analysis often involves diverse cross-functional teams, made up of employees from different functional areas within an organization that collaborate to reach a stated objective. The result of a SWOT analysis is a chart or list of subject characteristics, as shown in the following analysis example for an imaginary retail employee [31]:

- *Strengths*: Good communication skills, on time for shifts, handles customers well, gets along well with all departments, physical strength, good availability

- *Weaknesses*: Takes long smoke breaks, has low technical skills, very prone to spending time chatting
- *Opportunities*: Storefront worker, greeting customers and assisting them to find products, helping keep customers satisfied, assisting customer post-purchase and ensuring buying confidence, stocking shelves
- *Threats*: Occasionally missing time during peak business due to breaks, sometimes too much time spent per customer post-sale, too much time in interdepartmental chat

The SWOT analysis supports an organization to gain insight into its current and future position in business, technology, employee skills and others against a stated goal, e.g., cybersecurity awareness or cybersecurity strategy, and others. Therefore, a SWOT analysis may be used to assess and consider a range of goals and action plans such as [31]:

- Creation and development of products or services
- Evaluating and improving customer service, opportunities, and performance
- Hiring, promotion, or other human resources activities
- Making investments in technologies, geographical locations, or markets
- Setting business strategies to improve competitiveness or improve business performance

Against this background, the preparation for a SWOT analysis depends on four essential steps followed in order:

1. Clear statements of the purpose of the SWOT analysis
2. Definition of the objectives analyzed
3. Establishment of the SWOT analytical team members
4. Standardization of work methodology and motivation of SWOT team members

Every one of the SWOT analysis team members must agree on a particular procedure of SWOT analysis and adhere to it. Main motivation of team members knows the purpose of SWOT analysis and that their work could be supportive and useful. Furthermore, team members must have enough time and expertise to acquire essential and certain information in discussions with the staff responsible in the organization. Then consecutive steps carried out referring to

- Identification and assessment of the internal factors' strengths and weaknesses
- Identification and assessment of the external factors' opportunities and threats

Based on the knowledge gained, the SWOT analysis matrix is developed as follows [32]:

- *Determine Factors of Strategic Significance*: Internal factors' strengths and weaknesses are of high relevance as well as the external factors' opportunities and threats of high values, i.e., the benefits and risks of high levels are of strategic significance.
- *Generating of Alternative Strategies*: Based on combining strengths and weaknesses (internal factors) with identified opportunities and threats (external factors), which consist of four different strategic options:

  - *Weaknesses–Opportunities Strategy*: Strategy of searching
  - *Strengths–Opportunities Strategy*: Strategy of taking advantage
  - *Weaknesses–Threats Strategy*: Strategy of avoidance
  - *Strengths–Threats Strategy*: Strategy of confrontation

Objective outcomes of the SWOT analysis help organizations make a more precise overview of the prospects of further development. Thus every one of the SWOT analysis must ensure that the outcomes of assessments are the most objective.

7.5.2 SWOT Analysis Best Practice Examples

Cybersecurity incidents have increased in all industrial sectors, causing serious breach-in attacks. In the context of the NIST CSF (see Chap. 5), cybersecurity is a process that involves identification, protection, detection, response, and recovery from cyberattacks. Thus, cybersecurity includes cybersecurity awareness as well as cybersecurity strategies. Hence, cybersecurity maturity models used to gain insight into organizations' actual status in cybersecurity awareness and cybersecurity defense strategy. Therefore, cybersecurity maturity models are useful in guiding organizations in the development of processes leading to a state of maturity indicator level in the area for which the Maturity Model was developed. In the industrial sectors specific cybersecurity maturity model approaches are used, assessing compliance to cybersecurity standards and guidance or specific information systems components. Two examples of maturity level models are shown in Table 7.13, which make use of a questionnaire to test the achieved maturity level of maturity indicator level (see Table 7.4).

7.5.2.1 Company Analysis

Increasing digitization, as part of the digital transformation, faces great challenges to companies, examined as part of a SWOT company analysis. Distinctive factors for a SWOT analysis are branded products, core competences, fixed costs, and others. These factors are assigned to the four dimensions in the SWOT analysis matrix, the internal factors strength and opportunities, and risks as external factors. As a result, the companies' actual situation is available at a glance in the context of SWOT factors, as follows:

Table 7.19 Factor examples in SWOT business analysis

Strengths	Weaknesses
High reputation	Bad financial situation
High visibility on the internet	Harmful working atmosphere
Innovative technologies and processes	High personnel costs
Loyal customers	Low customer loyalty
Modern production facilities	Old fashioned technology and processes
Motivated employees	Poor marketing and sales
Recognized expertise	Poorly skilled employees
Opportunities	**Threats**
Changed customer expectations	New competitors come into market
Good financial situation for innovation	New regulations due to climate change
New regulations with positive impact	Product price goes down
New product trends	Risk of an economic recession
Rising product demand	Stagnant demand
Technological innovations	Uncertain supply chains

Internal Factors:

- *Strength*: High reputation, high visibility on the Internet, innovative technologies and processes, loyal customers, modern production facilities, motivated employees, recognized expertise, and others
- *Weaknesses*: Bad financial situation, harmful working atmosphere, high personnel costs, low customer loyalty, old fashioned technology and processes, poor marketing and sales activities, poorly skilled employees, and others

External Factors:

- *Opportunities*: Changed customer expectations, good financial situation for innovation, new regulations with positive impact, new product trends, rising product demand, technological innovations, and others
- *Threats*: New competitors come into market, new regulations due to climate change, product price going down, risk of an economic recession, stagnant demand, uncertain supply chains, and others

All factors are entered in the four quadrants' matrix, as shown in Table 7.19.

Proposals should follow the principle *strengthening strength and weakening weaknesses*, sorted according to the degree of difficulty in practical implementation in order to be able to set priorities. The strengths and weaknesses lie in the past and present. The measures required being aimed to the future. Thus a potential question is: "Which opportunities need to be seized and which risks are to be avoided?" It is important to note that the SWOT factors were backed up with reliable numbers, data, and facts.

7.5.2.2 Cybersecurity SWOT Analysis

The SWOT analysis is a useful approach analyzing internal factors (strengths and weaknesses) in organizations that cybersecurity can control and external factors (opportunities and threats) an organization cannot control. Therefore, the SWOT analysis type used is on an organization's functional basis, e.g., organizations cybersecurity strengths. Common organization's system strength protect against cyberattacks by firewalls; anti-malware, anti-virus programs, and others. Another strength is employees who respect using authorized passwords and use computer systems and apps only for work-related activities. Opportunities are external factors that help to increase an organization's strengths such as enhancing storage capacity through Cloud-as-a-Service backed up by the outside cloud provider. Weaknesses of cybersecurity can be difficult to overcome. A weakness against ransomware attacks (see Chap. 6) would cost an organization beside the ransom, downtime in production, and hence product selling cannot be achieved on time or is delayed, which may also cause loss in reputation. To overcome this threat potential, training of employees in cybersecurity awareness skills and following the organization's policies, as well as introducing an organization-wide cybersecurity strategy, installing new cybersecurity tools, and regularly updating software security packages, help to protect the organization against cybersecurity risk(s). Finally, threats are another external factor an organization has to face such as viruses, hacking, and phishing attacks, and others. Beside this, threats can be gained through social media by negative reports about the organization's business, product, services, quality, and others, which the organizations have to defend too.

Developing a cybersecurity SWOT analysis the organization should conduct, an audit on employees, identifying whether or not they follow the organization's policies. This also allows to identify whether or not employees have gained enhanced cybersecurity awareness skills and knowledge about the organization-wide cybersecurity strategy, introduced by business leaders. In this regard, strength in cybersecurity skills between an organization and its employees at its bests is symbiotic, meaning the more the employee is aware about the organization's cybersecurity risks and encouraged to understand and promote the cybersecurity-minded culture, the better protected the organization's corporate network becomes.

Due to easy handling of the SWOT analysis, however, it should be noted that the categorization of the factors (internal and external) within the four dimensions can be subjective, which is why a draft of the survey must be stringent and dedicated in order to enable the most objective data analysis possible. For this purpose, defining the essential quantitative parameters in a brainstorming session for both the internal and the external factors is required. Table 7.20 provides an example of questions and possible answers chosen from the cybersecurity domain.

Due to the qualitative nature of interviews in SWOT analysis, an evaluation is carried out, contributing to the clarification of contradictions and abnormalities by deepening certain questions about selected objectives of the SWOT analysis. The results of best practice examples on cybersecurity and new technologies are recently

Table 7.20 Questions and possible answers to cybersecurity

Strengths	Weaknesses
Cybersecurity is of great importance in all business processes for the organization to prevent cyberattacks. This usually requires the appropriate expertise, e.g., to recognize cyberattacks at the workplace and to fend them off as best as possible. In your opinion, which recognized methodological procedures are already in use? ☐ Regularly updated antivirus software ☐ Robust access policy for user accounts ☐ Robust data encryption practices ☐ Regularly updates about attempted cyberattack forms according to their threat potential ☐ Regularly updated and binding organization-wide cybersecurity strategy ☐ Using a Virtual Private Network (VPN) to control access to the network to reduce its exposure	Cybersecurity is of great importance in all business processes of organizations preventing cyberattacks. Sufficient cybersecurity awareness is required to be able to recognize cyberattacks at an early stage to ward them off. In your opinion, what could be the reason why the cyber-security strategy used is yet not sufficient? ☐ Lack of centralized tracking of cyberattacks ☐ Lack of organization-wide cybersecurity strategy by senior management (CIO) ☐ Lack of data encryption practices ☐ Missing access policy for user accounts ☐ Poor funding of cybersecurity activities ☐ Spotty update process for security patches
Opportunities	**Threats**
Cybersecurity is of great importance in organizations' business processes to prevent cyberattacks. Therefore, continuous improvement in cybersecurity is required to be able to react effectively to possible threat. What changes in your organization may gain benefits? ☐ Implementation of an intelligent edge platform to interact and learn with the changing threat landscape (watching the watchers) ☐ Keeping data secure and backed up off site ☐ Recognized priority and support from senior management ☐ Regular updating of expertise on new data storage technologies in order to keep them up-to-date with the latest technology and to save them ☐ Sustainable growth through a business model protected against cyberattacks	The importance of cybersecurity in the organizations' business processes is recognized. However, the necessary steps to be able to react appropriately to possible cyberattacks are unclear. Where do you see the greatest threats to the organization? ☐ Attacks introducing viruses ☐ Failure complying with legal requirements ☐ Hacker attacks injecting malware ☐ No binding organization-wide cyber-security strategy that regulates the uniform procedure by cyberattacks ☐ No centrally managed BYOD usage that potentially increases risks of information theft and risk of hacker attacks ☐ Non-efficient cybersecurity expertise built up, ransomware is everywhere ☐ Unprotected WLAN allows unauthorized persons access to the organization's (company) network

published in [25, 26]. The main limitation of the study reported in [25, 26] is the sub-participation of some sectors. Since cybersecurity is a sensitive topic, many companies do not want to discuss, fearing revealing significant weaknesses.

7.6 Exercises

7.6.1 Maturity Models

What is meant by the term *Maturity* Index?
Describe the main characteristics and capabilities of the Maturity Index.
What is meant by the term *Performance in Maturity Index*?
Describe the performance characteristics and capabilities of the Maturity Index.
What is meant by the term *Reporting in Maturity Index?*
Describe the main characteristics and capabilities of the Reporting Maturity Index.
What is meant by the term *Maturity* Model?
Describe the main characteristics and capabilities of at least four Maturity Models.
What is meant by the term *Maturity Level?*
Describe the main characteristics and capabilities of the regular six Maturity Levels.
What is meant by the term *ISO 9004:2008*?
Describe the features and capabilities of the ISO 9004:2008.
What is meant by the term *Features and Capabilities of the ISO 9004:2018?*
Describe the main characteristics, capabilities, and features of the ISO 9004:2018.
What is meant by the term *Application Security?*
Describe the characteristics and capabilities of the Application Security.
What is meant by the term *Information Security?*
Describe the characteristics and capabilities of the Information Security.
What is meant by the term *Network Security?*
Describe the characteristics and capabilities of the Network Security.
What is meant by the term *Operational Security?*
Describe the characteristics and capabilities of the Operational Security.
What is meant by the term *NIST CSF Maturity* Model?
Describe the specific focus of the NIST CSF Maturity Model.
What is meant by the term *ISO/IEC 27 K Maturity* Model?
Describe the specific focus of the ISO/IEC 27 K Maturity Model.
What is meant by the term *CIS20 Maturity* Model?
Describe the specific focus of the of the CIS20 Maturity Model.
What is meant by the term *Radar Chart*?
Describe the main characteristics and capabilities of the Radar Chart.
What is meant by the term *Fuzzy Elements in Maturity Models*?
Describe the main characteristics and capabilities of the Fuzzy Elements in Maturity
    Models.
What is meant by the term *Membership Function*?
Describe the main characteristics and capabilities of five Membership Functions.
What is meant by the term *Capability Maturity Model?*
Describe the main characteristics and capabilities of the Capability Maturity Model.
What is meant by the term *Cybersecurity Capability Maturity Model?*
Describe the main characteristics and capabilities of the Cybersecurity Capability
    Maturity Model.

7.6.2 SWOT Analysis

What is meant by the term *SWOT Analysis*?
Describe the main characteristics and capabilities of the SWOT Analysis.
What is meant by the term *SWOT Dimension*?
Describe the main characteristics and capabilities of the four SWOT Dimensions.
What is meant by the term *SWOT Analysis Matrix*?
Describe the main characteristics and capabilities of the SWOT Analysis Metrix.
What is meant by the term *Cybersecurity SWOT Analysis*?
Describe the main characteristics and capabilities of the Cybersecurity SWOT Analysis.
What is meant by the term *Internal Factors in SWOT Analysis*?
Describe the main characteristics and capabilities of the Internal Factors on SWOT Analysis.
What is meant by the term *External Factors in SWOT Analysis*?
Describe the main characteristics and capabilities of the External Factors on SWOT Analysis.
What is meant by the term *Company Analysis in SWOT Analysis*?
Describe the main characteristics and capabilities of the SWOT Company Analysis.

References

1. International Telecommunication Union. Index of Cybersecurity. 2017
2. Slapnicas, S., Marko, T.V., Drascek, M.: Effectiveness of Cybersecurity Audit. In: International Journal of Accounting Information Systems, Vol. 44, 2022
3. Kahyaoglu, S.B., Caliyurt, K.: Cybersecurity Assurance Process from the Internal Audit Perspective. In: Managerial Audit Journal, Vol 33, No. 4, pp. 360–376, 2018. https://doi.org/10.1108/MAJ-02-2018-1804 (Accessed 12.2022)
4. Deloitte's Cyber risk Capabilities, cyber strategy, secure, vigilant, and resilient. Deloitte Report, 2017. https://www2.deloitte.com/content/dam/Deloitte/at/Documents/risk/cyber-risk/Deloitte-Cyber-Risk-Capabilities-Broschuere.pdf (Accessed 12.2022)
5. Mutune, G.: 27 Top Cybersecurity Tools for 2020. 2020. https://cyberexperts.com/cybersecurity-tools/ (Accessed 12.2022)
6. Hackenbrack, K., Knechel, W.R.: Resource Allocation Decisions in Audit Engagements. In: Contemporary Accounting Research, Vol. 14, No. 3, pp. 481–499, 1997
7. Rothrock, R.A., Kaplan, J., Van Der Oord, F.: The board's role in managing cybersecurity risks. MIT Sloan Management Review Vol. 59, No. 2, pp. 12–15. 2014
8. Rife, R.: Planning for Success. 2004. https://iaonline.theiia.org/2006/Pages/Planning-for-Success.aspx (Accessed 12.2022)
9. Chambers, R.: From Good to Great: Strategic Planning Can Define an Internal Audit Function. 2014. https://iaonline.theiia.org/blogs/chambers/2014/Pages/From-Good-to-Great—Strategic-Planning-Can-Define-an-Internal-Audit-Function.aspx (Accessed 12.2022)
10. Alonso, J., Martinez de Soria, I., Orue-Echevarriai, L., Vergara, M: Enterprise Collaboration Maturity Model (ECMM): Preliminary Definition and Future Challenges. https://scholar.google.de/scholar_url?url=https://www.researchgate.net/profile/Leire-Orue-Echevarria Arrieta/publication/226088235_Enterprise_Collaboration_Maturity_Model_ECMM_

Preliminary_Definition_and_Future_Challenges/links/00b495319682d7473f000000/Enter
prise-Collaboration-Maturity-Model-ECMM-Preliminary-Definition-and-Future-Challenges.
pdf&hl=de&sa=X&ei=a_EiY4rCIPuSy9YPrsi4iAw&scisig=AAGBfm2P_n2IJTfYoA_J0
ej6K8J9AAZAJQ&oi=scholar (Accessed 12.2022)

11. Möller, D., Iffländer, L., Nord, M., Leppla, B., Krause, P., Czerkewsky, P., Lenski, N., Mühl,
K.: Cybersecurity in the German Railway Sector. Accepted Paper CRITIS 2022. Published in
Springer Computer Science Procedures, 2022

12. Becker, J., Knackstedt, D. and Pöppelbuß, J.: Developing Maturity Models for IT
Management – A Procedure Model and its Application. In: Business Information Systems
Engineering Vol. 1, pp. 213–222, 2009

13. Venkatraman, V.: The Digital Matrix: New Rules for Business Transformation through Technology. LifeTree Book Publ., 2017

14. Rogers, D. l.: The Digital Transformation Playbook: Rethink your Business for the Digital Age.
Columbia University Press, 2016

15. Rossmann, A.: Digital Maturity Models: Theoretical Foundations and Practical Applications
(in German). 2016. https://www.researchgate.net/publication/334509326_Digitale_
Reifegradmodelle_theoretische_Grundlagen_und_praktische_Anwendung (Accessed 12.2022)

16. NIST Cybersecurity Framework. https://www.nist.gov/cyberframework, last accessed 2022/02/
14 (Accessed 12.2022)

17. CIS Top 20 Critical Security Controls Solutions. https://www.rapid7.com/solutions/
compliance/critical-controls/ (Accessed 12.2022)

18. Akkasoglu, G.: Methodology for Conception and Application-specific Maturity Models considering Information Uncertainty (in German). PhD Thesis, University of Erlangen
Nuremberg, 2013

19. https://www.metaplan.com/wp-content/uploads/2021/04/Metaplan_Basiswissen_engl.pdf
(Accessed 12.2022)

20. https://www.usertesting.com/blog/affinity-mapping (Accessed 12.2022)

21. Czabanski, R., Jezewski, M., Leski, J.: Introduction to Fuzzy Systems. In: Prokopowicz, P.,
Czerniak, J., Mikilajewsi, D., Apiecionek, L., Slezak, D. (Eds.) Theory and Applications of
Ordered Fuzzy Numbers: Studies in Fuzziness and Soft Computing. Springer Publ. 2017

22. Zadeh, L.: The Concept of a Linguistic Variable and its Application to Approximate
Reasoning. In: Internat. Information Science, Vol. 8, pp. 199–249, 1975

23. Weckenmann, A., Akkasoglu, G.: Maturity Determination and Information Visualization of
New Forming Processes considering Uncertain Indicator Values. In: American Institute of
Physics Conference Proceedings, Vol. 1431, pp. 899 ff, 2012. https://doi.org/10.1063/1.
4707649 (Accessed 12.2022)

24. von Altrock, C., Krause, B.: Multi-criteria Decision Making in German Automotive Industry
Using Fuzzy Logic. In: Fuzzy Sets and Systems, Vol. 63, No. 3, pp. 375–380, 1994. https://doi.
org/10.1016/0165-0114(94)90223-2 (Accessed 12.2022)

25. Möller, D., Iffländer, L., Nord, M., Leppla, B., Krause, P., Czerkewsky, P., Lenski, N., Mühl,
K.: Cybersecurity in the Railway Sector. In: Proceedings 17th International Conference on
Critical Information Infrastructures Security.; will be published in LNCS, Springer Publ., 2022

26. Möller, D.P.F., Iffländer, L., Nord, M., Krause, P., Leppla, B., Mühl, K., Lenski, N.,
Czerkewski, P.: Emerging Technologies in the Era of Digital Transformation: State of the Art
in the Railway Sector. In Proceedings 19th International Conference on Informatics in Control,
Automation and Robotics, pp. 721–726. SCITEPRESS, 2022

27. Kour, R., Karim, R., Thadurii, A.: Cybersecurity for Railways – A Maturity Model. In: Proceedings of the Institution of Mechanical Engineers , Part F: Journal of Rail and Rapid Transit,
Vol. 234, No. 10, pp. 1129–1148, 2020. https://doi.org/10.1177/0954409719881849 (Accessed
12.2022)

28. Kour, R.: Cybersecurity in Railway: A Framework for Improvement of Digital Asset Security.
PhD Thesis, Lulea University of Technology, Sweden, 2020

29. Gümüsten, Ü.: What is SWOT Analysis? 2021. https://umitgumusten.com/what-is-swot-analysis/ (Accessed 12.2022)
30. Sarsby, A.: SWOT Analysis. Leadership Library Publ., 2016
31. Bigelow, S.J., Pratt, M.K.; Tucci, L.: SWOT Analysis. 2022. https://www.techtarget.com/searchcio/definition/SWOT-analysis-strengths-weaknesses-opportunities-and-threats-analysis (Accessed 12.2022)
32. Rehak, D., Grasseova, M.: The Ways of Assessing the Security of Organization Information System through SWOT Analysis. Chapter 7, pp. 162–184, 2011. https://doi.org/10.4018/978-1-61350-311-9.ch007 (Accessed 12.2022)

Chapter 8
Machine Learning and Deep Learning

Abstract Machine Learning is a sub-category of Artificial Intelligence enabling computers with the ability of pattern recognition, or to continuously learn from, making predictions based on data, and carry out decisions without being specifically programmed for doing so. In this context, Machine Learning is a broader category of algorithms being able to use datasets to identify patterns, discover insights, and enhance understanding and make decisions or predictions. Compared with Machine Learning, Deep Learning is a particular branch of Machine Learning that makes use of Machine Learning functionality, and moves beyond its capabilities. Deep Learning Algorithm is interpreted as a layered structure that tries to replicate the structure of the human brain. These capabilities enable Machine Learning and Deep Learning Algorithms usage in applications to identify and respond to cybercriminals manifold cyberattacks. This is achieved by analyzing Big Datasets of cybersecurity incidents to identify patterns of malicious activities. For this purpose, Machine Learning and Deep Learning compare known threat event attacks with detected threat event attacks to identify similarities they automatically dealt with trained Machine Learning or Deep Learning model for response. Against this background, this chapter seeks to offer a clear explanation of the classification of Machine Learning and Deep Learning and comparing them with regard to effectivity and efficiency in their specific application domains. This requires (i) discussing the methodological background of Machine Learning and Deep Learning; (ii) introducing relevant application areas of Machine Learning and Deep Learning like Intrusion Detection Systems; and (iii) use cases showing how to combat against threat event attacks based cybersecurity risks. In this context, this chapter provides, in Sect. 8.1, a brief introduction in classical Machine Learning, which consists of Supervised, Unsupervised, and Reinforcement Machine Learning. In this regard, Sect. 8.1.1.1 introduces Supervised Machine Learning, while Sect. 8.1.1.2 refers to Unsupervised Machine Learning, and Sect. 8.1.1.3 focuses on Reinforcement Machine Learning. Sect. 8.1.1.4 finally compares the different Machine Learning methods with regard to advantages and disadvantages. Based on this methodological introduction of classical Machine Learning, Sect. 8.2.1 introduces in Machine Learning and cybersecurity issues. Machine Learning-based intrusion detection in industrial application is therefore the topic of Sect. 8.2.1.1. Section 8.2.1.2 introduces Machine Learning-based intrusion detection based on feature learning, and Machine Learning-based intrusion detection of unknown cyberattacks is the topic of Sect. 8.2.1.3. In Section 8.3, the classification of Deep Learning methods is given which contains in Sect. 8.3.1 the topics Feedforward Deep Neural Networks, Convolutional Feedforward Deep Neural Networks, Recurrent Deep Neural Networks, Deep Beliefs Networks, and the Deep Bayesian Neural Network. Based on this methodological background of Deep Learning methods, Sect. 8.3.2 introduces Deep Bayesian Neural Networks, while Sect. 8.3.3 refers to Deep Learning-based intrusion detection. Finally, Sect. 8.4 refers to Deep Learning methods in cybersecurity applications. Section 8.5 contains comprehensive questions from the topics Machine Learning and Deep Learning, followed by "References" with references for further reading.

8.1 Introduction

With the ever-increasing amounts of data in electronic form, the need for automated methods for data analysis continues to grow which machine learning methods provide. Machine learning is a sub-category of artificial intelligence, enabling computer systems with the ability to automatically detect patterns in data, and then to use the uncovered pattern to predict future data or other outcomes of interest, or to perform other kinds of decision-making under uncertainty. In this context, machine learning drives many advances in industrial, public, and private organizations and in modern society, from web searches to content filtering, e.g., on social networks and others. Furthermore, machine learning is essentially a form of applied statistics with increased emphasis on the usage of computer systems to estimate statistically complex functions and less on proving confidence intervals around these functions. Thus, machine learning is used to identify objects in images, transcribe speech into text, match new items, selecting relevant results of search, and other applications. Nevertheless, every machine learning workflow begins with three general questions:

- What kind of data is used to work with?
- What insights are wanted to get from it?
- How and where those insights will be applied?

Answering these questions helps to decide about the machine learning model used. Classical machine learning consists of supervised, unsupervised, and reinforcement machine learning models. Machine learning in general trains a model on known input and output data so that it can predict future outputs, whereby, in case of unsupervised learning, hidden or intrinsic structures in input data may be found. Against this background, machine learning allows to tackle tasks that are too difficult to solve with traditional programming paradigms. In this regard, machine learning tasks described how to process to solve a problem. A problem is a collection of features quantitatively measured at the object under investigation, which use the machine learning algorithm. Typically, a problem can be represented as a vector

$$x \in \mathfrak{R}^n,$$

where each entry x_i of the vector is another feature. For example, the features of an image are usually the values of the pixels in the image. In this context, machine learning models may be assumed as sets of

$$n \; input \; values \; x_1, \; \ldots, x_n,$$

and associate them with an output y. These models learn a

$$set \; of \; weights \; w_1, \; \ldots, w_n,$$

and compute their output such as

$$f(x, w) = x_1 w_1 + \ldots + x_n w_n.$$

Thus, a huge number of problems are solved based on the respective machine learning algorithms. Some of the most common machine learning applications are shown in Table 8.1 [1].

Table 8.1 indicates that machine learning is used in case of solving complex tasks or problems involving a large amount of data and many variables, for which no traditional formula or equations exist, to solve the addressed problem. Therefore, machine learning is an adequate option to deal with these situations. In this regard, one of the earliest problems in threat event attacks solved by machine learning was spam detection making use of spam filters, which create detection rules, based on machine learning algorithms. The approach used to solve the spam problem was learning spam filters to recognize junk mails and phishing messages by analyzing rules across a huge number of computer systems. In addition to spam detection, social media websites are using machine learning as a way to identify and filter abuses.

8.1.1 Classical Machine Learning Techniques

Classical Machine Learning Techniques (CMLT) are divided into several types such as predictive or supervised machine learning, descriptive or unsupervised machine learning, as well as reinforcement machine learning, as shown in Fig. 8.1.

The first type of machine learning, the Predictive or Supervised Machine Learning, groups and interprets data based only on input data. Thus, its goal is to learn a mapping from inputs x to outputs y, given a labeled set of input–output

Table 8.1 Common machine learning-based applications with their characteristics

Machine learning application	Characteristics
Anomaly detection	Machine learning algorithm sifts through a set of events or objects, and flags some of them as being atypical or unusual (see Sect. 8.1.1). Hence, a machine learning algorithm is an algorithm that learns from data.
Classification	Machine learning algorithm specifies which of k-categories some inputs belong to. To solve this problem, machine learning algorithm generates a function $f: \mathfrak{R}^n \rightarrow \{1, \ldots, k\}$. For $y = f(x)$ the classification assigns an input described by vector x to a category identified by the numeric code y.
Denoising	Machine learning algorithm has as input a *corrupted problem vector* $x_c \in \mathfrak{R}^n$ obtained by an unknown corruption process from an *uncorrupted (regular) problem vector* $x_u \in \mathfrak{R}^n$. The machine learning algorithm predicts the uncorrupted (regular) problem vector x_u from its corrupted problem vector x_c, or more in general, it predicts the conditional probability distribution $p(x_u \mid x_c)$.
Density estimation	Machine learning algorithm has to learn a function $p: \mathfrak{R}^n \rightarrow \mathfrak{R}$, where $p(x)$ may be interpreted as a probability density function if x is continuous.
Imputation of missing values	Machine learning algorithm has to solve the *problem vector* $x \in \mathfrak{R}^n$, but some entries x_i of x are missing. Thus, the machine learning algorithm has to provide a prediction of the values of the missing entries.
Machine translation	Machine learning algorithm input consists of a sequence of symbols in some language that may be converted into a sequence of symbols in another language by the machine learning algorithm.
Regression	Machine learning algorithm has to output a function $f: \mathfrak{R}^n \rightarrow \mathfrak{R}$.
Synthesis and sampling	Machine learning algorithm generates new examples that are similar to those in the training data.
Transcription	Machine learning algorithm observes an unstructured representation of some kind of data and transcribes it into discrete, textual form.

$$D = \{x_1 . y_1\}_{i=1}^{N}$$

with D as training set, and N as the number of training samples. In the simplest setting, each training input x is a D-dimensional vector of numbers, called features, attributes, or covariates, which often is stored in an $N = D$ design matrix. However, x_i could be also a complex structured object, such as an image, a sentence, a time series, a graph, and many others. Output y in principle can be anything, but most techniques in machine learning assume that y_i is a categorical or nominal variable from some finite set

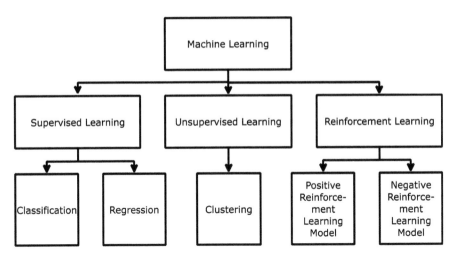

Fig. 8.1 Machine learning techniques

$$y_i \in \{I, \ldots, Z\}$$

or a real value scalar. In case y_i is categorical, the machine learning method is a classification or pattern recognition approach, and if y_i is real-valued, the machine learning method is termed a Regression Approach [2]. Thus, the scope of Supervised Machine Learning (SML) is to build a model that makes predictions based on evidence in the presence of uncertainty. In this context, an SML algorithm uses a known set of input data and known responses to the output data, and trains the model to generate reasonable predictions for the response to new data. SML uses classification techniques that predict discrete responses and regression techniques to predict continuous responses. Applications for classification technique are, for example, whether an email is genuine or spam and for regression technique are, for example, fluctuations in power demand [3].

The second type of machine learning, the Descriptive or Unsupervised Machine Learning (UML), is a predictive model based on both input and output data. The goal of this machine learning technique is finding interesting hidden patterns or intrinsic structures in the data, which is termed knowledge discovery. UML algorithms are used to draw inferences from datasets consisting of input data without labeled responses. Clustering is the most common UML technique used for exploratory data analysis to find hidden patterns of groupings in data. Applications for clustering include gene sequence analysis, market research, and object recognition [3].

The third type of machine learning is Reinforcement Machine Learning (RML). RML algorithms interact with an environment, so there is a feedback loop between the RML system and human experiences. This method is useful for learning how to act or behave for given occasional reward or punishment signals.

8.1.1.1   Supervised Machine Learning

Supervised Machine Learning (SML) is the most common form of machine learning, based on features associated with a label or target. Therefore, SML algorithms study the dataset of an object of interest to classify the object features into different classes, based on their measurements, which is termed as density estimation. In this context, y indicate in SML to a variable that can be predicted, which means that SML algorithms enable predictions. The advantage of SML algorithms is obvious, developing intelligent systems that precisely adapt to their specific task. Hence, they are not standard software, but rather highly specialized software tools for individual and specific tasks. However, their basic operational principle is always the same: a certain dataset (input x) runs through a previously trained algorithm of an SML technique, and at the end is the result (output y).

As shown in Fig. 8.1, SML algorithms are divided into classification and regression models. Both types make use of different learning methods, capable for learning under specific constraints. Some of the most important SML methods are

- *Classification Learning Methods*: These methods include:

  - *Linear Discriminant Analysis (LDA)*: Makes predictions to estimate the probability that new inputs belong to each class under investigation. Against this background, the class under investigation that received the highest probability is the output class identified by prediction. The LDA model uses the Bayes theorem to estimate the probabilities. The Bayes theorem is used to estimate the probability of the output class (k) given the input (x) using the probability of each class and the probability of the data belonging to each class:

$$P(Y = x | X = x) = \frac{(Pk * fk(x))}{\text{sum}(PI * fI(x))}$$

  where Pk refers to the base probability of each class (k) observed by training data. In the context of the Bayes' theorem, this is termed prior probability

$$Pk = \frac{nk}{n}$$

  The term $f(x)$ represents the estimated probability of x belonging to the class k. A Gaussian Distribution Function (GDF) is used for $f(x)$ and simplifying the result in the discriminate function $Dk(x)$ for class k and given input x.

  The typical implementation of LDA assumes a given complete dataset for training in advance and learning, carried out in one batch. This technique is widely used in face recognition, data mining, and knowledge discovery applications.

  - *Naive Bayes Method (NBM)*: Are a set of supervised learning algorithms based on applying Bayes' theorem with the naive assumption of conditional independence between every pair of features given the value of the class variable.

Bayes' theorem states the following relationship, given class variable and dependent feature vector x_1 through x_n as described in [4], which is followed:

$$P(y|x_1, \ldots, x_n) = \frac{(P(y)P(x_1, \ldots, x_2|y))}{(P(x_1, \ldots, x_n))}$$

Using the naive conditional independence assumption that

$$P(x_1|y, x_1, \ldots, x_{i-1}, x_{i+1}, \ldots, x_n) = P(x_i|y)$$

for all i, this relationship is simplified to

$$P(y|x_1, \ldots, x_n) = \frac{(P(y)\prod_{i=1}^{n} P(x_i|y))}{(P(x_1, \ldots, x_n))}$$

Since

$$P(x_1, \ldots, x_n)$$

is constant given the input, the following classification rule can be used:

$$P(y|x_1, \ldots, x_n) \propto P(y) \prod_{i=1}^{n} P(x_i|y)$$

that follows

$$y = \arg\max_{y} P(y) \prod_{i=1}^{n} P(x_i|y)$$

for which Maximum A Posteriori (MAP) estimation is used to estimate

$$P(y)$$

and

$$P(x_i|y)$$

the former is then the relative frequency of class y in the training set. The different naive Bayes classifiers differ mainly as described in [4] by the assumptions they make regarding the distribution of

$$P(x_i|y)$$

In [5], a Python module that integrates a wide range of machine learning algorithms for medium scale supervised and unsupervised classifier problems is introduced.

Against this background, NBM represents a classification method to the probability of assign class label C_{al} to problem instance I_p represented by a vector $x = \{x_1, \ldots, x_n\}$ assigned to the problem instance I_p probability

$$P(C_{al}|x_1, x_2, \ldots, x_n).$$

Let's assume that a particular event E_p investigated is equal to the assigned class label if instance I_p is equal to the target output, then the naïve Bayes assumption is proof P divided into attributes that are independent of each other as follows with probability p:

$$p(E_p|P) = \frac{p(P_1|E_p) \cdot p(P_2|E_p) \cdot \ldots \cdot p(P_n|E_p)p(PE_p)}{p(P)}$$

– *Nearest Neighbor Method (NNM)*: Optimization problem of finding the point in a given dataset that is closest to a given point, which means nearest neighbor, based on the following constraints [6, 7]

$$P = \{\bar{p}_i\}\,_1^N$$

is a set of n points in a d-dimensional space, and let

$$\vec{q_i}$$

be a query point, then the nearest neighbor point method calls for finding the point

$$\vec{p_c}$$

in P, which is the minimum distance from

$$\vec{q}$$

which finally results in [7]

$$\left\| \vec{q} - \vec{p_c} \right\| \leq \left\| \vec{q} - \vec{p_i} \right\| \forall \vec{p_i} \in P.$$

The nearest neighbor method is applied in industrial applications. In manufacturing, e.g., it can be important knowing the frequency at which objects required for a specific assembling process passing through under a monitoring camera, or monitoring whether required different assembling objects inter arrival times fit with the assembly sequence.

- *Regression Learning Methods (RLM)*: These methods include:

  - *Decision Tree Method (DTM)*: Representation of a flowchart-like structure in which each node of the tree characterizes a test on features for which a decision is taken after computing all features which represent the gained information of the corresponding leaf node class label. In this context, branches represent conjunctions of features that lead to those class labels. Hence, the paths from root to leaf represent the classification rules. Against this background, the decision tree method represents an approach to split datasets based on different conditions, which is most widely used for SML, used in data mining, statistics, and other applications. Maimon and Rokach published a book [8] that specially covers the data-mining topic. The datasets format used in the DTM contain the dependent variable y, which is the target variable to classify. The vector x is composed of the

$$fearures, x_1, x_2, \ldots, x_n.$$

  Let's assume a regression problem with the

$$fearure\ variables\ (x_1, x_2),$$

  and a numerical output y result in the general notation

$$(x, y) = (x_1, x_2, \ldots, x_n, Yy)$$

  Using the DTM requires a specific sequential procedure to execute this method, which contains the following steps [9]:

  (a) Determine datasets that take into consideration to execute the DTM recursively at each node.
  (b) Compute the uncertainty of datasets.
  (c) Determine questions which need to be asked at that node.
  (d) Partition rows split into true rows and false rows, based on each question asked.
  (e) Calculate information gain based on Gini Impurity and partition of data from previous step. The Gini Impurity is used to build Decision Trees to determine how the features of a dataset should split nodes to form the tree.
  (f) Update highest information gain based on each question asked.
  (g) Update best question based on information gain.
  (h) Divide the node on best question.
  (i) Repeat from top step until a pure node is identified.

  In [9], the code for these steps is given in detail and the code for calculating the required Gini Impurity.
  - *Ensemble Learning Methods (ELM)*: Build prediction models combining the strengths of a collection of simpler based models. They are meta-algorithms, combining individual trained classifiers whose predictions are evaluated to provide a single output. Some of the Ensemble methods include bagging and

boosting. Bagging is a bootstrap ensemble method that creates individual classifiers for its ensemble by training each classifier on a random redistribution of the training set [10], in which training datasets for each classifier is randomly generated from the original. One way to reduce the variance of an estimate is to average together multiple estimates. For example, by training M different trees on different subsets of the data that can be chosen randomly with replacement and compute the ensemble:

$$f(x) = \frac{1}{M} \sum_{m=1}^{M} f_m(x)$$

In this regard, the main goal for bagging is to reduce overfitting.

Boosting encapsulates a range of methods whose scope is to create a series of classifiers. The training set used for each predictor in a series based on the performance of the earlier classifiers [11]. The predictions then combined through a weighted majority vote, the classification, or a weighted sum, the regression, to create the final prediction. In a penalized regression with large basis expansion, all possible terminal node regression trees $T = \{T_k\}$, which could be realized on the training dataset, is basis function in $\Re^p$. The linear model is

$$f(x) = \sum_{k=1}^{K} (x)_k T_k(x)$$

where $K = card\,(T)$. Let's assume the coefficients are to be estimated by Least Squares.

The principal difference between boosting and methods such as bagging is that base learners are trained in sequence on a weighted version of the data. Therefore, the goal of boosting methods is to create classifiers that enable to correctly classify examples for which the previous ensemble predictions are poor. Similar to bagging, classifiers are separated based on data points but differ in the way chosen [12].

To conclude, Supervised Machine Learning Algorithms (SMLA) make predictions or statements. One of their typical tasks for example is spam prediction in the digital transformation era, meaning, is an email spam or not? Another more complex problem is to calculate the ideal sales price for an innovative digital product with regard to the sales price of a competitor's sales price for the same product category. During the training phase in SML, an intelligent algorithm tries to approach the correct result.

8.1.1.2 Unsupervised Machine Learning

Unsupervised Machine Learning (UML) studies how computer systems can learn to represent particular input patterns in a way that reflects the statistical structure of the overall collection of input patterns. In this context, the goal of UML is to find hidden patterns or intrinsic structures in data of objects of interest. Thus, UML is used to draw inferences from datasets consisting of input data without labeled responses. Clustering is the most common UML technique, used for exploratory data analysis to find hidden patterns or groupings in data. Applications for clustering include gene sequence analysis, market research, and object recognition [3]. By contrast with SML (see Sect. 8.1.1.1) or Reinforcement Machine Learning (see Sect. 8.1.1.3), in UNL there are no explicit target output associated with each input; rather the UML algorithm brings to bear prior biases as to what aspects of the structure of the input should be captured in the output. Therefore, in UML exists

$$N \; observations \; (x_1, x_2, \ldots, x_N)$$

of a

$$random \; p - Vector \; X$$

that has a

$$joint \; density \; P(X).$$

The goal of UML is to directly infer the properties of the probability density without the support of a supervisor. It's a fact that the dimension of X is sometimes higher than in SML, and the properties of interest are often more complicated than simple location estimates. Therefore, these factors are somewhat mitigated by the fact that represents all of the variables under consideration; one is not required to infer how the properties of $P(X)$ change, conditioned on the changing values of another set of variables [13]. In case of low-dimensional problems, $p \leq 3$, a variety of effective nonparametric methods for directly estimating the density $P(X)$ at all X-values exist, and graphically represented. Due to dimensionality, these methods fail in high dimensions. Therefore, one must settle for estimation by rather crude global models, such as Gaussian mixtures or various simple descriptive statistics that characterize $P(X)$ [13].

As shown in Fig. 4.1, UML algorithms are characterized as clustering models, capable for learning by the following methods:

- *Clustering Learning Methods (CLM)*: These methods include:

  - *Density-based Clustering Methods (DCM)*: Identify distinctive groups/clusters in data, based on the idea that a cluster in a data space is a contiguous region of high data point density, separated from other such clusters by contiguous regions and low data point density. The data points in the separated regions of low data point density are typically considered as noise/outliners [14].
  - *Hierarchical Clustering Methods (HCM)*: Approach that groups similar objects into groups, termed clusters. The method tries to achieve a set of clusters where each cluster is distinct from the other clusters, and the objects within each cluster are broadly similar to each other. This method can give important clues to the structure of datasets, and therefore suggests results and

hypothesis for the underlying problem to be solved [15]. Basis algorithms for hierarchical clustering are agglomerative and divisive algorithms;

- *Partitioning Clustering Methods (PCM)*: This method relocates instances by moving them from one cluster to another, starting from an initial partitioning. Typically, it requires that the number of clusters is preset by the user. To achieve global optimality in partitioned-based clustering, exhaustive enumeration process of all possible partitions is required [16]. There are several algorithms used in partitioning clustering such as:

    K-means Algorithm: Partitioning the data into K clusters

$$(C_1, C_2, \ldots, C_K),$$

    are represented by their centers of means. The k-means Algorithm finds locally optimal solutions with respect to the clustering error.

    K–medoids Method: Partitioning algorithm to minimize the Sum of Squared Error (SoSE), similar to the K-means Algorithm. It differs due to its representation of different clusters. Each cluster is represented by the most centric object in the cluster.

    K-mode Clustering: Corresponds to a two-way CANDULUS model, which decomposes each element in an N-way data array into a function of N sets of P clusters, when both ways corresponding to the data matrix modeled via discrete parameters and a L_{FCS}-norm-based fitting function, where *FCS* is infinitesimally small [17].

    K-medians Clustering: Special case of the two way CANDCLUs model when S—a matrix formed from the general elements given N-way data array—is constrained to be a partition and a Minimum Absolute Deviation (MAD) or an $L1$-norm-based fitting function is used [18].

    K-means in general: Have advantages and disadvantages.

    Advantages:

    (a) Easy to implement.
    (b) K-means computationally is faster compared to hierarchical clustering for large number of variable, if K is small.
    (c) K-means produces tighter clusters than hierarchical clustering.

    Disadvantages:

    (a) Predicting number of clusters (K-value) is difficult.
    (b) Initial seed has a strong impact on the final results.
    (c) Order of data has an impact on final results.

- *Probability-based Clustering Methods (PbCM)*: The basic idea is modeling clusters by means of probability distributions, presentation of algorithms, discussion of the differences to partitioning methods.
- *Fuzzy C-Means (FCM)*: Clustering method enables one piece of data to belong to two or more clusters. The method is based on minimization of a target function J.

$$J_m = \sum_{i=1}^{N} \sum_{j=1}^{C} \mu_{ij}^m \|x_i - c_j\|2, 1 \leq m < \infty$$

where m is any real number greater than 1, μ_{ij} is the degree of membership of x_i in the cluster j, x_i is the ith of d-dimensional measured data, c_j is the d-dimension center of the cluster, and $\|*\|$ is any norm expressing the similarity between any measured data and the center.

- *Gaussian Mixtures*: Statistical tool for modeling densities, widely used in SML and UML. The parameters of a mixture can be estimated using the Expectation-Maximization Algorithm (EMA) [19, 20], which converges to a local maximum of the likelihood that depends on the initial parameter values.
- *Hidden Markov-Model (HMM)*: Used to model and analyze temporal data sequences, e.g., continuous speech recognition, protein structure analysis in the sequencing domain, data clustering, and others. In this regard, the HMM is a non-deterministic stochastic finite state automata model that consists of a connected set of state,

$$S = (S_1, S_2, \ldots, S_n).$$

In case of a first order HMM, the state of a system at the immediate previous time point is [21]

$$P(St|St-1, \, St-2, \, \ldots, S1) = P(St|St-1).$$

A continuous density HMM with n states for data having m temporal features can be characterized in terms of three sets of probabilities [22]: (1) the initial state probabilities, (2) the transition probabilities, and (3) the emission probabilities.
- And others.

To conclude: Unsupervised Machine Learning Models (UMLM) include methods such as data analysis, factor analysis, hidden Markov models, K-means clustering, neural networks, and others. They are used in applications domains such as automation, classification, and maintenance. In [17], a good overview is given. Furthermore, UMLM is used for defending against cyberattacks with different degrees in sophistication and success.

8.1.1.3 Reinforcement Machine Learning

Reinforcement Machine Learning (RML) is the problem an investigator faces, to learn a behavior of interest through trial-and-error interactions within a dynamic environment, thought of as a class of problems, rather than as a set of techniques. There are two main strategies for solving RML problems: (a) search the space of behaviors in order to find one that performs well, which has been used in genetic algorithms and genetic programming; and (b) use statistical techniques and dynamic programming methods to estimate the utility of taking actions [23]. In a standard

Reinforcement Machine Learning Model (RMLM), an investigator, or a software agent is connected to its environment through perception and action. For each step, the investigator receives as input, *inp*, some indication of the current state, *cst,* of the environment. The investigator then chooses an action, *act*, to generate as output. The action changes that state of the environment of interest, and the value of this state transition communicated to the investigator through a scalar reinforcement signal *rs.* The investigator's behavior, *B,* should choose actions that tend to increase the long-run sum of values of the RML signal. It can learn doing this over time by systematic trial and error, guides by a wide variety of algorithms. Formally, the RMLM consists of [23]:

- A discrete set of environmental states *ES*.
- A discrete set of actions *ACT*.
- A set of scalar reinforcement signals *rs*, or the real numbers.
- An input function *inp* determines how the investigator views the environment state.

Against this background, RML is useful for learning how to act or behave when given occasional reward or punishment signals [2]. RML is also considered as part of Deep Learning that supports to maximize some portion of cumulative reward. However, UML is somewhat less commonly used. The two kinds of reinforcement machine learning are:

- *Positive Reinforcement Machine Learning (PRML)*: Takes action on a specific behavior. It increases the strength and the frequency of the behavior and impacts positively on the action taken by the investigator or software agent. PRML tries to maximize the performance and sustain change for a more extended period. However, too much reinforcement may lead to over-optimization of environmental state, which can affect the results.
- *Negative Reinforcement Machine Learning (NRML)*: Strengthening behavior that occurs because of a negative condition, which should have been stopped or avoided. It supports defining the minimum of performance. However, the drawback of this method is that it provides enough to meet up the minimum behavior.

Three Reinforcement Machine Learning Algorithms (RMLAs) are available:

- *Value-based Reinforcement Learning Algorithm (VRLA)*: Tries to maximize a Value Function *V(s)*. By this algorithm, the investigator or software agent is expecting a long-term return of the current environment states under a policy *π.*
- *Policy-based Reinforcement Learning Algorithm (PRLA)*: Tries to come up with a policy that the action performed in every environment state helps to gain a maximum reward in the future. There are two types of policy-based methods:
  - *Deterministic*: For any environment state, the same action is produced by the policy *π.*
  - *Stochastic*: Every action has a certain probability, which is determined by the following stochastic policy equation

$$n\{act\backslash es) = P\backslash ACT, = act\backslash ES, = ES]$$

- *Model-based Reinforcement Learning Algorithm (MRLA)*: Creates a virtual model for each environment state for which the investigator or software agent learns to perform in that specific environment state. There are two learning models available,

  - *Markov Decision Process (MDP)*: Mathematical approach mapping a solution termed Markov Decision Process (MDP), for which several parameters are used to generate a solution:

    Set of actions *ACT*
    Set of environmental states *ES*
    Reward signal *RS*
    Policy *PY*
    Value *VA*

  - *Q Learning (QL)*: Value-based method that supplies information to inform which action an investigator or software agent should take. The method can be explained by the following example:

    There are *N* rooms in a building, connected by doors.
    Each room is numbered *1* to *N*.
    Door numbers *1* and *N-1* lead into the building from room *N*.

  The characteristics of RML are:

- There is no supervisor, only a real number or reward signal.
- Sequential decision-making.
- Time plays a crucial role in solving reinforcement problems.
- Feedback is always delayed, not instantaneous.
- Investigator or software agent's actions determine the subsequent data it receives.

8.1.1.4 Comparison of Machine Learning Methods

The different types of machine learning methods, partially described in Sects. 8.1.1.1, 8.1.1.2, and 8.1.1.3, perform complex processing tasks based on different kinds of machine learning algorithms, whereby each of which have advantages and disadvantages with regard to their specific application domains.

Table 8.2 summarizes the advantages and disadvantages in SML.

In Table 8.3, the advantages and disadvantages in UML are summarized.

The advantages and disadvantages in RML are summarized in Table 8.4.

Tables 8.2, 8.3, and 8.4 illustrate the manifold differences between SML, UML, and RML. In this context, Table 8.5 introduces a different aspect by comparing SML against RML with regard to specific parameters.

Table 8.6 finally shows a comparison of SML versus UML versus RML, based on the same evaluation criteria to identify the respective pros and cons of the different classical machine learning models.

Table 8.2 Advantages and disadvantages in supervised machine learning

Supervised machine learning	
Advantages	Disadvantages
Specific about class definition; trains classifier for perfect decision boundary; distinguishes different classes accurately	Decision boundary might be over trained; training set may not include examples in class if no correct class labels are achieved after training
Specifically determine how many classes are required.	If input is not from any of the classes, classification shows wrong class label.
Keep decision boundary as mathematical formula to classify future inputs.	Selecting good examples from each class for training classifier is required.
	Training needs a lot of computation time to run the classification.

Table 8.3 Advantages and disadvantages in unsupervised machine learning

Unsupervised machine learning (K-means)	
Advantages	Disadvantages
Easy to implement	Difficult to predict the number of clusters (K-values)
For a larger number of variables, K-means computes fast.	Initial variables have strong impact on final outputs.
K-means generates tight clusters.	Order of data has impact on final output.
If centroids are recomputed, instance change cluster.	Sensitive to scale.

Table 8.4 Advantages and disadvantages in reinforcement machine learning

Reinforcement machine learning	
Advantages	Disadvantages
Solve very complex problems that cannot be solved by conventional techniques	Not preferable to use for solving simple problems
Learning model similar to human learning; achieving better perfection	Can lead to an overload of states which can diminish output
Create perfect model to solve a particular problem	Needs lots of data and lots of computation
In absence of training data, it is bound to learn from experience	Dimensionality limits usage heavily for real physical systems
Achieve long-term outputs, which are hard to achieve	Achieve long-term outputs, which are hard to achieve

The conclusion of Table 8.6 is:

- In SML, the learning model learns from labeled datasets with guidance.
- In UML, the learning model is based on training, based on unlabeled data without any guidance.
- In RML, an investigator or software agent interacts with its environment performing actions and learns from errors or rewards.

Table 8.5 Comparison of supervised and reinforcement machine learning for specific parameters

Specific parameters	Supervised machine learning	Reinforcement machine learning
Best suited for	Mostly operated with an interactive software system or applications	Supports and works better in AI, where human interaction is prevalent
Decision style	Decision made on the input given at the beginning	Support to take decisions sequentially
Dependency of use decision	Decisions that are independent of each other, so labels given for every decision	Learning decision is dependent. Thus, give labels to all dependent decisions
Example	Object recognition	Chess game
Works good on	Examples or given sample data	Interacting with the environment

Table 8.6 Comparison of supervised machine learning versus unsupervised machine learning versus reinforcement machine learning for same evaluation criteria

Evaluation criteria	Supervised machine learning	Unsupervised machine learning	Reinforcement machine learning
Machine learning model	Learn on labeled datasets with guidance	Trained on unla-bled data without any guidance	Agent interacts with environment-performing actions and learns from errors or rewards
Solved problem types	Regression and classification;	Association and clustering	Reward based
Required data type	Labeled data	Unlabeled data	No predefined data
Method of training	External supervision	No supervision	No supervision
Method of machine learning	Maps labeled inputs to known outputs	Understands pattern and discovers output	Follow trial and error method
Computational complexity	Computationally simple	Computationally complex	Depends on usefulness of samples [17]
Accuracy	Highly accurate and trustworthy method	Less accurate and trustworthy method	Uncertainty reduces predictability
Best ssuited for	Interactive software system or applications	Automation and Classification ...	Supports and works in AI, human interaction
Cybersecurity applications	Find function or model that explains completely labeled datasets to detect intrusion, anomalies, and malware.	Find patterns, structures, know-ledge in unlabeled datasets to detect intrusion, anomalies, DDoS attacks, and unauthorized access	Labeling datasets during acquisition to detect intrusions of threat event attacks

In this regard, machine learning methods are applied to defend cyberattacks and doing some action required to cybersecurity such as quick scanning large amounts of data and statistical analysis. Therefore, machine learning is a powerful method used to cyber-secure computer systems, networks, infrastructure resources, and others, as seen from Table 8.6. So it's no wonder that machine learning is a useful and powerful tool in the domain of cybersecurity.

8.2 Machine Learning and Cybersecurity

Machine learning is a set of methods including SML, UML, and RML that use mathematical methods for building behavior models. These models are used for prediction(s) of new patterns in input data. In this regard, machine learning can automatically detect pattern in input data to compare them with measured data, and as a result, decide whether these data are usual as expected or unusual, indicating that an intrusion incident happened or not. An identified intrusion incident that has a negative impact is interpreted as a cybersecurity risk in the context of the usual operational behavior of computer systems, networks, and infrastructure resources and others. Cybersecurity means applying security prevention to provide confidentiality, integrity, and availability of data, as described in the CIA Triad of information security (see Sect. 1.6.2). In order to be on the defensive site of cybersecurity, it is specialized as Application Security, Computer Security, Data Security, End-Point Security, Information Security, Network Security, Operational Security, and others (see Sect. 1.2.1). Each of these has, at a minimum, a Firewall, Antivirus Software, and an Intrusion Detection System (IDS). The IDS (see Sect. 3.1) discovers, determines, and identifies unauthorized use, duplication, alteration, and destruction of information systems [24]. There are three main types supporting cybersecurity analytics by IDS: Anomaly-based, Misuse-based, and Specification-based Intrusion Detection Systems (see Sects. 3.1.1, 3.1.2, and 3.1.4). A novel approach supporting cybersecurity in IDS incorporates machine learning. In this context, machine learning leverages in various domains and risks scenarios within cybersecurity. In this regard, cybersecurity processes are enhanced to make it easier for cybersecurity analysts to quickly identify and deal with and remediate new cyberattacks. Therefore, machine learning is applied to a variety of cybersecurity risks as discussed in [18]:

- Anomaly detection
- Biometric recognition
- Denial of Service (DoS) and Distributed Denial of Service (DDoS) Attack Detection
- Detection of Advanced Persistent Threats (APTs)
- Detection of hidden channels
- Detection of identity theft(s)
- Detection of information leakage(s)

- Detection of software vulnerabilities
- Malware detection and identification
- Spam mail and phishing page detection
- Social media analysis
- User identification and authentication

Thus, machine learning acts as a method to deal with uncertainty, enabling a clear decision-making process, as required in an IDS application of identifying and detecting cyberattack incidents. From a practical perspective, this process consists of two phases, training and testing, with the following intrinsic steps [19]:

- Identify class attributes and classes from training data or patterns.
- Identify a subset of the attributes necessary for classification.
- Learn model using training data or patterns.
- Use trained model to classify the unknown data or patterns.

Thus, cybersecurity measures used to identify and detect and then prevent cyber-attack incidents in data or patterns. This concept categorizes in anomaly intrusion detection and misuse intrusion detection (see Sects. 3.1.1 and 3.1.2). Misuse intrusion detection methods are intended to recognize known patterns, described by rules; anomaly intrusion detection focusses on detecting unusual activities in data or patterns [20–22].

Considering the training phase in misuse intrusion detection, each misuse class is learned by using appropriate scenarios from the training set. In the test phase, new data or patterns are run through the model and a new classified scenario belongs to one of the misuse classes. If the scenario does not belong to any of the misuse classes, the scenario is classified as normal.

In anomaly intrusion detection, the normal data or traffic pattern of the object under test is defined in the training phase. In the testing phase, the learned model is applied to the input data or patterns of the actual scenario against the testing set and any scenario classified as either normal or anomalous.

For most machine learning methods, three phases are used instead of two. These are: training, validation, and testing [25]. The majority of cyberattack Intrusion Identification and Detection methods are misuse intrusion detection, due to their reliance on rule sets. Rule-based solutions divided into Blacklist- and Whitelist-based methods. A Blacklist method is a list of discrete entities that have been previously determined to be associated with a malicious activity. A Whitelist method is a list of discrete entities, such as hosts, email addresses, network port numbers, runtime processes, or applications authorized to be present or active on computer systems, networks, and infrastructure resources and others, according to a well-defined baseline. Besides Blacklist and Whitelist methods, Graylist methods also exist. A Graylist method is a list of discrete entities that has not yet been established as benign or malicious and that requires more information to move the Graylist method items to the Blacklist or Whitelist method. Blacklist-based methods can be refined into signature-based and heuristic-based approaches, comparing whether a match was found or not, indicating that a cyberattack incident happened or not.

Signature-based approaches enable cyberattack incident identification and detection based on specific cyberattack incident patterns, like malicious byte sequences and others. Compared to that, heuristic methods enable identification and detection of unknown cyberattack incidents by using expert-system probabilistic rule sets to describe malicious indicators. Heuristic approaches often complement signature-based cybersecurity solutions; a major drawback is their susceptibility to high false positive rates.

Blacklist intrusion detection requires an automatic daily update for intrusion detection in real-time. Whitelist intrusion detection usually includes policies, which enable cyberattack incident detection based on the deviation from a pre-defined negative baseline configuration, e.g., an IP whitelist. Therefore, the application of whitelist software prevents installation or execution of any application not specifically authorized for use on a particular host. This mitigates multiple categories of cyberattack incidents, including malware and other unauthorized software-based cyberattacks. A good overview of misuse and anomaly-based intrusion detection methods is provided by the work published by Modi [26] and Mitchell [27]. In this context, machine learning provides methods to automatically infer generalized data models, based on suspicious patterns identified in data. In supervised machine learning, the categorized and labeled training data feed into classification or regression models during a training phase. As described in Sect. 8.1.1.1, SML techniques include classification learning methods and regression learning methods. In the UML model, training samples are used with no knowledge of corresponding category labels. As described in Sect. 8.1.1.2, UML techniques include clustering and auto-encoder learning methods. In RML, as described in Sect. 8.1.1.3, the learning model optimizes behavior strategies through a trial-and-error method, which is different from the learning methods described in Sects. 8.1.1.1 and 8.1.1.2. Against this background, machine learning is becoming important for IDS because intrusion detection is the process enabling dynamic monitoring whether cyberattack incidents occur or not by analyzing for signs of potentially suspicious incidents and interdicting the unauthorized access [28]. Thereafter, the suspicious incident(s) is analyzed for possible cybersecurity solutions, making the cyberattacked object cyber secure against cyberattack incidents by use of machine learning-based methods.

In this regard, cybersecurity is based on a set of technologies and processes designed to protect data or patterns against cyberattack intrusions, unauthorized access, change, destruction, and others. In this context, cybersecurity devices used have as minimum a Firewall, Antivirus Software, or an Intrusion Detection System (IDS) to support cybersecurity analytics in misuse detection, anomaly detection, or hybrid detection methods. The different IDS types and their features are described in detail in Sect. 3.1.

8.2.1 Machine Learning Examples in Cybersecurity

8.2.1.1 Machine Learning-Based Intrusion Detection in Industrial Applications

With the exponential growth of computer systems with their enormous variety of different applications, the potential cybersecurity risks also come to the fore. The gradual increase in the number of successful complicated cybersecurity attacks, developed by skilled cyberattackers, against industrial applications of advanced computer technological systems, are hard to comprehend and even harder to mitigate. In this regard, the increased amount of cybersecurity issues of relevant data and business assets, processed in the manifold industrial systems, enhances their cybersecurity risk to deal with unprecedented cyberattacks on a real-time basis. This has led to an urgent need to create cybersecurity defense strategies for accurate and timely intrusion detection of potentially resulting Industrial Automation and Control System anomalies. Therefore, a machine learning-based framework is required for anomaly intrusion detection. This framework takes data, learn what is normal (regular), and then apply a statistical test to determine whether any data point for the same time series in the future is normal (regular) or anomalous (irregular). Deploying this metric in computer systems, networks, infrastructure resources, and others conquer their potential cybersecurity risks [29, 30]. However, every set of data comes in and goes through a classification process, whereby the human expert has to decide what type of model fits best.

Let's consider a machine learning framework for learning normal (regular) behavior consisting of three steps, as shown in Table 8.7.

Let's further assume that the intrusion detection approach, based on machine learning, follows a normal distribution, represented by an average standard deviation. This means that for a given large number of data points, most data points submitted should fall within the average, plus or minus three times the standard deviation. With the additional assumption that the data points have a known normal distribution, then the data fall within the previously mentioned bounds, meaning they have a normal (regular) behavior. If data points fall outside these bounds, their behavior is anomalous (irregular) because the data points are outside the average. This results in the mathematical constraints, as shown in Table 8.8.

However, choosing just one model for machine learning-based intrusion detection does not work well because the available models may measure different metrics

Table 8.7 Process steps required learning normal (regular) behavior

Process step	Actions taken
1	Model the normal (regular) behavior of the metrics using a statistical model.
2	Devise a statistical test to determine if samples are explained by the model.
3	Apply the test for each sample. Flag as anomalous (irregular) if it did not pass the test.

Table 8.8 Mathematical constraints

Constraint	Constraints for average standard deviation model
1	Assumed normal behavior is normal distribution.
2	Estimate the average, standard deviation over all samples.
3	Testing any sample for *x-average	> 3 * standard deviation* as anomalous (irregular).

Table 8.9 Characteristics of SML and UML and a hybrid method

Constraints to be fulfilled before using a machine learning-based method		
When cyberattack incidents are well-defined, SML.	When cyberattack incidents are not well-defined, UML.	Hybrid method
Requires a well-defined set of cyberattack incidents to identify Learning a model to classify data points as normal (regular) or anomalous (irregular) Requires labeled examples of anomalies Cannot detect new types of cyberattack incidents	Learning a normal model only Requires statistical test to detect anomalies Can detect any type of anomaly, known or unknown	Use a few labeled examples to improve detection of unsupervised methods OR Use unsupervised detection for unknown cases, or supervised detection to classify already known cases.

that behave differently. Thus, other approaches for intrusion detection have to be taken into account besides the technique for modeling the normal behavior, by using the normal distribution approach. In this regard, different constraints may have to be fulfilled before using machine learning-based concepts and applying them for intrusion detection issues. The research paper in [31] summarizes the characteristics of SML and UML and a hybrid method while applying them for intrusion detection purposes, as shown in Table 8.9.

In Fig. 8.2, a machine learning approach developed for intrusion detection illustrates the capability to monitor and protect Industry 4.0 (I4.0) Industrial Automation and Control Systems (IACS) against potential cyberattacks. In this approach, machine learning based on data exploration is used to learn about normal (regular) and anomalous (irregular) behaviors, according to interaction of IACS with each other for the prediction of malicious cyberattack incidents, achieved at early stages to output alerts in case of a cyberattack incident. Another option is to detect malicious cyberattack incidents and decide how to prevent them. Against this background, the chosen machine learning-based model, illustrated in Fig. 8.2, is a machine learning technique based on an experience approach.

Developing a cybersecurity model for intrusion detection of anomalous cyberattack incident(s) requires enumerating unique characteristics of and requirements for I4.0 IACS [32, 33], as shown in Table 8.10, following [29]:

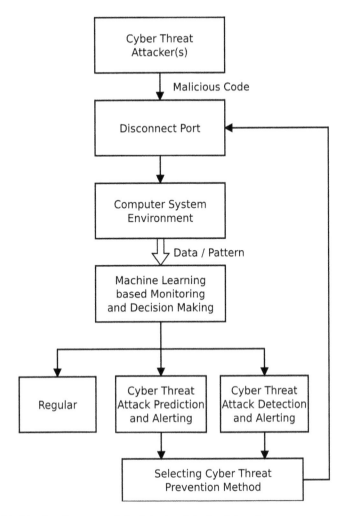

Fig. 8.2 Machine learning approach for intrusion detection interaction

In the context of the characteristics of IACS, as summarized in Table 8.10, it is important to test not only the accuracy of the chosen machine learning-based model for intrusion detection but it is also essential to test the robustness against cyberattackers' capability, manipulating data input(s) to the intrusion detection method. This means that the cyberattack model not only focuses on simple attack scenarios executed by a cyberattacker, while there are more advanced stealthy attacks available. Thus, to defend against advanced cyberattackers capable of learning the defense process, some solutions are proposed inspired by the classical challenge–response paradigm to ensure the non-deterministic behavior in the data [37, 38]. Moreover, in [29], recommendations are made that may be useful for researchers and practitioners in the design of cybersecurity for IACS. The key findings in [29] are summarized in a reduced form in Table 8.11.

Table 8.10 Requirements and characteristics for IACS

Requirements	Characteristics
1	IACS is often required to operate uninterrupted for long periods, without any downtime for activities such as controllers' code patching. Components in an IACS require deterministic responses with an acceptable level of jitter or delay, whereas IT systems can tolerate a higher level of delay in network traffic without noticeable impact on the system performance.
2	Physical process controlled by an IACS is continuous and hence unexpected outage of the systems that monitor and control it is unacceptable. In IT systems, rebooting or temporary shutdown of the systems occurs much more often than it does in a physical plant that provides continuous critical services.
3	In typical IT systems, the CIA Triad (see Sect. 1.2.2) includes the primary concerns to ensure availability. On the contrary, in IACS the CIA Triad is perceived as AIC, wherein priority is accorded to the availability of data followed by integrity and confidentiality. For example, it might be desired to ensure the integrity of sensor data, while the confidentiality of data itself might not be of major concern.
4	In IT systems, security focus is on safeguarding IT assets through which data transfer takes place. In IACS, the primary focus is safeguarding the edge clients such as the programmable logic controllers, sensors, and actuators.
5	A successful attack on an IACS may have a severer impact than one on an IT system. The damage in the physical components of an IACS could lead to service disruption and may even affect human life. In contrast, a successful attack on an IT system generally leads to information loss.
6	The behavior of an IT system is highly uncertain and varied, whereas that of IACS components is much more stable and predictive.
7	Payload of IACS data is shorter than IT data due to delay-tolerance requirements. Unlike in IT systems, data generated in an IACS are highly correlated and obey the system design specifications.
8	A typical IACS operates in a significantly resource-constrained environment and usage of third-party applications is restricted. On the contrary, the computational efficiency of the IT system may be frequently updated based on user requirements.
9	Communication protocols used for data transfer among the IACS components are proprietary and different from the well-known protocols used in traditional IT environments [34–36].

8.2.1.2   Machine Learning-Based Intrusion Detection Using Feature Learning

In [41], a machine learning-based IDS is introduced, building a combination of two typical methods commonly used for IDS, learning or training and classification. The system consists of the following functional blocks: input dataset block, pre-processing block, and the core IDS block, based on Feature Learning (FL), which contains a learning or training component and a classification element. The outcome of this machine learning-based Intrusion Detection System differentiates between benign and malicious cyberattack types. Feature Learning is an approach to improve the learning process of a machine learning algorithm, which consists of feature construction, extraction, and selection. As described in detail in [41], feature

Table 8.11 Recommendations and key findings for developing IADS

Recommendations	Key findings
1	Improve transparency of anomaly-based intrusion detection: From a plant operator's point of view, most of the intrusion detection approaches created and deployed in operational plants behave like a block-box that inputs the current state of the plant and generates alert(s), indicating a process anomaly, the intrusion detection.
2	Are intrusion detection and false alarm rates adequate for evaluating anomaly-based intrusion detectors of IACS?
3	Is design of anomaly-based intrusion detection based on domain constraints? Research reported in [39, 40] focuses on the design of a generic system of anomaly-based intrusion detection of IACS, operating in different domains.
4	Anomaly-based detection should be capable of distinguishing faults from the cyberattacks: A physical process could enter an anomalous-based state due to one or more reasons: human error, component fault, misconfiguration, cyberattack, and others.

construction expands the original features to enhance their expressiveness, whereas feature extraction transforms the original features into a new form and feature selection eliminates unnecessary features [41, 42]. The classification task used in [41] based on an SML method to distinguish benign and malicious traffics, based on the data provided, which usually comes from the previous steps in the blocks of input dataset and pre-processing. The pre-processing block commonly consists of normalization and balancing steps (see Sect. 3.1.7). Data normalization is a process to output same value ranges of each attribute, which is essential for effective learning by machine learning algorithms [43]. As described in [41], the frequency of benign traffic is larger than the one of malign traffic. Against this background, this property makes it difficult for the core IDS element to learn the underlying patterns correctly [44]. Therefore, the balancing process that creates the dataset with an equal ratio for both benign and malicious possibilities is an essential prerequisite step for the training process.

8.2.1.3   Machine Learning-Based Intrusion Detection of Unknown Cyberattacks

Machine learning is capable to detect known and new cyberattack types. The later, introduced as *Unknowns*, is introduced in Sect. 2.2.1. In this context, Machine Learning Algorithms compare whether identified cyberattacks belong to the attack category *Known-Knowns* or not. In case of not *Known* cyberattack types, based on the cyberattack category *Known-Unknowns*, as illustrated in Table 2.10, the detected new cyberattack automatically is output of the cyberattack category *Known-Unknowns* cybersecurity risk type. Thus, the machine learning model is based on machine learning on experience. This requires that the environment must have, on the one hand, a cyber-secure communication between its system devices and networks and, on the other hand, a cybersecurity-based intelligence with focus on the capabilities of machine learning methods, especially in the cyberattack category of *Known-Unknown* cyberattacks. This is often the case in the era of digital transformation with its complex, connected, and intelligent environments, which work in diverse surroundings to accomplish the required different process goals. Therefore, maintaining the security requirements with their manifold cyberattack surfaces in the cyber and physical domains is challenging, meeting with the comprehensive cybersecurity requirements [45, 46]. Satisfying the desired security requirement requires that the chosen cybersecurity solution must include holistic considerations, with regard to the existing distributed access points. Therefore, cybersecurity becomes a complex and challenging task due to anytime accessibility of the connected and accessible IACS, which enables cyberattackers to easily gain access to critical and crucial computer systems, networks, infrastructure resources, and others. Thus, active cyberattacks such as eavesdropping, as well as passive cyberattacks, such as malware, man-in-the-middle, or denial of service attacks, may happen that affect the overall cybersecurity measures. In this sense, the cyber-attacks affect especially the main security requirements of the CIA Triad (see Sect. 1.6.2).

With regard to practicability, for instance, in misuse-based intrusion detection, a rule based approach can be used based on association rules in the simplest form

IF A AND B THEN C,

which describes the relationship that when *A and B* are present *Then C* is present too. For an IACS problem this association rule can be:

IF service request AND data set AND port address THEN attack type XY

Therefore, an essential initial step is finding associative rules with high confidence. Against this background, association rule mining support to discover that the rules explain the relationships clearly. This intrusion detection approach is helpful to develop machine learning-based algorithms for detecting cyberattack signatures by comparing each dataset against all rules that describe the respective signatures, to detect the respective cyberattack. Besides signature-based machine learning, clustering of rules is another option, whereby association rule clustering minimizes the number of comparisons necessary to determine which rules are triggered by given input datasets [19]. In case of an association rule algorithm for machine learning IDS, attack category *Known-Unknowns* is the underlying cybersecurity approach to investigate the relationship among the various variables in the training datasets.

Let U, V, W be variables in a dataset. The association rule algorithm investigates the relationship between the variables to identify their correlations and hence build a model, used to predict the class of new samples that co-exist to cyberattacks [45]. In [46], an association rule algorithm is discussed which shows a good performance in cyberattack intrusion detection.

Another promising approach in machine learning and Intrusion Detection combines the outputs of classification methods to generate a collective output to enhance classification performance. In this context, Ensemble Learning is a learning method,

Table 8.12 Machine learning-based models for securing IACS

Machine learning model	Working principle	Advantages	Disadvantages	Application(s)
Associated rule-based learning	Studies relationships between variables in trainings' datasets	Simple algorithms, easy to use	Time complexity of algorithm is high	Intrusion detection
Ensemble learning	Combine concepts of different classification methods	Robust algorithm, adapt better as single classifier methods	Time complexity of algorithm is high	Anomaly intrusion detection malware intrusion detection

trying to combine heterogeneous and homogeneous multi-classifiers to obtain successful classification results [47, 48]. The Ensemble Learning Method (ELM) uses several machine learning-based methods to reduce the variance and is robust to overfitting [49]. In Table 8.12, the potential for two machine learning-based methods for securing IACS is summarized.

8.3   Introduction to Deep Learning

Machine learning is a specific type of practically implemented artificial intelligence (AI) with the objective of giving a computer system, network, infrastructure device, and others access to data storage devices, and enabling it to learn from it. In this context, the techniques of machine learning process data in their raw form to discover automatically representations for identification or classification purposes. Machine learning features developed by humans and the developed machine learning algorithms then learns automatically how to map the required features to an output. Machine learning approaches are characterized by distinguishing between SML, UML, and RML.

By comparing machine learning with Deep Learning, the latter method originally comes from the advancements of Neural Network Algorithms. To overcome the limitations of hidden layers in neural networks, different methods can be applied. As described in [41], the methods applied employ hierarchical cascaded consecutive hidden layers. Due to a variety of models belonging to Deep Learning, Aminanto et al. [44] classified several Deep Learning Models (DLM) based on approaches introduced by [50, 51], which differentiate Deep Learning in three subgroups: generative, discriminative, and hybrid models. Against this background, Deep Learning is an important element in Data Science, which includes statistics and predictive modeling. Predictive modeling, on the one hand, is a mathematical method used to predict future events or outcomes by analyzing patterns in given datasets, and on the other, a method of predictive analytics, a type of data analytics which uses current and historical data to forecast activities, behaviors, and trends. In this context, deep learning relies on a multi-layered representation of output data and performs feature selection automatically through a process defined as representation learning. This enables deep learning to obtain computing simpler, by transforming non-linear models one level representation into a representation at a higher, slightly abstract level, to support solving real world problems with more accuracy. Moreover, deep learning extracts patterns from datasets using neural networks. Neural networks date back decades based on neurons. In Deep Learning, a neuron is a perceptron, the structural building block of Deep Learning. The perceptron is a forward propagation, described as follows:

$$y = g\left(\sum_{1=1}^{m} x_i w_i\right)$$

where y is the output of the perceptron, g is a non-linear activation function of the perceptron, x_i are the perceptron inputs, and w_i are the perceptron weights which can be rewritten as

$$y = g\left(X^T W\right)$$

with

$$X = \begin{bmatrix} x_1 \\ \dots \\ x_m \end{bmatrix}$$

and

$$W = \begin{bmatrix} w_1 \\ \dots \\ w_m \end{bmatrix}.$$

Activation function g can be sigmoid functions, hyperbolic tangent, as well as the rectified linear units. The purpose of activation functions is to introduce non-linearity into the Deep Learning Neural Network. Linear activation functions generate linear decisions, no matter the neural network size. In contrast, non-linear activation functions enable to approximate arbitrarily complex functions. Assume that

$$W = \begin{bmatrix} 6 \\ -4 \end{bmatrix}$$

then

$$y = g\left(X^T W\right) = g\left(\begin{bmatrix} x_1 \\ x_2 \end{bmatrix}^T \begin{bmatrix} 6 \\ -4 \end{bmatrix}\right) = g(6x_1 - 4x_2)$$

which represent a graph in a 2D space.

Deep Learning models have a huge number of parameters, but acquiring enough labeled data to train because Deep Learning models can become difficult. To overcome the problem of labeled training data, one can focus on an unsupervised learning approach. The most natural way to perform this is to use generative models such as directed, undirected, and mixed models [52]. Deep Learning also eliminates some of the pre-processing data that is typically involved with machine learning. Moreover, Deep Learning models are capable of different types of learning.

8.3.1   Classification of Deep Learning Methods

Like machine learning, Deep Learning methods are characterized by distinguishing between Supervised and Unsupervised Deep Learning Methods. The former techniques require a training process with a large and representative set of data, previously classified by a human expert or through other means. The latter approaches do not require a pre-labelled training dataset.

Supervised Deep Learning (SDL) methods used for cybersecurity intrusion detection are:

- *Feedforward Deep Neural Networks (FFDNN)*: A variant of a Deep Neural Network (DNN) where every neuron is connected to all neurons in the previous layer. FFDNN do not make any assumption on the input data and provide a flexible and general-purpose solution for classification at the expense of high computational costs.
- *Convolutional Feedforward Deep Neural Networks (CFFDNN)*: A variant of a DNN where each neuron receives its input only from a subset of neurons of the previous layer. This characteristic makes CFFDNN effective in analyzing spatial data, but their performance decreases when applied to non-spatial data. CFFDNN have lower computation cost than FFDNN.
- *Recurrent Deep Neural Networks (RDNN)*: A variant of a DNN whose neurons can send their output also to previous layers. This design makes them harder to train than FFDNN. They excel as sequence generators, especially their recent variant, the long short-term memory.

Unsupervised Deep Learning (UDL) methods used for cybersecurity intrusion detection are [53]:

- *Deep Belief Networks (DBN)*: A model that is partially directed and partially undirected. The DBN are based on an unsupervised fast greedy layer-by-layer training algorithm. Upper layers are supposed to represent more abstract concepts that explain the input observation x, and lower layers extract low level features from x.

  Based on complimentary priors, a fast Greedy algorithm, which build up a solution piece by piece always choosing the next piece that offers the most obvious and immediate benefits, that can learn a Deep Belief Directed Network (DBDN). The DBDN contains at the bottom level the observed patterns, and the remaining hidden layers. DBDN are trained in three steps: Pertaining one layer in time in a Greedy way, using unsupervised learning at each layer in a way that preserves information from the input and disentangles factors of variation, fine-tuning the whole network with regard to the criterion of interest [54].

DBN supports in solving the optimization problem of Deep Multi-Layer Structure Learning Algorithm that uses a space relative relationship to reduce the number of parameters to improve the training performance. The model is defined in three hidden layers as follows [52]:

$$p(h_1, h_2, h_3, v|\theta)$$

where h_i are the hidden layers, v is the visible mode, and

$$\theta = (w_0, w_1, w_2, w_3)$$

are the parameters in the model.

DBN is also modelled through a composition of Restricted Boltzmann Machines (RBM), a class of neural networks with no output layer. The RBM is a special topological structure of a Boltzmann Machine (BoM), which originates from statistical physics as a modeling method, based on an energy function that describes high-order interactions between variables. In this regard, the BM is a symmetric coupled random feedback binary unit neural network, composed of a visible layer and a plurality of hidden layers. The network node is divided in a visible and a hidden unit, whereby both units are used to express a random network and a random environment. The learning model expresses the correlation between units by weighting [55]. In this regard, the RBM defines a probability distribution p on data vectors v as follows:

$$p(v) = \sum_h \frac{e^{-E(v,h)}}{\sum_{u,g} e^{-E(u,g)}}$$

where v is the input variable, and the variable h corresponds to observed features, which can be thought of as hidden causes not available in the original dataset [56]. Thus, a Restricted Boltzmann Machine describes a joint probability on the observed and unobserved variables, which refers to visible and hidden units, respectively. The distribution is marginalized over the hidden units to give a distribution over the visible units. The probability distribution is defined by an Energy Function (EF) [57].

DBN is successfully used for pre-training tasks because they excel in the function of feature extraction. They require a training phase, but with unlabeled datasets;

- *Stacked Auto-Encoders (SAE)*: Composed by multiple Auto-Encoders, a class of neural networks where the number of input and output neurons is the same. SAE excel at pre-training tasks similarly to a DBN, and achieve better results on small datasets.

8.3.2 Deep Bayesian Neural Network

Deep Bayesian Neural Network (DBNN) is a model to add uncertainty handling in models. Instead of having deterministic weights to learn, it is all to learn the parameters of a random variable used to sample weights during forward propagation. To learn the parameters, backpropagation is used which sometimes requires adoption to make parameters differentiable. In this context, in a DBNN, the probability of incident I if the evidence E is given can be written as follows:

$$p(I|E) = \frac{p(E|I)p(I)}{p(E)}$$

with $p(I)$ as a priori probability of I, which means a probability of a threat event attack incident before the evidence is known, and $p(I|E)$ as a posteriori probability of I, which means a probability of a threat event attack incident after the evidence is known.

8.3.3 Deep Learning-Based Intrusion Detection System

Three different categories of Intrusion Detection Systems (IDS) are used in Deep Learning, namely, the generative, the discriminative, and the hybrid models. As described in [40], the Generative Models consist of IDS that use Deep Learning for Feature extraction only and use shallow methods for the classification task. For classification, the DNN uses a comprising multilayer FFNN. The Discriminative Model contains IDS that uses a single Deep Learning method for both feature extraction and classification, using a Software-Defined Network (SDN) (see Sect. 1.2.16) in the DNN. The hybrid model includes IDS that uses more than one Deep Learning methods for generative and discriminative purposes, which combines Auto-Encoder (AE) [57] and Generative Adversarial Networks (GAN). Moreover, a survey is given in [40], referring to 10 malware detection studies based on deep learning due to the increasing number of malware incidents.

Deep Learning-based solutions are used to make network-based IDS (NIDS) efficient in detecting malicious cyberattacks. However, the massive increase in the network traffic and the resulting security threats has posed many challenges for the NIDS systems to detect malicious intrusions efficiently. As reported in [41], the research on using Deep Learning methods for NIDS is in its early stages and there is still enormous room to explore this technology within NIDS for efficient detection of cyberattack intruders within the network. The key idea is to furnish up-to-date information on recent Deep Learning-based NIDS approaches to provide a baseline for new researchers who want to start exploring this important domain. Main contributions described in [64] that

- Conduct a systematic study to select recent journal articles focusing on various ML- and DL-based NIDS published during the last three years (2017–2020).
- Review each article extensively and discuss its various features such as its proposed methodology, strength, weakness, evaluation metrics, and the used datasets.

- Highlight the recent trends of using AI methods for NIDS in ML- and DL-based NIDs, which provided different future directions of the important defense mechanism in networks against intrusions.

8.4   Deep Learning Method Example in Cybersecurity

Deep Learning methods used for cybersecurity Intrusion are (1) Deep Neural Network (DNN); (2) Feedforward Deep Neural Network (FFDNN); (3) Recurrent Neural Network (RNN); (4) Convolutional Neural Network (CNN); (5) Restricted Boltzmann Machine (RBM); (6) Deep Belief Network (DBN); (7) Deep Auto-Encoder (DAE); 8) Deep Migration Learning (DML); (9) Self-Taught Learning (STL); (10) Replicator Neural Network (ReNN); and others. In this regard, a comparative study of Deep Learning approaches and datasets for cybersecurity intrusion detection is published in [58], and a survey of Deep Learning Methods used for cybersecurity applications is published in [59]. The manifold cyberattack types, including false data injection, insider threat event attacks, malicious domain names used by botnets, malware, network intrusion, spam, and others, which result in different approaches used for cybersecurity of Industrial Control System Architectures (ICSA).

With regard to Sect. 8.2.1, one type of Deep Learning used in cybersecurity intrusion detection is DBN, a Probabilistic Generative Model (PGM) consisting of multiple layers of stochastic and hidden variables. In the study published in [60], the application of DBN in malware detection is described. Furthermore, DBN is modeled through a composition of Restricted Boltzmann Machines (RBM). The RBM enables the hidden layer to train datasets effectively through activation for further training stages relevant to network anomaly intrusion detection, based on experiments showing the feasibility of the Deep Learning Approach to network traffic analysis [61]. For Deep Learning Application in cybersecurity, based on a DNN, several variants exist, whereby one of which is the Generative Adversarial Nets Framework Approach (GANFA), in which two neural networks compete against each other in a zero-sum approach to outsmart each other, whereby one neural network acts as a generator and the other neural network acts as a discriminator [62]. The generator takes input datasets and generates output datasets with the same characteristics as real datasets. The discriminator takes real datasets and datasets from the generator and tries to distinguish whether the input is real or fake. When training has finished, the generator is capable of generating new datasets that are not distinguishable from real datasets [63]. For this purpose, the Generative Adversarial Network (GAN) represents a framework to estimate generative models through an adversarial process, which simultaneously trains the two models, the generative model that captures the dataset distribution, and the discriminative model that estimates the probability that a sample came from the training datasets rather than the generative model. The model framework corresponds to a minimax two-player approach. In the space of arbitrary functions of the generative model and the discriminative model, a unique solution exists, with the generative model recovering the training dataset distribution and the discriminative model to 0.5 everywhere. In case the generative model and the discriminative model are defined by multilayer perceptron's, the entire system is trained by a Back-Propagation Algorithm (BPA). A BPN uses a layered hierarchical architecture of simple neurons (nodes), employing a high degree of connectivity between layers. The algorithm back propagates derivatives through generative processes using the observation

$$\min_{\sigma \to 0} \nabla_x E_{0 \sim N(0, \sigma_2 I)} f(x+) = \nabla_x f(x).$$

When training has finished, the generator is capable of generating new datasets that are not distinguishable [63]. The architectural GAN concept, as shown in Fig. 8.3.

Since developed, GANs have shown wide applicability, especially to images. Examples include image enhancement [65], caption generation [66], and optical flow estimation [67], as reported in [63].

A recently published book that bridges the areas of Deep Learning and cyberse-curity, providing Deep Learning tools and frameworks to allow users to quickly develop workable and advanced prototypes, is referenced in [68]. It also introduces into recently made advances in the fields of cybersecurity intrusion detection methods, malicious code analysis, and forensic identification approaches, showing how Deep Learning methods can be used to advance cybersecurity objectives, including detection, modeling, monitoring, and analysis as well as defense against various cyberattack incidents. The performance evaluation of DNN for cybersecurity issues, including Android malware classification, incident detection, and Fraud Detection (FD) cases, is published in [69]. *Fraud Detection* using *machine learning* to enable automated transaction processing on an example dataset or an own dataset. In the DNN, the traffic matrix is a nonlinear function applied to a weighted sum of the units of the previous layer. There are a number of different nonlinear functions used whereby the most common are [63]:

- Sigmoid functions to understand how a DNN learns complex problems.
- Softmax function that transforms input values into values between 0 and 1, so that they can be interpreted as probability.
- Hyperbolic tangent function as an alternative of the sigmoid function to be used as activation function for transforming the summed weighted input from the node into the activation of the node or output for that input.

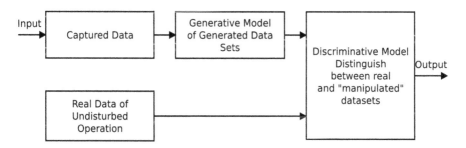

Fig. 8.3 Generative adversarial network

- Rectified linear unit is a piecewise linear function that outputs the input directly if it is positive, otherwise, it outputs zero. It has become the default activation function for many types of neural networks because a model that uses it is easier to train and often achieves better performance.

8.5   Exercises

8.5.1   Machine Learning

What is meant by the term *Machine Learning?*
Describe the main characteristics of Machine Learning.
What is meant by the term *Anomaly Detection?*
Describe the main characteristics of Anomaly Detection.
What is mean by the term *Misuse Detection?*
Describe the main characteristics of Misuse Detection.
What is meant by the term *Denoising?*
Describe the main characteristics of Denoising.
What is meant by the term *Supervised Learning?*
Describe the main characteristics of Supervised Learning.
What is meant by the term *Classification?*
Describe the main characteristics of Classification.
What is meant by the term *Classification Learning Method?*
Describe the main characteristics of Classification Learning Methods and give examples of them.
What is meant by the term *Regression?*
Describe the main characteristics of Regression.
What is meant by the term *Regression Learning Method?*
Describe the main characteristics of Regression Learning Methods and give examples of them.
What is meant by the term *Unsupervised Learning?*
Describe the main characteristics of Unsupervised Learning.
What is meant by the term *Clustering?*
Describe the main characteristics of Clustering.
What is meant by the term *Clustering Learning Method?*
Describe the main characteristics of the Clustering Learning Methods and give examples of them.
What is meant by the term *Reinforcement Learning?*
Describe the main characteristics of Reinforcement Learning and give examples of it.
What is meant by the term *Machine Learning in Intrusion Detection?*
Describe the main characteristics of Machine Learning Methods in Intrusion Detection.

8.5.2   *Deep Learning*

What is meant by the term *Deep Learning?*
Describe the main characteristics of Deep Learning.
What is meant by the Term *Neural Network?*
Describe the main characteristics of a Neural Network.
What is meant by the term *Feedforward Deep Neural Network?*
Describe the main characteristics of Feedforward Deep Neural Network.
What is meant by the term *Convolutional Feedforward Deep Neural Network?*
Describe the main characteristics of Convolutional Feedforward Deep Neural
    Network.
What is meant by the term *Recurrent Deep Neural Network?*
Describe the main characteristics of Recurrent Deep Neural Network.
What is meant by the term *Deep Belief Network?*
Describe the main characteristics of Deep Belief Network.
What is meant by the term *Deep Bayesian Neural Network?*
Describe the main characteristics of Deep Bayesian Neural Network.
What is meant by the term *Deep Learning in Cybersecurity?*
Describe the main characteristics of Deep Learning in Cybersecurity.
What is meant by the term *Deep Learning-based Intrusion Detection?*
Describe the main characteristics of Deep Learning-based Intrusion Detection.
What is meant by the term *Generative Adversarial Network?*
Describe the main characteristics of a Generic Adversary Network.
What is meant by the term *Stacked Auto Encoder?*
Describe the main characteristics of Stacked Auto Encoder.
What is meant by the term *Restricted Boltzmann Machine?*
Describe the main characteristics of a Restricted Boltzmann Machine.
What is meant by the term *Deep Neural Network?*
Describe the main characteristics of a Deep Neural Network.

References

1. Goodfellow, I., Bengio, Y., Courvill, A.: Deep Learning Book. WorldCat.Org, 2016
2. Murphy, K.P.: Machine Learning – A Probabilistic Perspective. MIT Press, 2012
3. Machine Learning with MATLAB, 2022. https://se.mathworks.com/campaigns/offers/next/machine-learning-with-matlab.html (Accessed 12.2022)
4. Naïve Bayes. In: scikit learn 1.2.9, 2022. https://scrikit-learn.org/stable/modules/naïve_bayes.html (Accessed 12.2022)
5. Pedregosa F., Varoquaux, G., Gamfort, Michel, V., Thiron, B., Grisel, O., Blondel, M., Prettenhofer, P., Weiss, R., Dubourg, V., Vanderplas, J., Passos, A., Cournapeau, D., Brucher, M., Perrot, M., Durchesnay, E Scikit-learn: Machine Learning in Python. In: Journal of Machine Learning Research, Vol. 12, pp. 2825–2830, 2011
6. Biau, G., Devroy, L.: Lectures on the Nearest Neighbor Method. Springer Publ. 2015

7. Greenspan, M., Godin, G.: A Nearest Neighbor Method for Efficient ICP. 2001. http://citeseerx. ist.psu.edu/viewdoc/download?doi=10.1.1.494.8938&rep=rep1&type=pdf   (Accessed 12.2022)
8. Maimon, O., Rokach, L. (Eds,): Data Mining, and Knowledge Discovery Handbook. Springer Publ. 2010
9. Yadav, P.: Decision Tree in Machine Learning. Medium Publications, 2018. https:// towardsdatascience.com/decision-tree-in-machine-learning-e380942a4c96 (Accessed 12.2022)
10. Opitz, D., Maclin, R.: Popular Ensemble Methods: An Empirical Study. In: Journal of Artificial Intelligence Research, Vol. 11, pp. 169–198, 1999
11. Gopika, D., Azhagusundan, B.: An Analysis on Ensemble Methods in Classification Tasks. In: International Journal of Advanced Research in Computer and Communication Engineering, Vol. 3, No. 7, pp. 7423–7427, 2014
12. Miller, S., Busby-Earle, C.C.R.: Multi-Perspective Machine Learning – A Classifier Ensemble Method for Intrusion Detection. In: Proceedings International Conference on Machine Learning and Soft Computing, 2017. https://doi.org/10.1145/3036290.3036303 (Accessed 12.2022)
13. Hastie, T., Tibshirani, R., Friedman, J.: The Elements of Statistical Learning, Chapter Unsupervised Learning, pp. 485-585. Springer Series in Statistics, 2009. https://doi. org/10.1007/978-0-387-84858-7_14 (Accessed 12.2022)
14. Sander, J.: Density-based Clustering. In: C. Sammut, G. I. Webb (Eds.) Encyclopedia of Machine Learning, Springer Publ. 2010
15. Carlsson, G., Memoli, F.: Characterization, Stability, and Convergence of Hierarchical Clustering Methods. In: Journal of Machine Learning Research, Vol. 11, pp. 1425–1470, 2010
16. Rokach, L., Maimon, O.: Clustering Methods. Chapter 15 in O. Maimon, L. Rokach (Eds,): Data Mining, and Knowledge Discovery Handbook. Springer Publ. 2010
17. Carroll, J.D., Chaturvedi, A., Green, P.E.: K-Means, K-Medians, and K-Modes: Special Cases of Partitioning Multiway Data. In: Proceedings Annual Meeting of the Classification Society North America, Vol. 11, 1994
18. Khanum, M., Mahboob, T., Imtiaz, W., Ghafoor, H.A., Sehar, R.: A Survey on Unsupervised Machine Learning Algorithms for Automation, Classification, and Maintenance. In: International Journal of Computer Applications, Vol. 119, No. 13, pp. 34–39, 2015
19. Dempster, A., Laird, N., Rubin, D.: Maximum Likelihood Estimation from Incomplete Data via the EM Algorithm. In: Journal Royal Statistical Society, Series B, Vol. 39, No. 1, pp. 1-38, 1977
20. McLachlan, G., Krishnan, T.: The EM Algorithm and Extensions. Jon Wiley & Sons Publ., 1997
21. Li, C., Biswas, G.: Applying the Hidden Markov Model Methodology for Unsupervised Learning of Temporal Data. In: International Journal of Knowledge Based Intelligent Engineering Systems, 2008
22. Rabiner, L.R.: A Tutorial on Hidden Markov Models and Selected Applications in Speech Recognition. In Proceedings of the IEEE Vol. 77, No. 2, pp. 257–285, 1989
23. Kaebling, L.P., Littman, M.L., Moore, A.W.: Reinforcement Learning: A Survey. In: Journal of Artificial Intelligence Research, Vol. 4, pp. 237–285, 1996
24. Mukkamala, A., Sung, A., Abraham, A.: Cyber Security Challenges: Designing efficient Intrusion Detection Systems and Antivirus Tools. In: Enhancing Computer Security with Smart Technology, V. R. Vemuri (Ed.), pp. 125–163, Auerbach Publ. 2005
25. Buczak, A.L., Guven, E.: A Survey of Data Mining and Machine Learning Methods for Cybersecurity Intrusion Detection. In: IEEE Communications Surveys & Tutorial, Vo. 18, No. 2, pp. 1153–1175, 2016
26. Modi, Patel, D.J., Borisaniya, B., Patel, H., Patel, A., Rajarajan, M.: A Survey of Intrusion Detection Techniques in Cloud. In: Journal of Network and Computer Applications, Vol. 36, No. 1, pp. 42–57, 2013
27. Mitchell, R., Chen, I.R.: A Survey of Intrusion Detection Techniques for Cyber-Physical Systems. In: ACM Computer Survey Vol. 46, No. 4, pp. 55.1–55.29, 2014
28. Mell, P., Scarfone, M.; Guide to Intrusion Detection and Prevention Systems. National Institute of Standards and Technology. NIST Sp-800-94, 2007. http://www.nist.gov/customcf/get_pdf. cfm?pub_id=50951 (Accessed 12.2022)

29. Gauthama Raman, M.R., Ahmed, C.M., Mathur, A.: Machine Learning for Intrusion Detection in Industrial Control Systems: Challenges and Lessons from Experimental Evalluation. In: Cybersecurity Vol. 4, pp. 27 ff, 2021. https://doi.org/10.1186/s42400-021-00095-5 (Accessed 12.2022)
30. Ganapathi, P., Shanmugapriya, D.: Handbook of Research on Machine and Deep Learning Applications for Cyber Security, Hershey, PA, IGI Global, 2020
31. Ultimate Goals to Building a Machine Learning Anomaly Detection Systems: Design Principles. Anodot, 2017
32. Stoufer, K. Guide to Industrial Control Systems. NIST Special Publication 800–882, Revision 2, 2014
33. Wang, Q., Chi, H., Li, Y., Vucetic, B.: Recent Advances in Machine Learning-based Anomaly Detection for Industrial Control Networks. In: Proceedings 1st International Conference on Industrial Artificial Intelligence, pp. 1–6, 2019
34. Drias, Z., Serhrouchni, A., Vogel, O.: Taxonomy of Attacks on Industrial Control Protocols. In: Proceedings IEEE International Conference on Protocol Engineering (ICPE) and International Conference on New Technologies of Distributed Systems (NTDS), 2015
35. Feng, X., Li, Q., Wang, H., Sun, L.: Characterizing Industrial Control System Devices on the Internet. In: Proceedings IEEE 24th International Conference on Network Protocols (ICNP), pp. 1–10, 2016. https://doi.org/10.1109/ICNP.2016.7784407 (Accessed 12.2022)
36. Miran, A., Ma, Z., Adrian, D., Tischer, M., Chuenchujit, T., Yardley, T., Berthier, R., Masión, J., Durumeric, Z., Halderman, J.A., Bailey, M.: An Internet-wide View of Industrial Control System Devices. In: Proceedings 14th Annual Conference on Privacy and Trust (PST), pp. 96–103, 2016. https://doi.org/10.1109/PST.2016.7906943 (Accessed 12.2022)
37. Ahmed, C.M., Gauthama Raman, M.R., Mathur, A.P.: Challenges in Machine-Learning-based for Real-Ime Anomaly Detection in Industrial Control Systems. In: Proceeding 6th ACM on Cyber-Physical Security Workshop, 2020
38. Ahmed, C.M., Ocha, M., Zhou, J., Marthur A., Svanning the Cycle: Timing-based Authentication o Programmable Logic Controllers, 2021 arXiv e-prints. Feb:arXiv-2102 (Accessed 12.2022)
39. Priyanga, S., Gauthama Raman, M.R., Jagtap, S.S., Aswin, N., Kirthivasab, K., Shankar Sriram, V.: An improved rough Set Theory-basec Feature Selection Approach for Intrusion Detection in SCADA Systems. In: Journal Intelligent Fuzzy Systems, Vol. 36, pp. 1–11. 2019
40. Krithivasan, K., Priyanga, S., Shankar Sriram, V.S.: Detection of Cyberattacks in Industrial Control Systems Using Enhanced Principal Component Analysis and Hypergraph-based Convolution in Neural Network. In: IEEE Transactions on Industrial Applications, Vol. 56, pp. 4394–4404, 2020
41. Kim, K., Aminanto, M.E., Tanuwidjaja, H.C.: Network Intrusion Detection usinging – A Feature Learning Approach. In: Springer Briefs on Cyber Security Systems and Networks, 2018
42. Motoda, H., Liu, H.: Feature Selection, Extraction and Construction. In: Communication of Institute of Information and Computing Machines, Vol. 5. Pp. 67–72, 2002
43. Sabhnani, M., Serpen, M.: Application of Machine Learning Algorithm to KDD Intrusion Detection Dataset within Misuse Detection Context. In: Proceedings International Conference Machine Learning Model, Technologies and Applications, pp. 209–215, 2003
44. Aminanto, M.E., Kim, K.: Deep Learning in Intrusion Detection Syste,: An Overview. Ín: Proceedings International Research Conference on Engineering and Technology, pp. 28–30, 2016
45. Al-Garadi, M.A., Mohamed, A., Al-Ali, A.A., Du, X., Guizari, M.: A Survey of Machine and Deep Learning Methods for Internet of Things Security. Cornell University, 2018 arXiv:1807.11023 (Accessed 12.2022)
46. Banerjee, A., Venkatasubramanian, V.V., Mukherjee, T., Gupta, S.K.S.: Ensuring Safety, Security, and Sustainability of Mission-Critical Cyber Physical Systems. In: Proceedings of IEEE, Vol. 100, No. 1, pp. 283–299, 2012
47. Wozniak, M., Grana, M., Corchado, S.: A Survey of Multiple Classifier Systems as Hybrid Systems. In: Information Fusion, Vol. 16, pp. 3–17, 2014

48. Domingos, P.: A Few Useful Things to Know About Machine Learning. In: Communications of the ACM, Vol. 55, No. 10, pp 78–87, 2012
49. Zhang, C., Ma, Y.: Ensemble Machine Learning Methods and Applications. Springer Publ. 2012
50. Deng, L.: A Tutorial Survey of Architectures, Algorithms and Applications for Deep Learning. In: APSIP Transaction on Signal and Information Processing, Vol. 3, 2014
51. Deng, L., Yu, D.: Deep Learning Methods and Applications. In: Foundations and Trends in Signal Processing, Vol. 7, No., 3-4, pp. 197–387, 2014
52. K. P. Murphy, "Machine Learning – A Probabilistic Perspective", MIT Press 2012
53. Apruzzese, G., Colajanni, M., Ferretti, L., Guido, A., Marchetti. M.: On the Effectiveness of Machine and Deep Learning for Cybersecurity. In: Proceedings 10th International Conference of Cyber Conflicts, pp. 371–390, 2018. https://ccdcoe.org/uploads/2018/10/Art-19-On-the-Effectiveness-of-Machine-and-Deep-Learning-for-Cyber-Security.pdf (Accessed 12.2022)
54. Du, K.L., Swamy, M.N.S: Neural Networks and Statistical Learning. Springer Publ., 2020
55. Xin, Y., Kong, L., Liu, Z., Chen, Y., Li, Y., Zhu, H., Gao, M., Hou, H., Wang, C. Machine Learning and Deep Learning Methods for Cybersecurity. In: IEEE Access, Vol. 6, pp. 35365–35381, 2018
56. Ghahramani, Z.: Unsupervised Learning. In: Advanced Lectures in Machine Learning, pp. 72–112, 2004
57. Arnold, L., Rebecchi, S., Chevallier, S., Paugam-Moisy, H. An Introduction to Deep Learning. In: Proceeding European Symposium on Artificial Networks, 2011
58. Ferrag, M.A., Maglaras, L., Moschoyiannis, S., Janick, H.: Deep Learning for Cybersecurity Intrusion Detection: Approaches, Datasets, and Comparative Study. In: Journal of Information Security and Applications, Vol. 50, 2020. https://doi.org/10.1016/j.jisa.2019.102419 (Accessed 12.2022)
59. Dimokranitou, A.: Adversarial Autoencoder for Anomalous Event Detection in Images. PhD dissertation, Purdue Iniversity, 2017
60. Ding, Y., Chen, S., Xu, J.: Application of Deep Belief Networks, for Opcode based Malware Detection. In: International Joint Conference of Neural Networks, pp. 3901–3908, 2016
61. Kwon, D., Kim, H., Kim. J., Suh, S.C., Kim, I., Kim, K.J.: A Survey of Deep Learning-based Network Anomaly Detection. In: Clustering Computation, Vol. 4, No. 3, pp. 1–13, 2017
62. Goodfellow, I., Pouget-Abadie, J., Mirza, M., Xu, B., Warde-Farley, D., Ozair, S., Couville, A., Bengio, Y.: Generative Adversarial Nets. In: Advances in Neural Information Processing Systems, MIT Press, pp. 2672–2680, 2014
63. Berman, D.S., Buczah, A.L., Chavis, J.S., Corbet, C.L.: A Survey of Deep Learning Methods for Cyber Security. Information 2019, 10, 122
64. Ahmad, Z., Khan, A.S., Shang, C.W., Abdullah, J., Ahmad, F.: Network Intrusion Detection Systems: A Systemic Study of Machine Learning and Deep Learning Approaches. In: Transactions on Emerging Telecommunications Technologies, pp. 1–29, Wiley & Sons, 2021 (Accessed 12.2022)
65. Ledig, C., Theis, L., Huszar, F, Caballero, J., Cunningham, A., Acosta, A., Aitken, A., Tejani, A., Trotz, J., Wang, Z.: Phota Realistic Single Image Super-Resolution using a Generative Adversarial Network. arXiv 2016, arXiv:1609.04802 (Accessed 12.2022)
66. Reed, S., Akata, Z., Yan, X., Logeswaran, L., Schiele, B., Lee, H.: Generated Adversarial Text to Image Synthesis. arXiv 2016, arXiv:1605.05396 (Accessed 12.2022)
67. Dosovitskiy, A., Fischer, P., Ilg, E., Hausser, P., Hazirbas, C., Golkov, V., van der Smagt, P., Cremers, D., Brox, T.: Flownet: Learning Optical Flow with Convolutional Networks. In: Proceeding IEEE International Conference on Computer C'Vision (ICCV), pp. 2758–2766, 2015
68. Alazab, M., Tang, M.J. (Eds.): Deep Learning Applications for Cybersecurity. Springer Publ. 2019
69. Vinayakumar, R., Barathi Ganesh, H.B., Prabaharan, P., Anand Kumar, M., Soman, K.P.; Deep-Net: Deep Neural Network for Cybersecurity Use Cases. 2018. https://arxiv.org/ftp/arxiv/papers/1812/1812.03519.pdf (Accessed 12.2022)

Glossary

A

AA Application Awareness—term for systems that have built-in information or awareness about individual applications, in order to better interact with these applications.

AABM Adversary Attack Behavior Model—formalization of an attacker behavior to a computer or networked system.

AAR Autonomous Acting Robots—theory and applications of robotic systems capable of some degree of self-sufficiency.

AIPS Anomaly-based Intrusion Prevention System—detecting both network and computer intrusions and misuse by monitoring system activity and classifying it as either normal (regular) or anomalous (irregular).

AIPS Anomaly-based Intrusion Prevention System—takes samples of network traffic at random and compares them to a pre-calculated baseline performance level.

AE Auto Encoder—type of artificial neural network used to learn efficient codings of unlabeled data.

AESK Advanced Encryption Standard Key—symmetric key encryption, which involves the use of only one secret key to cipher and decipher information.

AI Artificial Intelligence.

AID Anomaly-based Intrusion Detection—utilizes machine learning to train the detection system to recognize a normalized baseline.

AM Additive Manufacturing—technologies that grow three-dimensional objects one superfine layer at a time.

ANN Artificial Neural Network—consists of input, hidden, and output layers with connected neurons (nodes) to simulate the human brain.

API Application Programming Interface—software intermediary that allows two applications to talk to each other.

APLCM Analytics Project Life Cycle Management—benefits from the introduction of data-based knowledge, since these insights help to make the right decisions about product strategy.

APT Advanced Persistent Threat—cyberattack carried out using complex attack technologies and background information.

AR Augmented Reality—integration of digital information with the user's environment in real time.

ASI Accenture Security Index—based on a comprehensive model measuring 33 specific *cybersecurity* capabilities.

ATT&CK Adversarial Tactics, Technics, & Common Knowledge—guideline for classifying and describing cyberattacks and intrusion.

AuC Area under Curve—between two points is found out by doing a definite integral between the two points.

B

B2B Business-to-Business—type of electronic commerce (e-commerce), is the exchange of products, services, or information between businesses.

BCP Business Continuity Plan—document describing how an industrial organization continues to operate in the event of an unplanned business interruption BIA: Business Impact Analysis.

BIA Business Impact Analysis—systematic process to determine and evaluate the potential effects of an interruption to critical business operations as a result of a disaster, accident, or emergency.

BoM Boltzmann Machine—neural network of symmetrically connected nodes that make their own decisions whether to activate.

BPA Back Propagation Algorithm—based on generalizing the Widrow-Hoff learning rule. It uses supervised learning, which means that the algorithm is provided with examples of the inputs and outputs that the network should compute, and then the error is calculated.

BPaaS Business Process-as-a-Service—business process outsourcing (BPO) delivered based on a cloud services model.

BPM Business Process Management—methodology for managing the processes in an organization.

C

CaaS Cloud-as-a-Service—*services consumed and paid for on a subscription or pay-per-use basis.*

CAAP Cyber Attackers' Attack Policy—living document, continuously updated about attackers' behavior requirements to successful cyberattack types and technological developments.

CABM Cyber Attack Behavior Model—approximation of adversarial cyberattack behavior against a computer system and others.

CACS Cyber Attackers' Capability Skills—*a*bilities that enable cyber attackers to successfully attack computer systems and others.

CADKAC Cyber Attack Defenders Knowledge of Cyber Attackers' Capabilities—*knowledge and skills to identify attack capabilities to successfully defend an organization.*

CADM Cyber Attackers Defending Model—a specific defense model (algorithm) to defend against well-defined cyberattacks.

CAM Critical Access Management—practice to protect valuable organization's assets by securing the access points that lead to them.

CAPEC Common Attack Pattern Enumeration and Classification—*list of software weaknesses.*

CAPA Cyber Attackers' Profile to Attack—refers to the different profiles of cyber attackers such as cybercriminal, insider, hacktivist, nation state, terrorist, and others.

CC Correlation Coefficient—statistical measure of the degree to which changes to the value of one variable predict change to the value of another.

CCMM Cybersecurity Capability Maturity Model—tool to support organizations to evaluate their cybersecurity capabilities and optimize security investments.

C2M2 Cybersecurity Capability Maturity Model—tool to support organizations to evaluate their cybersecurity capabilities and optimize security investments.

CDT Cognitive-based Detection Techniques—make use of audit classification technique.

CER Critical Entity Resilience—EU Directive 2022/2557.

CERT Critical Emergency Readiness Team—group of information security experts responsible for the protection against, detection of, and response to an organization's cybersecurity incidents.

CFA Confirmatory Factor Analysis—statistical technique used to verify the factor structure of a set of observed variables.

CFFDNN Convolutional Feedforward Deep Neural Network—variant of a DNN where each neuron receives its input only from a subset of neurons of the previous layer.

CFoRA Cost Factors of Ransomware Attack—cost of ransom payment, cost of disruption and downtime, cost of forensic and recovery, cost of data losses, cost of reputation loss.

CIP Continuous Improvement Process—ongoing effort to improve products, services, or processes.

CISA Cybersecurity and Infrastructure Security Agency.

CISCSC Center for Internet Security Critical Security Control—community-driven nonprofit organization, responsible for the CIS Controls® and CIS Benchmarks™, globally recognized best practices for securing IT systems and data.

CITA Cyber Threat Intelligence Algorithms—provides better insight into the threat landscape and threat actors, along with their latest tactics.

CLASP Comprehensive Lightweight Application Security Process—role-based set of process components guided by formalized best practice.

CLM Clustering Learning Methods—*algorithm for grouping data points into different clusters, consisting of similar data points.*

CM Cognitive Model—deals with simulating human problem-solving and mental processing in a computerized model.

CMLT Classical Machine Learning Technique—algorithms such as linear regression, logistic regression, SVMs, nearest neighbor, decision trees, PCA, naive Bayes classifier, and k-means clustering.

CMM Capability Maturity Model—*methodology used to develop and refine organizations' software development processes.*

CM2 Capability Maturity Model—*methodology used to develop and refine organizations software development processes.*

CMMI Capability Maturity Model Integration—incorporates best components of individual disciplines of CMM like Software CMM, Systems Engineering CMM, and others.

CMP Control Message Protocol—reporting protocol that network devices such as routers use to generate error messages to the source IP address.

CoA Center of Area—point of a plane figure that coincides with the center of mass of a thin uniform distribution of matter over the area of the figure.

COG Center of Gravity—point from which the weight of a body or system may be considered to act.

CPM Critical Process Management—practices to identify critical processes maturity level.

CPS Cyber-Physical Systems—integrate sensing, computation, control, and networking into physical objects and infrastructure, connecting them to the Internet and to each other.

CPT Conditional Probability Table—set of discrete and mutually dependentrandom variables to display conditional probabilities of a single variable with respect to the others.

CRI Cyber Readiness Index—method to evaluate and measure the preparedness level for certain cybersecurity risks.

CRM Customer Relationship Management—technology for managing all organizations' relationships and interactions with customers and potential customers.

CPPS Cyber Physical Production System—consists of autonomous and co-operative elements and subsystems that are connected based on the context within and across all levels of production, from processes through machines up to production and logistics networks.

CPT Conditional Programmable Table—has multiple *tables*, some of which collect continuously, but others on demand only.

CRI Cyber Readiness Index—represents a way of examining the cyber-readiness problem.

CSA Cybersecurity Audit—provides a comprehensive assessment of information systems to evaluate compliance and identify gaps in security policy implementation.

CSAI Cybersecurity Audit Index—combining indicators into one benchmark measure to monitor and compare the level of cybersecurity.

CSF Cybersecurity Framework—set of guidelines for mitigating organizationalcybersecurity risks, based on existing standards, guidelines, and practices.

CSMM Cybersecurity Maturity Model—current capabilities in cybersecurity workforce planning, establishing a foundation for consistent evaluation.

CSMMA Cybersecurity Maturity Model Assessment—regulatory requirements managing information maturity appropriately.

CSML Cybersecurity Maturity Level—identifies and evaluates practical ways to evaluate and/or enhance the maturity level.

CSMS Cybersecurity Management System—basis for the protection of networked systems.

CSRF Cross-Site Request Forgery—attack that forces an end user to execute unwanted actions on a web application in which they are currently authenticated.

CSV Comma Separated Value—used to refer to a computer file containing tabular data that is presented in plain text.

CTA Cyber Threat Attacker—malicious actor that seeks to damage data, steal data, disrupt digital operations, and others.

CTI Cyber Threat Intelligence—process of identifying and analyzing cyber threats.

CTIA Cyber Threat Intelligence Algorithm—detecting, analyzing, and defending against *cyber-threats.*

C2C Command and Control—exercise of authority and direction by a properly designated responsible actor.

CVE Common Vulnerabilities and Exposures—*provides a reference-method for publicly known information-security vulnerabilities and exposures.*

CVSS Common Vulnerability Scoring System—*free and open industry standard for assessing the severity of vulnerabilities of computer system security.*

CWE Common Weakness Enumeration—*category system for hardware and software weaknesses and vulnerabilities.*

D

DAE Deep Auto Encoder—composed of two symmetrical deep-belief networks that typically have four or five shallow layers representing the encoding half of the net and a second set of four or five layers that makes up the decoding half.

DAIDS Dynamic Anomaly Intrusion Detection System—run partially ordered sequence of distinct incidents.

DCM Density-based Clustering Methods—refers to unsupervised learning methodologies used in model building and machine learning algorithms.

DBDN Deep Belief Directed Network—directed acyclic graph composed of stochastic variables.

DBN Deep Belief Network—sophisticated generative model that employs a deep architecture.

DBNN Deep Bayesian Neural Network—type of artificial intelligence based on Bayes' theorem with the ability to learn from data.

DDoS Distributed Denial of Service—subclass of denial of service (DoS) attacks.

DLM Deep Learning Method—machine learning technique that teaches computers to do what comes naturally to humans: learn by example.

DNS Domain Name System—Internet service that manages the name space on the Internet and translates the name of a domain into numerical IP addresses.

DoB Degree of Belief—represents the strength with which the truth of various propositions is believed.

DoS Denial of Service—shut down a machine or network, making it inaccessible to its intended users.

DPI Deep Packet Inspection—enable high performance filtering.

DRP Disaster Recovery Plan—documented, structured approach that describes how an organization can quickly resume work after an unplanned incident.

DS Digital Sovereignty—refers to the ability to have control over own digital destiny—the data, hardware, and software that you rely on and create.

DTE Digital Twin Environment—integrated multi-domain physics application space for operating digital twins for a variety of purposes.

DTM Decision Tree Method—non-parametric supervised learning algorithm, which is utilized for both classification and regression tasks.

DTM Digital Twin Model—virtual representation of an object or system that spans its lifecycle, is updated from real-time data, and uses simulation, machine learning, and reasoning to support decision-making.

DTSF Digital Twin Shop Floor—physical manufacturing shop floor.

E

EaaS Edge-as-a-Service—platform for realizing distributed cloud architectures and integrating the edge of the network in the computing ecosystem.

EDR Endpoint Detection and Response—integrated endpoint security solution that combines real-time continuous monitoring and collection of endpoint data with rules-based automated response and analysis capabilities.

EGAN Evolutionary Generative Adversary Network—effective for learning generative models for real-world data.

ELM Ensemble Learning Method—machine learning technique that combines several base models in order to produce one optimal predictive model.

EMA Expectation-Maximization Algorithm—approach that underlies many machine learning algorithms, although it requires that the training dataset is complete.

ERP Enterprise Resource Planning—type of software system that helps organizations automate and manage core business processes for optimal performance.

F

FaaS Function-as-a-Service—cloud-computing service that allows customers to execute code in response to events, without managing the complex infrastructure typically associated with building and launching applications.

FAR False Alarm Rate—statistic measure defined to take action when there is a measure of the background trigger rate.

FCM Fuzzy C-Means—data clustering technique in which a data set is grouped into N clusters with every data point in the dataset belonging to every cluster to a certain degree.

FD Fraud Detection—set of activities undertaken to prevent assets from being obtained through false pretenses.

FFANN Feedforward Artificial Neural Network—simplest form of a single layer perceptron.

FFDNN Feedforward Deep Neural Network—*Neural Network* with one hidden layer as a "shallow" *network* or simply *a Feed-Forward network*.

FFNN FeedForward Neural Network—artificial neural network in which the connections between nodes do not form a cycle.

5G Fifth Generation of Wireless Broadband Cellular Technology—mobile communications standard that has been gaining popularity since 2019.

FIPS Firewall Integrated Intrusion Detection System—next-generation firewall (NGFW) or unified threat management (UTM) solution.

FL Feature Learning—representation learning is a set of techniques that allows a system to automatically discover the representations needed for feature detection or classification from raw data.

FM F-Measure—combines both precision and recall into a single measure that captures both properties.

FNPR False Negative Prediction Rate—probability that a true positive missed by the test.

FNR False Negative Rate—condition tested for is not present when condition is actually present.

FPPR False Positive Prediction Rate—probability that a true negative missed by the test.

FPR False Positive Rate—— condition tested for is not present when condition is actually present.

G

GAN Generative Adversarial Network—generate new, synthetic instances of data that can pass for real data.

GOA Government Accountability Office—legislative branch government agency that provides auditing, evaluative, and investigative services for the United States Congress.

H

HCM Hierarchical Clustering Methods—groups similar objects into groups, termed clusters.

HIDS Host-based Intrusion Detection System—monitors the infrastructure system on which it is installed, analyzing traffic and logging malicious behavior.

HIPS Host-based Intrusion Prevention System—analyzes all received and sent data or packet traffic streams.

HLPD High Level Policy Document—describes the various activities that organizations performs to manage cybersecurity and cyber-robustness.

HMI Human Machine Interface—dashboard that allows users to communicate with machines, computer programs, or systems.

HMM Hidden Markov Model—stochastic model in which a system is modeled by a Markov chain with unobserved states.

HTTP Hyper Text Transfer Protocol—communications protocol on the World Wide Web with which data can be transmitted.

HTTPS Hyper Text Transfer Protocol Secure—communications protocol on the World Wide Web with which data can be transmitted securely.

I

IaaS Infrastructure-as-a-Service—a type of cloud computing service that offers essential compute, storage, and networking resources on demand, on a pay-as-you-go basis.

IACS Industrial Automation and Control Systems—use of control systems, such as computers or robots, and information technologies for handling different processes and machineries in an industry to replace a human being.

ICMP Internet Control Message Protocol—network layer protocol used by network devices to diagnose network communication issues.

ICSS Industrial Control System Security—protection of industrial control systems against threats from cyber attackers.

ICT Information and Communication Technology (IDC) is a diverse set of technological tools and resources that enable to transmit, store, create, share or exchange information.

IDC Intrusion Detection Capability is the ratio of mutual information between the IDS input and output to the entropy of the input.

IDDM Intelligence Driven Defense Model—supports the intent to stop offensive maneuvers during a cyberattack while maintaining a defensive posture.

IDPS Intrusion Detection and Intrusion Prevention System—network security technology and key part of any organization's security system that continuously monitors network traffic for suspicious activity and takes steps to prevent it.

IDPSA Intrusion Detection and Prevention System Architecture—one of the most critical considerations in intrusion detection and prevention.

IDS Intrusion Detection System—detects suspicious activities and generates alerts when they are detected.

IDSA Intrusion Detection System Architecture—software and/or hardware components that monitor computer systems and analyze events occurring in them for signs of intrusions.

IEC International Electrotechnical Commission—worldwide used IEC International Standards and Conformity Assessment systems to ensure safety.

IEEE Institute of Electrical and Electronic Engineers—world's largest technical professional society, promoting the development and application of electrotechnology and allied sciences for the benefit of humanity, the advancement of the profession, and the well-being of members.

IIoT Industrial Internet of Things—extension and use of the IoT in industrial sectors and applications.

IoC Indicators of Compromise—piece of digital forensics that suggests that an endpoint or network may have been breached.

IoCL Indicator of Compromise Lifecycle—time during which it preserves its malicious activity with a high probability.

IoTDS Internet of Things and Data Services—consists of Enterprise, Business Area, and Data Warehouse logical data models developed for companies providing internet and broadband products and services.

IoT Internet of things—system of interrelated computing devices, digital components, objects and others that are provided with unique identifiers and the ability to transfer data over a network without requiring human–human or human-to-computer interaction.

IP Internet Protocol—sent data from one computer to another on the Internet.

IPS Intrusion Prevention System—security tool, which can be a hardware device or software that continuously monitors a computer system or network for malicious activity and takes action to prevent it, including reporting, blocking, or dropping it, when it does occur.

IPSA Intrusion Prevention System Architecture—data collection, analysis, and response module.

IPv6 Internet Protocol Version 6—set of specifications from the Internet Engineering Task Force (IETF) that is essentially an upgrade of IP version 4 (IPv4), a category of IP addresses in IPv4-based routing.

IRMI Information Risk Maturity Index—based on a weighted risk maturity excellence model comprising four separate elements: strategy, people, communications, and security.

ISA International Society for Automation—non-profit professional association of engineers, technicians, and management engaged in industrial automation.

ISMS Information System Management System—applying computer-base for managing information in organizations for management roles such as interpersonal roles, informational roles, and decision-based roles.

ISO International Organization for Standardization—worldwide federation of national standards bodies.

IT Information Technology—use of any computers, storage, networking, and other physical devices, infrastructure and processes to create, process, store, secure, and exchange all forms of electronic data.

ITS Information Technology Systems—means all communications systems and computer systems used by a Group Company, including all hardware, software, and websites, but excluding networks generally available to the public.

I4.0 Industry 4.0—refers to the fourth industrial revolution, is the cyber-physical transformation of manufacturing.

K

KCSI Kaspersky Cybersecurity Index—set of indicators that allow the evaluation of the level of risk for Internet users worldwide.

KPA Key Process Area—building blocks that indicate the areas an organization should focus on to improve its software process.

KPI Key Performance Indicator—measurable value that demonstrates how effectively an organization is achieving key business objectives.

L

LAN Local Area Network—collection of devices connected together in one physical location, such as a building, office, or home.

LDA Linear Discriminant Analysis—used as dimensionality reduction technique in the pre-processing step for pattern-classification and machine learning applications.

LoR Loss of Reputation—negative publicity if anyone acting on its behalf is implicated in any fraud, bribery, or corruption issue.

LSTM Long Short-Term Memory—type of recurrent neural network capable of learning order dependence in sequence prediction problems.

M

MAD Maximum Absolute Deviation—is the maximum of the absolute deviations of a sample from that point.

MAEC Malware Attribute Enumeration and Characterization—community-developed structured language for encoding and sharing high-fidelity information about malware based upon attributes such as behaviors, artifacts, and relationships between malware samples.

MANTIS Model-based Analysis of Threat Intelligence Sources—consists of several Django Apps that, in combination, support the management of CTI expressed in standards such as STIX, CybOX, OpenIOC, IODEF (RFC 5070), and others.

MDP Markov Decision Process—*model decision making in discrete, stochastic, sequential environments.*

MFA Multi-F Authentication—security technology that requires multiple methods of authentication from independent categories of credentials to verify a user's identity for a login or other transaction.

MID Misuse-based Intrusion Detection—match computer activity to stored signatures of known exploits or attacks.

MFT Master File Table—database in which information about every file and directory on an NT File System (NTFS) volume is kept.

MIT Massachusetts Institute of Technology—private land-grant research university in Cambridge, Massachusetts. Established in 1861.

MOM Mean of Maximum—defuzzification method for pattern recognition applications.

MRLA Model-based Reinforcement Learning Algorithm—refers to learning optimal behavior indirectly by learning a model of the environment by taking actions and observing the outcomes that include the next state and the immediate reward.

MVKC Minimum Viable Kill Chain—Intelligence Driven Defense® model for identification and prevention of cyber intrusions activity.

M2M Machine-to-Machine—term used to describe any technology that enables networked devices to exchange information and perform actions without the manual assistance of humans.

MWF Malware Filtering—instead of just flagging potential *malware*, it will instead remove it from the system.

N

NaaS Network-as-a-Service—enables users to easily operate the network and achieve the outcomes they expect from it without owning, building, or maintaining their own infrastructure.

NBM Naïve Bayes Method—classification technique based on Bayes' Theorem with an assumption of independence among predictors.

NIDS Network-based Intrusion Detection System—designed to help organizations monitor their cloud, on-premise, and hybrid environments for suspicious events that could indicate a compromise.

NBM Naïve Bayes Method—set of supervised learning algorithms based on applying Bayes' theorem with the naive assumption of conditional independence between every pair of features given the value of the class variable.

NCCIC National Cybersecurity and Communication Integration Center—division of the Office of Cybersecurity and Communications within the National Protection and Programs Directorate (NPPD), is the operational arm of NPPD and responsible for providing full-time monitoring, information sharing, analysis, and incident response capabilities to protect Federal agencies' networks and critical infrastructure and key resources, such as industrial control systems.

NCI National Cybersecurity Index—focuses on measurable *cybersecurity* aspects that are implemented by the central government, and aims to identify which gaps in policies and strategies.

NCSA National Cyber Security Alliance—non-profit organization on a mission to create a more secure, interconnected world.

NCSI National Cybersecurity Index—global index, which measures the preparedness of countries to prevent cyber threats and manage cyber incidents.

NGAV Next Generation Antivirus—takes traditional antivirus software to a new, advanced level of endpoint security protection.

NGFW Next Generation Firewall—part of the third generation of firewall technology, combining a traditional firewall with other network device filtering functions, such as an application firewall using in-line deep packet inspection (DPI), an intrusion prevention system (IPS).

NFR Network Flight Recorder—programmable traffic analysis/intrusion detection engine that can be instantly updated when a new attack is discovered.

NFV Network Function Virtualization—improve the flexibility of network services provisioning and reduce the time to market of new services.

NIDS Network-based Intrusion Detection System—designed to help organizations monitor their cloud, on-premise, and hybrid environments for suspicious events that could indicate a compromise.

NIPS Network-d Intrusion Prevention System—examines the transmitted data or packet traffic stream at the protocol or application level.

NIS Network and Information Security—first piece of EU-wide cybersecurity legislation (EU 2016).

NIST National Institute of Standards and Technology—promotes US innovation and industrial competitiveness by advancing measurement science, standards, and technology in ways that enhance economic security and improve quality of life.

NIST-CSF NIST Cybersecurity Framework—consists of standards, guidelines, and best practices to manage *cybersecurity* risk.

NNM Nearest Neighborhood Method—classifies a sample based on the category of its nearest neighbor.

NoSQL No Standard Query Language—*group of database systems which are not based on structured query language.*

NRML Negative Reinforcement Machine Learning—method used to help teach specific behaviors.

M2M Machine-to-Machine: describes any technology that enables networked devices to exchange data and perform actions without manual assistance of humans.

NVD National Vulnerability Database—US government repository of standards-based vulnerability management data represented using the Security Content Automation Protocol (SCAP).

O

OCS&C Office of Cybersecurity and Communications—official website of the US *Department* of Homeland Security about DHS, Accessibility, Budget and Performance.

OEM Original Equipment Manufacturer—term refers to any company that produces parts or products designed to be incorporated into end products produced by other companies.

OpenIoC Open Indicators of Compromise—open framework that currently exists for organizations that want to share threat event attack information internally and externally in a machine-digestible format.

OpEX Operational Expenditure—money a company spends on an ongoing, day-to-day basis in order to run a business or system.

OS Operating System—program that, after being initially loaded into the computer by a boot program, manages all of the other application programs in the computer.

OSS Operating's System Scheduler—special system software, which handles process scheduling in various ways.

OT Operational Technology—hardware and software that detects or causes a change, through the direct monitoring and/or control of industrial equipment, assets, processes, and events.

OTA Over the Air—wireless delivery of new software, firmware, or other data to mobile devices.

OTSEC Operational Technology Security—enables cybersecurity teams to fine-tune traditional, digitized, and new technology-based OT processes.

OWASP Open Web Application Security Project—nonprofit foundation that works to improve the security of software.

OWL Web Ontology Language—designed for use by applications that need to process the content of information instead of just presenting information to humans.

P

PaaS Platform-as-a-Service—cloud-computing model where a third-party provider delivers hardware and software tools to users over the Internet.

PbCM Probability-based Clustering Method—form of a finite mixture model, where each component in the mixture is a multivariate probability density or distribution function for a particular group (cluster).

PCM Partitioning Clustering Methods—relocate instances by moving them from one cluster to another, starting from an initial partitioning.

PDM Proactive Design for Manufacturability—process of proactively designing products to optimize all the manufacturing functions: fabrication, assembly, test, procurement, shipping, service, and repair; and ensure the best cost, quality, reliability, regulatory compliance, safety, time-to-market, and customer satisfaction.

PERT Program Evaluation and Review Technique—visual tool used in project planning.

PFF Packet Filtering Firewalls—network security feature that controls the flow of incoming and outgoing network data.

PGM Probabilistic Generative Model—statistical model of the joint *probability* distribution $P(X, Y)$.

PIMS Privacy Information Management System—framework for data privacy that builds on ISO 27001.

PIN Personal Identification Number—numeric or alphanumeric string that is used to authenticate a person to a system.

PIPSA Policy-based Intrusion Prevention System Architecture—implies predefined action pattern that is repeated by an entity (IPSA) whenever certain conditions occur.

PGM Probabilistic Generative Model—statistical model of the joint *probability* distribution $P(X, Y)$.

PIMS Privacy Information Management System —framework for data privacy that builds on ISO 27001.

PIN Personal Identification Number —numeric or alphanumeric string that is used to authenticate a person to a system.

PLC Programmable Logic Controller—type of tiny computer that can receive data through its inputs and send operating instructions through its outputs.

PLCS Product Life Cycle Sustainability—measures a company's total environmental impact from raw materials to production, distribution, consumer use, and disposal of the product by the consumer.

POTS Plain Old Telephone Service—refers to the traditional, analog voice transmission phone system implemented over physical copper wires (twisted pair).

PRLA Policy-based Reinforcement Learning Algorithm—a branch of machine learning dedicated to training agents to operate in an environment, in order to maximize their utility in the pursuit of some goals.

PRML Positive Reinforcement Machine Learning—when an event occurs due to a particular behavior, PRML increases the strength and the frequency of the behavior.

P2P Peer-to-Peer—bilateral communication between two or more equal partners.

Q

QoS Quality of Service—set of technologies that work on a network to guarantee its ability to dependably run high-priority applications and traffic under limited network capacity.

QL Q-Learning—efficient way for an agent to learn how the environment works.

QMS Quality Management System—formalized system that documents processes, procedures, and responsibilities for achieving quality policies and objectives.

R

RaaS Ransomware-as-a-Service—offer of malware that can be used for a fee.

RAT Remote Access Trojan—malware an attacker uses to gain full administrative privileges and remote control of a target computer.

RIDS Rule-based Intrusion Detection System—attack can either be detected if a rule is found in the rule base or goes undetected if not found.

RBM Restricted Boltzmann Machine—are shallow, two-layer neural nets that constitute the building blocks of deep-belief networks.

R-CCMM Railway Cybersecurity Capability Maturity Model—allow *railway* organizations to improve their *capability* to reduce the impacts of *cyberattacks.*

R-C2M2 Railway Cybersecurity Capability Maturity Model—allow *railway* organizations to improve their *capability* to reduce the impacts of *cyberattacks.*

RCE Resilience of Critical Entities—*carry out risk assessments of their own, take technical and organizational measures to enhance their resilience.*

RCO Recovery Point Objective—defined as the maximum amount of data—as measured by time—that can be lost after a recovery from a disaster, failure, and others.

RDF Resource Description Framework—represents the semantics of an entity as a set of objects or concepts.

RDNN Recurrent Deep Neural Network—type of artificial neural network commonly used in speech recognition and natural language processing.

RDP Remote Desktop Protocol—secure network communication protocol offered by Microsoft, allows users to execute remote operations on other computers.

RIDS Rule-based Intrusion Detection System—attack can either be detected if a rule is found in the rule base or goes undetected if not found.

RFA Random First Algorithm—*generate a random, completely solved, instance, and then remove numbers as long* as there's still a unique solution.

RFID Radio Frequency Identification—form of wireless communication that incorporates the use of electromagnetic or electrostatic coupling in the radio frequency portion of the electromagnetic spectrum to identify an object.

RIPE Robust Industrial Control Systems Planning and Evaluation—process-driven approach towards effective and sustainable industrial control system security.

RLM Regression Learning Methods—supervised machine learning algorithm with a continuous and constant slope-expected performance.

RMF Risk Management Framework—United States federal government guideline, standard, and process for risk management to secure information systems (computers and networks) developed by NIST.

RML Reinforcement Machine Learning—machine learning training method based on rewarding desired behaviors and/or punishing undesired ones.

RMLA Reinforcement Machine Learning Algorithm—*machine learning* method that is concerned with how software agents should take actions in an environment.

RMLM Reinforcement Machine Learning Model—area of machine learning concerned with how intelligent agents ought to take actions in an environment in order to maximize the notion of cumulative reward.

RMP Reputation-based Malware Protection—performed by attempting to identify communication between friendly hosts on the network that are protected, and hosts on the Internet that are believed to be malicious based upon a reputation for malicious actions.

RMPG Risk Management Principles and Guideline—business risk plan to avoid risks and minimize the fallout from issues based on several essential principles.

RMSE Root Mean Square Error—*standard deviation of the residuals (prediction errors)*.

ReNN Replicator Neural Network—used for anomaly detection.

ROI Return on Investment—calculation of the monetary value of an investment versus its cost.

RPO Recovery Point Objective—defined as the maximum amount of data—as measured by time—that can be lost after a recovery from a disaster, failure, or comparable event before data loss will exceed what is acceptable to an organization.

RPTS Railway and Public Transportation System—covers the entire range of *railway and public transportation* passenger *systems.*

RSIDS Rule Set-based Intrusion Detection System—attack can either be detected if a rule is found in the rule base or goes undetected if not found.

RTO Recovery Time Objective—maximum tolerable length of time that a computer, system, network, or application can be down after a failure or disaster occurs.

S

SaaS Software-as-a-Service—software distribution model in which a cloud provider hosts applications and makes them available to end users over the Internet.

SAE Stacked Auto-Encoder—extract important features from data using deep learning.

SAIDS Static Anomaly Intrusion Detection—detect if a portion of the system monitored does not act constant (static).

SbD Security by Design—approach to software and hardware development that seeks to make systems as free of vulnerabilities and impervious to attack as possible through such measures as continuous testing, authentication safeguards, and adherence to best programming practices.

SCADA Supervisory Control And Data Acquisition—a category of software applications for controlling industrial processes, which is the gathering of data in real time from remote locations in order to control equipment and conditions.

SCAP Security Content Automation Protocol—synthesis of interoperable specifications derived from community ideas.

SiIDS Signature-based Intrusion Detection System—detects attacks based of specific signatures such as number of bytes or number of 1s or number of 0s in the network traffic.

SCS Supply Chain Sustainability—uses environmentally and socially sustainable practices at every stage to protect the people and environments across the whole chain.

SDL Supervised Deep Learning—subcategory of machine learning and artificial intelligence.

SDN Software Defined Network—approach to networking that uses software-based controllers or APIs to communicate with underlying hardware infrastructure and direct traffic on a network.

SD-WAN Software Defined Wide Area Network—virtual WAN architecture that allows organizations to leverage any combination of transport services, including MPLS, LTE, and broadband Internet services to connect users securely to applications.

SEM Structural Equation Modeling—set of statistical techniques used to measure and analyze the relationships of observed and latent variables.

SIDS Specification-based Intrusion Detection System—use specified program behavioral specifications as a basis to detect attacks.

SEI Software Engineering Institute—American research and development center headquartered in Pittsburgh, Pennsylvania.

SID Signature-based Intrusion Detection—detects attacks based of specific signatures such as number of bytes or number of 1's or number of 0's in the network traffic.

SIMM Service Integration Maturity Model—standardized model for organizations to guide their transformation to a service based business model.

SIM2 Service Integration Maturity Model—standardized model for organizations to guide their transformation to a service based business model.

SM Statistical Model—mathematical model that embodies a set of statistical assumptions concerning the generation of sample data.

SMB Small and Medium Business—businesses whose personal and revenue numbers fall below certain limits.

SML Supervised Machine Learning—uses training datasets to achieve desired results.

SMLA Supervised Machine Learning Algorithm—*model relationships and dependencies between the target prediction output and the input features that can be predicted.*

SMLC Smart Manufacturing Leadership Coalition—accelerating the digital transformation of manufacturers and their supply chains through its membership.

SNI Server Name Indication—extension for the TLS protocol to indicate a hostname in the TLS handshake.

SOAR Security Orchestration, Automation, and Response—refers to a collection of software solutions and tools that allow organizations to streamline security operations in three key areas: threat and vulnerability management, incident response, and security operations automation.

SOC Security Operations Center—centralized function within an organization employing people, processes, and technology to continuously monitor and improve an organization's security posture while preventing, detecting, analyzing, and responding to cybersecurity incidents.

SOF Standardized Output Format—including the description of a malware instance, malware intrusion set, or malware families.

SOLTRA EDGE Soltra Edge Platform for Automated Standard Threat Intelligence Sharing—industry-driven threat intelligence sharing platform.

SOM Self Organizing Maps—unsupervised Machine Learning technique used to produce a low-dimensional (typically two-dimensional) representation of a higher dimensional dataset while preserving the topological structure of the data.

SoSE Sum of Squared Error—difference between the observed value and the predicted value.

SpID Specification-based Intrusion Detection—relies on pre-specific signatures of threat event attacks as basis to detect attacks.

SPT Security Penetration Test—targeted, controlled, and planned attack on organization's systems.

SQL Standard Query Language—standardized programming language used to manage relational databases and perform various operations on the data in them.

SSE-CMM Systems Security Engineering Capability Maturity Model—de facto standard for *security engineering* practices.

SSE-CM2 Systems Security Engineering Capability Maturity Model—de facto standard for *security engineering* practices.

SSH Secure Shell—enables secure system administration and file transfers over insecure networks.

SSI Snapshot Interval—select the desired interval to change the interval between snapshots.

SSL Secure Socket Layer—networking protocol designed for securing connections between web clients and web servers over an insecure network, such as the Internet.

STE Self-Taught Learning—use unlabeled data in supervised classification tasks.

STIX Structured Threat Information eXpression—standardized XML programming language for conveying data about cybersecurity threats in a common language, easily understood by humans and security technologies.

SW-CMM Software Capability Maturity Model—framework that lays out five maturity levels for continual process improvement.

SW-CM2 Software Capability Maturity Model—framework that lays out five maturity levels for continual process improvement.

SWOT Strength, Weaknesses, Opportunities, Threats—framework for identifying and analyzing an organization's strengths, weaknesses, opportunities, and threats, e.g., in cybersecurity.

T

TAXII^{TM} Transport Protocol Automated eXchange of Indicator Information—*free and open transport mechanism* that standardizes the automated exchange of cyber threat information.

TCO Total Cost of Ownership—overall cost of a product or service throughout its life cycle.

TCP Transmission Control Protocol—standard that defines how to establish and maintain a network conversation by which applications can exchange data.

TIDS Threat Intelligence Defense System—important for organizations looking to stay up-to-date in their defense planning.

TEA Threat Event Attack—noticed incident with a malicious intent.

TIM Threat Intelligence Model—overall picture of the intent and capabilities of malicious cyber threats, including the actors, tools, and TTPs through the identification of trends, patterns, and emerging threats and risks, in order to inform decision- and policymakers or to provide timely warnings.

TIMP Threat Intelligence Management Platform—collects, aggregates, and organizes threat intel data from multiple sources and formats.

TLS Transport Layer Security—designed to operate on top of a reliable transport protocol such as TCP.

TNPR True Negative Prediction Rate—proportion of samples that are genuinely negative that give a negative result using the prediction test.

TOE Target of Evaluation—in accordance with common criteria, an information system, part of a system or product, and all associated documentation that is the subject of a security evaluation.

TPPR True Positive Prediction Rate—*gives the proportion of correct predictions in predictions of positive class.*

TPR True Positive Rate—gives the proportion of correct predictions in predictions of positive class.

TROT Threat Response Operation Team—*group* of IT professionals in charge of preparing for and reacting to any type of organizational emergency.

TtD Time to Data—describes the required time to retrieve the backup data and deliver it to the restore location.

TTD Time to Detect—one of the main key performance indicators in incident management.

TTE Techniques, Tactics, or Exploits—*variety of exploits and tools* an APT possesses.

TTP Tactics, Technologies, and Procedures—methods, tools, and *strategies* that threat event actors use to develop and execute cyberattacks.

U

UCO Unified Cybersecurity Ontology—incorporates and integrates heterogeneous data and knowledge schemas from different cybersecurity systems and most commonly used cybersecurity standards for information sharing and exchange.

UDL Unsupervised Deep Learning—uses machine learning algorithms to analyze and cluster unlabeled dataset.

ULO Upper Level Ontology—provide semantic interoperability of ontologies across multiple domains.

UML Unsupervised Machine Learning—machine learning algorithms to analyze and cluster unlabeled datasets.

USP Unique Selling Point—also called unique selling proposition, is a marketing statement that differentiates a product or brand from its competitors.

V

VERA Violent Extremist Risk Assessment—indicators, categorized into five domains: beliefs and attitudes, context and intent, history and capability, commitment and motivation, and protective items.

VM Virtual Machine—virtual representation, or emulation, of a physical computer.

VoIP Voice over Internet Protocol—technology that allows you to make voice calls using a broadband Internet connection instead of a regular (or analog) phone line.

VPN Virtual Private Network—network which is self-contained and, above all, purely virtual.

VR Virtual Reality—computer-generated, interactive reality that is represented by 3D images and mostly sound.

VRLA Value-based Reinforcement Learning Algorithm—value-based reinforcement learning algorithm is *to optimize a value function V.*

W

Wi-Fi Wireless Fidelity—frequently but incorrectly thought to be the full version of the term Wi-Fi, which refers to the IEEE (Institute of Electrical and Electronics Engineers) 802.11 wireless LAN standards.

WM Whitelist Model—cybersecurity strategy under which a user can only take actions on their computer that an administrator has explicitly allowed in advance.

WSH Windows Script Host—COM-based runtime environment for scripting languages in Windows operating systems.

WSN Wireless Sensor Network—refer to networks of spatially dispersed and dedicated sensors that monitor and record the physical conditions of the environment and forward the collected data to a central location.

X

XaaS Everything as a Service—collective term that refers to the delivery *of anything as a service.*

XML Extensible Markup Language—"eXtensible" means that users may use XML to define their own data formats.

Z

ZTNA Zero Trust Network Access—IT security solution that provides secure remote access to an organization's applications, data, and services based on clearly defined access control policies.

Index

A

Accelerators, 37

Accenture Security Index (ASI), 308

Access
control, x, 78, 86, 106, 119, 160, 163, 165,
183, 186, 234, 237, 247, 274, 281, 282,
296–299, 301
governance, 284, 296–298, 301
policy, x, 103, 143, 274, 296, 297, 335, 342
rights, 77, 237, 296–298

Account monitoring and control, 324

Accuracy
mapping, 145

Activation function, 152, 374, 379, 380

Adjacent location, 140, 145

Advanced persistent threat (APT), x, 76, 77, 81,
82, 85, 91, 103, 111, 112, 114, 125, 267,
279, 364

Adversarial Tactics Techniques & Common
Knowledge (ATT&CK), 243–244, 258,
260, 267, 268

Adversary Attack Behavior Model (AABM), 197

Agent
rational, 75

Alert, 74, 99, 113, 120, 121, 134, 136–138, 140,
141, 148–150, 155–159, 163, 171, 191

Algorithm
hashing, 105
intelligent, 167
meta, 355
random first, 20
Welford, 19
Analytical Construction of Optimal
Regulation, 315

Android malware classification, 379

Anomalous
activity, 134, 139, 149

Anomaly-based Intrusion Detection (AbID),
137, 139–143, 145, 175, 366, 371

Application
programming interface (API), 106, 118
software security, 47, 324
specific maturity model, 314, 324
whitelisting & control, 283

Area under curve (AuC), 172

Artificial intelligence (AI), 4, 9–11, 60, 65, 75,
99, 100, 103, 146, 161, 255, 348, 373

Artificial neural network (ANN), 142, 145,
150–153, 167, 175, 185

As-is-state, 251, 253, 256

Assembly, 15, 16, 63, 354

Assessment criteria, 253, 254, 312

Asset management, 247, 248

Association
rule algorithm, 372
rules, 198

Associative rules, 372

Attack
active threat event, 78
Bayesian model, 197, 198, 200
brandjacking threat event, 79
capabilities, 72, 75, 76, 98, 108, 126, 140,
145, 155, 157, 160, 188, 196, 198, 199,
202–205, 207, 223, 224, 282
clickjacking threat event, 79
cross site, 86
crypto jacking, 73, 77
diamond model, 197, 198, 200

© The Editor(s) (if applicable) and The Author(s), under exclusive license to
Springer Nature Switzerland AG 2023
D. P. Möller, *Guide to Cybersecurity in Digital Transformation*, Advances in
Information Security 103 , https://doi.org/10.1007/978-3-031-26845-8

Printed in the USA
CPSIA information can be obtained
at www.ICGtesting.com
CBHW052213041124
16925CB00003B/21